Financial optimization

The use of formal mathematical models and optimization in finance has become common practice in the 1980s and 1990s. This book presents the exciting symbiosis between the fields of finance and management science/operations research. Prominent researchers from North American academic institutions (including economics Nobel Laureate Harry M. Markowitz) present the state-of-the-art in financial optimization, while analysts from industry discuss the latest business techniques practiced by financial firms in New York, London and Tokyo.

The book covers a wide range of topics: portfolio management of equities and fixed-income investments; the pricing of complex insurance, mortgage and other asset-backed products; and models for risk-management and diversification. Standard models are presented and analyzed, and the problem of portfolio management under uncertainty is examined. Key issues are analyzed, mathematical models proposed, and current practices are reviewed and evaluated.

Financial optimization

Edited by

STAVROS A. ZENIOS

Published by the Press Syndicate of the University of Cambridge
The Pitt Building, Trumpington Street, Cambridge CB2 1RP
40 West 20th Street, New York, NY 10011–4211, USA
10 Stamford Road, Oakleigh, Melbourne 3166, Australia

First published 1993
Reprinted 1995, 1997

Printed in Great Britain at the University Press, Cambridge

A catalogue record for this book is available from the British Library

Library of Congress cataloguing in publication data applied for

ISBN 0 521 41905 0 hardback

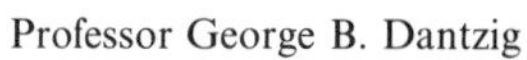

Professor George B. Dantzig

Professor Harry M. Markowitz

This volume is dedicated to Professors George B. Dantzig and Harry M. Markowitz, for their pioneering contributions to the fields of optimization and finance.

Contents

Contributors

Dr Heinrik Dahl *Vollmond Securities, Copenhagen*
Dr Alexander Meeraus *GAMS Development Corporation, Washington*
Professor Stavros Zenios *Decision Sciences Department, The Wharton School, Philadelphia*
Professor David F. Babbel *The Wharton School, Philadelphia*
Professor Peter Muller *BARRA, Berkeley*
Professor Harry M. Markowitz *Department of Economics and Finance, Baruch College, City University of New York*
Professor Michael Asay *Goldman Sachs and Co., New York*
Peter J. Bouyoucos *Goldman Sachs and Co., New York*
Anthony M. Marciano *Goldman Sachs and Co., New York*
Evdokia Adamiou *Goldman Sachs and Co., New York*
Yosi Ben-Dov *Vice President, Prudential-Bache Securities, New York*
Lisa Prendergast *Senior Associate, Prudential-Bache Securities, New York*
Vincent Pica *Managing Director, Prudential-Bache Securities, New York*
Professor Mordecai Avriel *A Tulin Professor of Operations Research, Technion-Israel Institute of Technology, Haifa*
Professor William T. Ziemba *Faculty of Commerce, University of British Columbia*
Dr Yuko Beppu *Seikei University, Tokyo*
Professor John Mulvey *Department of Civil Engineering and Operations Research, School of Engineering and Applied Science, Princeton University*
Professor Robert Nauss *School of Business Administration, University of Missouri at St. Louis*
Dr Ronald S. Dembo *Algorithmics Inc. Toronto*
Professor Paul Zipkin *Columbia University, New York*

Foreword

The service sector in the United States constitutes approximately 60% of GNP and over 70% of employment; the situation in Japan and Western Europe is similar. In addition, this sector consumes over 75% of all information technology investments. However, this investment in the largest component of the economy has not produced large, measurable improvements in productivity (causing some authors to dub this investment the "service sector sinkhole"). While productivity and quality in manufacturing remain a major concern for academia, industry, and government, the sheer size of and major productivity issues in the service sector demand attention.

The purpose of the Fishman-Davidson Center for the Study of the Service Sector at The Wharton School is to attack the above mentioned issues through a truly interdisciplinary approach. Founded in 1984, the Center is the only organization of its kind which is devoted to the full range of issues facing the service sector. The conference from which the current volume arose is an excellent example of the type of interdisciplinary approach which must be pursued if the "sinkhole" is to be avoided. As financial institutions move into a new era defined by both technological and regulatory advances, the best and brightest from finance, management science, computer science, and other fields are needed to improve the productivity of this sector as well as the ultimate value which is provided to the consumer.

The current volume is an excellent example of the research and scholarly dialogue which the Center has supported and will support in the future. Professors Dantzig and Markowitz are indeed the pioneers in their respective fields. The younger scholars who have contributed to this book are, along with the leading researchers in their fields, the best hope in breaking free of the "sinkhole." Dr Stavros Zenios and the staff at the Fishman-Davidson Center put on a first-class conference; I hope that you find the results of this conference presented herein to be as stimulating as the event itself.

PATRICK T. HARKER
Director, Fishman-Davidson Center for the Study of the Service Sector
The Wharton School
Philadelphia, May 1991

Preface

In November 1989 a Conference took place at The Wharton School, University of Pennsylvania on the topic of *financial optimization*. Several distinguished speakers from academia and industry presented that state-of-the-art in modeling and solving a variety of problems that are central to the financial services industry.

The objectives of the conference are best described by introducing the two keynote speakers: Professor George B. Dantzig from Stanford University and Professor Harry M. Markowitz from Baruch College of the City University of New York. These are the founding fathers of two diverse disciplines. Dantzig is credited with the origins of management science through his pioneering work with linear programming in the 1940s. This is a discipline that has touched many endeavors of our society: transportation, manufacturing, the military, economics, and financial services. Markowitz is credited with the foundations of modern portfolio theory for his seminal contribution in risk management through mean-variance models in 1952. His contribution has touched – over the years – most practitioner and academic studies for portfolio management and provided the basis for the derivation of equilibrium models for the financial markets. It was recognized by a Nobel prize in economics in 1990.

The aim of the conference was to bring together these two disciplines, and survey the current state of interaction between them. A second, but equally important objective, was to bring together academic researchers and leading practitioners. Representatives from industry included heads of research or fixed-income groups from Bear-Stearns, Goldman Sachs, Merill Lynch, Prudential-Bache, Salomon Brothers, and the World Bank. Speakers from academia included faculty from some of the US and Canadian leading Business Schools, as well as faculty from Operations Research departments.

The two key questions that were in my mind when I organized the conference were: First, "What are the important problem areas that face the financial services industry where management science is playing, or could play, a key role?" Second, "What tools has the management science community developed over the years to address problems of the financial industries?" Although these questions were never formally posed to the speakers partial answers were offered by the eighteen papers that were presented. A panel discussion, taking place in front of a full audience at the end of the conference – a Saturday afternoon! – summarized the general conclusions of the two-day deliberations:

Formal, quantitative, models for planning under uncertain and volatile interest-rate scenarios are particularly challenging in today's financial market place. The complexity and variety of novel financial instruments renders older and well-entrenched methodologies – especially those based on static models – obsolete. The global market place necessitates the development of methodologies for monitoring currency exchange risk, as well as the risk exposure in the individual markets.

Simplifying assumptions, that are often of great value in academic studies, find little sympathy among practitioners. It is better to provide an approximate solution to as complete a model as possible, than to obtain elaborate solutions to an oversimplified, approximate model.

In this environment tools like simulation, various optimization models, and novel computing technologies are of critical value. Optimization models for multiperiod planning in the face of noisy and uncertain future data are of particular importance. It was a welcome observation that such models, under the realm of *stochastic optimization*, do exist and have been applied with great – even if not widespread – success in several applications.

The consensus among the speakers from both industry and academia was that synergies exist between the fields of financial modeling and management science. Obstacles of course remain: the industry is rather reluctant to reveal its challenging questions or current practices. The academic community is often absorbed with the elegance and aesthetic value of solution methodologies that may have little to do with the realities of the application. Overall, however, a very healthy dialogue is taking place between the finance and modeling practitioners (i.e., the so-called *rocket scientists*). Equally active has been the interaction between practitioners and academics. The interaction has been extremely positive and beneficial for all parties: financial and management science analysts, practitioners, and academics.

Contents of the volume

The papers collected in this volume reflect both current applications of management science techniques by practitioners, as well as the development of new, but practical, methodologies by academic researchers. It is also my pleasure, on behalf of the contributors in this volume, to dedicate it to Professors Dantzig and Markowitz.

The volume is organized in three parts: In Part I I grouped papers that give a general overview of the use of optimization methodologies in financial planning and set the stage for subsequent chapters. In particular the two chapters by Dahl, Meeraus, and Zenios provide a characterization of financial risks and explain the use of optimization models for hedging against different risk dimensions. Markowitz adds some new results to his long list of contributions on market equilibrium and discusses their implications for the practical use of portfolio optimization. Muller provides an empirical evaluation of the mean-variance portfolio optimization models that are used frequently in practice.

Part II contains chapters on models that have been – or are currently being – used by practitioners. The emphasis here is on the use of methodologies that are well understood in addressing a problem of immediate and widespread practical concern. Asay, Bouyoucos and Marciano develop a simulation methodology for the valuation of single-premium deferred annuities and they discuss its use at Goldman Sachs. Adamidou, Ben-Dov, Pendergast, and Pica present Prudential-Bache's system for optimal portfolio selection under varying interest-rate scenarios. Avriel describes optimization-based systems that were developed for two distinct banking applications. Both systems have been in active use by a US and an Israeli bank respectively. Dahl describes a variety of factor models for interest-rate risk management and discusses experiences with the use of these models in the Danish market. Ziemba discusses strategies for currency hedging between the Japanese and US equities markets.

Part III contains chapters that are more methodological in nature. Nevertheless, all the chapters in this part present methodologies that have been used in practical applications. Mulvey describes models for optimal asset allocation that do not ignore – as is often the case – the impact of transaction costs. He proposes a novel network-based methodology for the solution of these models. Nauss discusses the use of integer programming formulations for managing bond portfolios. Dembo introduces the notion of scenario immunization: a portfolio immunization strategy that overcomes the severely restrictive assumption of "parallel shifts" that dominates current practices. Zipkin gives an exposition of valuation methodologies based on a Markov-chain description of the short-term interest rates. He illustrates the use of this methodology with a model for the valuation of mortgage-backed securities. Zenios presents the standard option adjusted analysis for mortgage-backed securities, and goes on to present novel computing techniques for execution of this computer-intensive application. In particular he uses alternative parallel computers.

STAVROS A. ZENIOS

The Wharton School, University of Pennsylvania
Philadelphia, March 1991.

Acknowledgments

The Conference on Financial Optimization was sponsored by the *HERMES Laboratory for Financial Modeling and Simulation*, and the *Fishman-Davidson Center for the Study of the Service Sector*, both at The Wharton School, and by Digital Equipment Corporation.

I would like to acknowledge several individuals who assisted me in the preparation of this volume: Henrik Dahl, Arjun Direcha, Randall Hiller, Ross Miller, and Torben Visholm provided editorial assistance. The support of the Fishman-Davidson Center, and its Director Patrick Harker – financial and otherwise – was instrumental in both organizing the Conference and publishing the proceedings. I was also fortunate to have received partial funding for this project from the National Science Foundation under grant SES–91-00216 and the Air-Force Office of Scientific Research under contract AFOSR–91-0169.

Part I

General overview

1 Some financial optimization models: I Risk management

HENRIK DAHL, ALEXANDER MEERAUS and
STAVROS A. ZENIOS

1 Introduction

Since the early seventies the domain of financial operations witnessed a significant transformation. The breakdown of the Bretton Woods Agreement, coupled with a liberalization of the financial markets and the inflation and oil crisis of the same time, led to increased volatility of interest rates. The environment of fixed-income securities, where private and corporate investors, insurance, and pension fund managers would turn for secure investments, became more volatile than the stock market. The fluctuation of bonds increased sharply after October 1979 when the Federal Reserve Bank adopted a policy allowing wider moves in short-term interest rates. According to the volatility indexes, compiled by Shearson Lehman Economics, bonds were more volatile than stocks by a factor of seven in the early eighties.

Uncertainty breeds creativity, but so does a dynamic market where intelligent answers to complex problems are rewarded immediately. As a result we have seen an increased use of advanced analytic techniques in the form of optimization models for many diverse aspects of financial operations. Several theoretical developments provided the building blocks on which an analyst could base a comprehensive planning model. Models for the estimation of the term structure of interest rates, the celebrated Black–Scholes formula for valuating options, and other complex instruments, were added to the long list of contributions since Markowitz's seminal work on mean-variance analysis for stock returns in 1952.

During the same period tools from management science/operations research reached a stage of maturity and sophistication that gained the attention of practitioners in this dynamic environment. Operations research analysts found a very exciting problem domain where their tools could have a significant impact. Developments in computing technology, with the advent of workstations, facilitated the easy development and validation of models. As a result optimization models are becoming indispensable tools in several domains of financial operations. It is probably too early to pass judgment but *financial optimization* promises to be an area of applications comparable to the use of management science models in logistics, transportation, and manufacturing.

In this chapter we hope to demystify several widely used optimization models. The field of financial optimization has reached a stage where the appropriate modeling techniques are well understood for most applications. Nevertheless, problems remain where either new modeling approaches are required or existing solution techniques are not appropriate. Even in the better understood models, however, knowledge of the management science and the finance communities about them remains anecdotal. We provide here a description of some key optimization models. Financial optimization models are classified into two broad areas of application: (1) *risk management* and (2) *financial engineering*. Within each class we discuss several models and point out how these models have a common underlying theme (often, but not always). For each application we provide a brief description of the problem with references to related literature, we define the underlying optimization model in its most basic form, and we discuss important variants or extensions. Quite often we point at open problems and difficulties either in modeling the problem or in solving the model. Our goal is to help analysts understand the applications, by removing much of the jargon which is usually encountered in this field. We also aim at helping users understand the models and remove the "black-box" syndrome from some very important analytic techniques.

This chapter is organized as follows: section 2 classifies the problem domains of financial optimization and section 3 discusses several models that relate to risk management. The companion chapter in this book presents models of financial engineering and provides a brief survey of current solution methodologies. A technical report by the same authors, Dahl, Meeraus, and Zenios (1989) provides a library of well-documented financial optimization models and data, developed in the general algebraic modeling system GAMS of Brooke, Kendrick, and Meeraus (1988).

2 Problem domains

Most financial optimization models may be classified in two broad classes according to their primary objective. These classes we call here: (1) *risk management* and (2) *financial engineering*. Risk-management models are used to select portfolios with specified exposure to different risks. Financial-engineering models are used to structure new financial instruments in order to target specific investor preferences, or to take advantage of arbitrage opportunities. Below we shall characterize both classes and describe the scope of optimization within each one.

2.1 Risk management

One primary function of financial markets is to transfer risk. The transfer mechanism assigns market prices to each type of risk – called *risk premia* – at which supplies and demands are equalized. Although this means that in equilibrium all risks are priced fairly, so all securities have the same expected instantaneous rate of return, one cannot infer that all securities are equally good for all investors. Thus, due to the nature of their business, some investors prefer current income over future performance, while others are concerned with always staying fully funded. Investors who are willing to take a bet on their market views can do so by taking on certain risks, while those who are uncertain or risk averse can hedge their positions.

Risk management is concerned, firstly, with selecting which risks one is to be exposed to and which risks to be immunized against. Secondly, it is concerned with assessing the risks of different securities, and, thirdly, with the construction and maintenance of portfolios with the specified risk-return characteristics. The focus of optimization models is primarily on the third activity, but all three are integrated and interdependent. We give here a classification of different financial risks and examine methods of risk control. Emphasis is placed on the role of optimization models for risk management.

2.1.1 Financial risks

Financial risk is multidimensional. Therefore, a prerequisite to the selection of risk exposure is the identification of the risk forms that are present. The following list serves as the general framework for our discussion:

1 Market risk,
2 Shape risk,
3 Volatility risk,
4 Sector risk,
5 Currency risk,
6 Credit risk,
7 Liquidity risk,
8 Residual risk.

Market risk has slightly different interpretations, depending on which market is analyzed. In the stock market this form of risk is associated with the movements in the market index of portfolio returns. According to the Capital Asset Pricing Model (CAPM), see for example Sharpe (1985), all securities must be priced so that their expected returns at equilibrium are a linear combination of the risk-free return and the market index portfolio return. The weight of the latter for a particular security is the security beta (β) which indicates the relative marginal variation of the returns of that security with respect to the market portfolio. Hence, when the market moves by 1% the expected return on a security will move by β%.

In the fixed-income market, the traditional measure of market risk is interest-rate risk. In general terms this is risk caused by movements in the overall level of interest rates on straight, default-free securities. More specifically, it is the risk associated with a uniform increase in all default-free interest rates. When interest rates rise marginally the price of a regular bond drops by the bond's dollar duration, viz. the first-order derivative of the yield–price relation. Measured as percentage changes, the price effect is given by the bond duration, viz. the elasticity of the yield–price relation for that bond.

CAPM in the stock market, and duration models in the fixed-income market, are single-factor models of security returns (i.e., both assume a single source of risk: the market). Security attributes, like bond cashflows, determine the security sensitivity to movements in this risk factor. Hence, when a risk premium for the factor is determined by general equilibrium conditions, security returns are also determined as the product of price (risk premium) and quantity (contents of the factor).

In more general models one encounters several independent risk factors. In this context security attributes determine the sensitivity of the security to each factor. At equilibrium, the total supply and demand for each risk must be equal. Each risk factor has a risk premium associated with it and, at equilibrium, security returns are determined as the sum over all factors of the total value of that factor in the security (premium times quantity of the factor). This hypothesis is termed the Arbitrage Pricing Theory, Sharpe (1985). Under this hypothesis, market risk is merely the effect of one out of many risk factors, the effect of which is measured by the conventional beta (for stocks) or duration (for bonds).

Shape risk is applicable to the fixed-income market. It is the risk caused by non-parallel shifts of interest rates on straight, default-free securities (i.e., changes in the shape of the term structure of interest rates). To see the effect of shape risk assume that the yield curve is initially flat at 10%. The prices of two zero coupon bonds, one maturing in one year and the other maturing in ten years, are therefore 90.91 and 38.55, respectively. The two securities have durations of 1 and 10, respectively, so for parallel shifts of the term structure, the price of the second security is expected to move roughly ten times as much as the price of the first one, and in the same direction. An investor who expects interest rates to decline would therefore prefer the ten-year bond which promises the highest return. However, if the yield curve tilts upward so the one-year rate decreases to 9% and the long rate increases to 11%, the investor will experience a loss of 9.56% relative to investing in the one-year bond, being subjected to shape risk.

Shape risk can be quantified. Empirical analyses of the returns on U.S. treasuries indicate that three independent factors are sufficient to explain 98% of the variations of the term structure; see Garbade (1986) and Litterman *et al.* (1988). These factors are characterized by their shape impact: changes in the first factor imply almost parallel shifts to the term structure, so this could be thought of as the market risk factor. Changes in the second factor steepens the curve, while changes in the third factor implies changes in the overall curvature of the curve. Furthermore, these factors have been historically stable. This implies that one can compute the return sensitivity of different instruments to each factor and be reasonably certain to quantify shape risk. Having quantified the risk allows investors to select securities that expose them to risk according to their views as to parallel movements, steepening of the term structure, and so on.

Volatility risk is most clearly displayed in options. These instruments are characterized by highly asymmetric returns. If the underlying instrument is worth more than the strike price of a call option on the expiration date, the option is exercised returning the difference. But, if the underlying security is worth less than the strike price, the option expires worthless. Thus, an option resembles an insurance policy: it has value only if chances are that something might happen. The more volatile the markets are, the higher is the price of an option. Therefore, volatility changes have a major impact on options and securities embedding options (for example callable bonds, Mortgage-Backed Securities (MBS) which include prepayment options, etc.), even in an environment that is unchanged in all other respects.

Volatility risk is not only present in options. Regular straight bonds are also subject to volatility risk. The reason is that the yield–price relation is convex. This property implies that bond prices are affected more by a unit yield drop than by a unit yield increase. Therefore, the higher the volatility of yields around a common expected value, the higher the bond expected return. The effect is more important the more convex the yield–price relation is, i.e., the longer the bond maturity and the more dispersed the bond cashflows are. Similarly, the sensitivity of options to volatility is derived from their high positive convexity. Thus, sensitivity to volatility may be approximated by convexity.

Sector risk stems from events affecting the performance of a group of securities as a whole. A sector is a set of securities sharing some common characteristics. Thus, the treasury sector consists of those securities issued by the US Treasury. The agency sector covers bonds issued by various government agencies like the Federal Home Loan Bank, the Government National Mortgage Association, and so on. Mortgage-backed securities may also be defined as a sector.

Since sectors share some common attributes, they are likely to be influenced by common risk factors. Consider for instance the mortgage sector: when interest rates decline, volatility increases, and macroeconomic factors are "favorable", the likelihood of prepayment increases. An investor who expects this scenario to happen may choose to take a bet on his views by selling MBS.

Currency risk is the risk caused by exchange-rate fluctuations. Investors who own portfolios in foreign currency denominated securities will lose when exchange rates depreciate and gain when they appreciate. Another type of risk in international investment is political, or country, risk. Governments may change tax policy, trade policy, or even expropriate foreign investments.

Credit risk covers risks due to up- or down-grading of a borrower's credit worthiness. These changes are caused by changing prospects on the issuer's ability to meet all future obligations. Thus, if a borrower is more likely to default on some or all future payments, his credit worthiness deteriorates, and investors in turn will demand a higher premium for holding the debt. This in turn implies a price drop, hence risk. Credit risk is of importance when considering corporate bonds, but it is also a major influence on corporate money market instruments, bonds issued by sovereigns, etc.

Bonds exposed to credit risk can be thought of as contingent claims. Thus credit risk is related to the shape and volatility risk factors. However, other factors relating to the management and activities of the issuing firm contribute to credit risk (e.g., variations in earnings, age, debt/equity ratio, and so on). Bond ratings are designed to indicate credit worthiness, and controlling portfolio composition across ratings may therefore be used as a means to partial credit risk protection. A comprehensive approach demands analysis of the individual firm to see if its particular circumstances justify taking a bet on its credit worthiness.

Liquidity risk is due to the possibility that the bid–ask spread on security transactions may change. This type of risk is especially important for actively managed portfolios which depend on frequent trading. If the liquidity of a particular instrument worsens (i.e., the bid–ask spread widens) losses will materialize when selling the security if all other market conditions are unaltered. There exist many and diverse reasons for liquidity drops. For instance, due to prepayment on mortgages the circulating volume of a MBS may drop, making it more difficult to match buy and sell orders. Another reason could be institutional changes in the market place. For instance one of the side effects of the recent Exchange Reform in Denmark has been a marked reduction in liquidity for large groups of securities, as can be witnessed from the daily published bond price list from the Copenhagen Exchange. Typical measures of the liquidity of a security are circulating volume and trade volume over a period.

Residual – or specific – risk is, as the name indicates, all other risk. In as much as the previous list accounts for systematic influences, residual risk is security specific and non-systematic. Much of the activity to "beat the market" lies in buying securities which are expected to be underpriced, thus representing relative value, and selling seemingly overpriced securities. In effect, this means taking a position on residual risk.

2.1.2 Risk control techniques

An important question for controlling risk is "How systematic is a particular type of risk (i.e., to what degree does it affect all securities in a given sector)?" Non-systematic risks, that result in returns with correlation close to zero across instruments, can be reduced by diversification. Diversification, however, only leads to risk averaging for highly correlated risks. In the latter case hedging strategies are required. Distinguishing systematic from non-systematic risk is important in order to develop the appropriate investment strategy.

To see how hedging works, consider the following single-factor model. Assume that the risk factor evolves according to some stochastic differential equation (i.e., an Itô process, Ritchken (1987)). Let:

F be the factor level,

t represent time,

μ and σ be two deterministic functions depending on time and the factor level, and ω represent a Gauss–Wiener process.

Then the factor evolves according to:

$$dF = \mu dt + \sigma d\omega \tag{1}$$

Now, assume that the price P of a given security is a twice continuously differentiable function of the single factor and of time. Then, by Itô's lemma, the security price evolves according to:

$$dP = \frac{\partial P}{\partial F} dF + \frac{\partial P}{\partial t} dt + \frac{1}{2} \frac{\partial^2 P}{\partial F^2} dF^2 \tag{2}$$

Substituting (1) into (2) and collecting terms yields:

$$dP = \left(\mu \frac{\partial P}{\partial F} + \frac{\partial P}{\partial t} + \frac{1}{2} \sigma^2 \frac{\partial^2 P}{\partial F^2} \right) dt + \sigma \frac{\partial P}{\partial F} d\omega \tag{3}$$

The first term at the right-hand side of equation (3) is deterministic and depends only on time. It is therefore risk free and represents the time value of the security. The second term, however, is stochastic. It represents the impact of random shocks to the underlying factor on the security price.

To develop a simple hedge, thus eliminating the factor risk, consider two securities exposed to the same factor with prices P and Q respectively. Choose nominal values in the two securities, x and y respectively, such that:

$$x\sigma \frac{\partial P}{\partial F} d\omega + y\sigma \frac{\partial Q}{\partial F} d\omega = 0$$

i.e.:

$$x \frac{\partial P}{\partial F} + y \frac{\partial Q}{\partial F} = 0 \tag{4}$$

Then the overall position is risk free. Equation (4) states that the factor dollar duration of the position must be zero to eliminate factor risk: $\frac{\partial P}{\partial F}$ is the sensitivity of the security price to marginal changes in the factor level, or the security's *factor loading*. This hedging arrangement works out only if the factor is common to both securities (i.e., it is systematic), so that marginal returns are perfectly correlated. It is easily seen that, if each of the two securities were also exposed to residual factors, the hedging arrangement in (4) would have eliminated the common factor risk but left the residual risks in the portfolio.

The investment strategy $\{x,y\}$ derived from (4) is only a local hedge. In general $\frac{\partial P}{\partial F}$ is time dependent and to ensure that the portfolio remains riskless the position must be continuously adjusted. In practice, this is not feasible, and the problem is solved by discrete portfolio rebalancing. This calls for stabilizing conditions on the hedge. Having eliminated first-order effects by (4) stabilization is done by forcing second-order, or convexity, restrictions to improve tracking. We could also view this problem from a different angle: By not adjusting the hedge continuously, the portfolio is subject to volatility risk. As described earlier volatility risk is controlled by convexity conditions.

Note that the principles behind the single-factor hedge in equations (1)–(4) apply to multifactor models as well. For instance, in the presence of two common factors three securities would suffice to eliminate risk. Two would be used to eliminate the first factor and the third would be combined with the other two to eliminate the second factor.

Dropping the non-descript term "factor" and assuming that F is bond yield then (4) is the classical duration matching strategy, see Zipkin (1989). Immunization against interest-rate risk is achieved when dollar durations on assets and liabilities are equal, (i.e., net dollar duration is zero). Another interpretation is reached by letting F represent the market. Then, in the single-factor world of CAPM a risk-free position is achieved when net sensitivity to the market is zero (i.e., the net portfolio beta is zero). In other words, equation (4) describes the principles of hedging market risk.

2.1.3 The scope for optimization in risk management

Viewed in the context of the previous section, risk management is nothing more than taking positions in generic or specific attributes of the securities. An important problem, however, is that generic attributes are not traded in the market! Rather, one can invest in securities which are in effect packages of attributes. This complicates risk management. For instance, suppose that we wish to target a particular exposure to interest-rate risk because we expect a parallel downward shift of the yield curve. If we could simply buy a pure duration bond (i.e., one exposed to interest-rate risk and nothing else) risk management would be simple. But such a bond is not traded. Instead we can buy a real bond which is simultaneously exposed to interest-rate risk and other factors. Targeting duration using one such bond may inadvertently increase shape risk, volatility risk, credit risk and so on. This means that comprehensive risk management is faced with the problem of simultaneously controlling the interaction of many securities and their attributes in shaping overall portfolio exposure.

The problem is complicated even further when realizing that most investors face several institutional requirements in setting up a portfolio. Thus, regulations or firm policy may limit holdings in particular sectors, accounting rules may imply bounds on holdings in securities traded at a discount or premium, and the resources allocated to risk management may favor more conservative approaches over active management, putting yet other constraints on the portfolio composition.

The setting in which the problems are addressed – as outlined above – explains the important role of optimization models in risk management. Mathematical programming techniques can effectively identify the solution to complex portfolio planning problems with many constraints. Using mathematical programming we can find feasible solutions to the problem, or demonstrate that a risk exposure target is unattainable.

An important unresolved issue is: "What to optimize?" On the surface, this question seems trivial: minimize the cost of setting up the portfolio, or maximize its expected return. However, modern portfolio theory dictates that systematic risks are rewarded in a way that makes expected instantaneous returns equal on all securities in equilibrium, and all securities with the same attributes should have the same expected return to eliminate arbitrage. Therefore, if all systematic risk is eliminated, we are left with a risk-free position plus non-systematic risk. The expected rate of return on this position must be equal to the risk-free rate to eliminate arbitrage. The consequence is that high expected returns imply exposure to high levels of risk. Thus, maximizing returns or minimizing costs has the hidden property of maximizing uncontrolled risk.

Consider as an example the standard bond portfolio immunization model (explained in detail in section 3.1). The objective is to maximize net portfolio yield subject to present value and dollar duration constraints. The results of this model depend on the selection universe. If, for instance, low credit corporate bonds are included, the model will pick such bonds to achieve immunization, since they have higher yields than straight, default-free bonds. An apparent gain is realized by speculating in credit and sector risk without a systematic measure of the exposure to these two risk factors.

When the universe is restricted to current US Treasuries (i.e., bonds with no default risk, liquidity risk, etc.) the optimization results in a two-bond portfolio, consisting of a very long and a very short bond (a barbell). The overall portfolio is dollar duration matched, hence free of interest-rate risk. However, the portfolio is maximally exposed to shape risk. Also, to maintain dollar duration matching, the portfolio must be rebalanced relatively frequently, incurring high transaction costs and liquidity risks which are not accounted for in the optimization model.

To achieve higher stability and reduce shape risk, second-order constraints may be imposed on the model. The result will be less dispersed cashflows across the horizon and a reduction of risk. The position will still be exposed to volatility risk, however, and to correct, one may continue to add higher-order constraints. In the limit, the result will be a cashflow matched portfolio! This could have been obtained right away by using a cashflow matching model – for example, the dedication model given in section 3.3 – which apparently results in a more costly portfolio than the immunization model. However, as the example shows, the reason for this cost differential is that the dedicated portfolio is less risky than the immunized portfolio. Therefore, there is reason to expect that, over a long period, a dedicated portfolio will yield approximately the same as an immunized portfolio. Empirical confirmation of this observation is given in Maloney and Logue (1989).

The lesson drawn from the examples is that optimization can not be used blindly. Instead it must be coupled with a careful analysis of which risks the model is buying when maximizing returns. Thus, the role of the portfolio manager becomes crucial. The manager must be able to formulate clear objectives, state the economic and institutional constraints, choose a suitable selection universe, and analyze risks and returns. In all this, optimization is a tool which simplifies the shaping process. If all systematic risks are monitored, optimization can be used to pick underpriced securities and thus enhance performance. If risks are not monitored, however, optimization will maximize exposure to them. Optimization will make a good portfolio manager better, and a bad one worse!

2.2 Financial engineering

With the continued deregulation of markets, the increased volatility, and the intensified competition in the financial industry, financial innovation with the engineering of new instruments has accelerated rapidly. Well-known new products include standardized financial futures and options, floating-rate instruments, caps and floors, interest-rate and currency swaps, mortgage-backed securities, adjustable rate preferred stock, as well as derivatives of all these (for a general reference, see Walmsley (1988)).

Most of these products represent new packages of old attributes. It is well known, for example, that options can be replicated by a continuously rebalanced portfolio of a risk-free security and the underlying instrument. Nevertheless the new products have been successful additions to the financial markets. The main reason is that the new instruments typically carve out a few generic risk attributes from the initial building blocks, making it easier to control risk. Options, for example, are primarily volatility instruments but they are also exposed to interest-rate risk. A position where a call option is bought and a put option is sold, and where both options have common underlying instrument, expiration date, and strike price, is equivalent to a forward contract. However, breaking up the forward into options, and buying or selling these, enables investors to take direct positions in volatility, a possibility which the forward itself does not offer.

In this sense, financial engineering is one more aspect of risk management. As discussed above, risk management is a process of creating a portfolio of securities with certain attributes, from existing attribute packages. However, the resulting portfolio is not sold as a standardized product, while the objective of financial engineering is to design products that are added to the market.

A functional categorization of financial engineering products follows the list of financial risks of section 2.1.1. Thus, interest-rate futures and forwards, floating-rate notes and inverse floaters are targeted at isolating *market risk*. Interest-rate swaps can be thought of as *shape* instruments, since their return depends on changes to the shape of the term structure. Options are *volatility* products. Index-linked securities often carry *sector* attributes. Currency swaps, options, futures, and rolling forwards are examples of *currency risk* instruments. Other instruments are primarily credit enhancing. These include a long list of securitized assets (MBS, securitized bank loans, etc.) and asset swaps. Liquidity enhancement is inherent in put bonds and secondary-market mortgage products. Finally, stock and bond indices can be thought of as stripping out residual risk.

Reviewing this list, it becomes apparent that financial engineering is a mixture of service activity and arbitrage. By repackaging and stripping risk attributes to fit investor preferences and needs, the financial engineer improves the marketability of the products. In effect, he takes over some of the responsibilities of decentralized portfolio managers, and should be rewarded with a service fee. At the same time, however, the intention is to utilize mispricings in the market to capture risk premia. Thus, by stripping a US Treasury in zero coupon bonds the financial engineer disaggregates shape risk. If he can sell the resulting portfolio for more than the price of the treasury, he captures a riskless arbitrage profit which is due to mispriced shape risk.

In order to standardize products, certain guidelines must normally be obeyed. Thus, to obtain an AAA rating on an issue, rating agencies may require proof that the structure is default free under both best-case and worst-case scenarios. To simplify this task, standard procedures and scenarios have been developed, which need not be completely in line with reality. For instance, when issuing Collateralized Mortgage Obligations (CMOs), one requirement is that the collateral must sustain bond retirements, both in the very unlikely case of an immediate complete prepayment of all mortgages in the collateral and in the equally unlikely case of no prepayment, Sykes (1987).

As in the case of risk management, the mere existence of such constraints make optimization a key tool in financial engineering. Mathematical programming efficiently ensures feasibility and identifies the difficulties when feasibility cannot be attained. However, there also appears to be genuine opportunities for optimization in financial engineering. The reason is that there is generally some flexibility in structuring constraints, whether rating-agency or market dictated. For example, when structuring CMOs, the expected life of a tranche determines its price. A three-year tranche is priced using a spread over three-year treasuries. However, three years in market terms means any time period in the interval (3.0, 3.4) years. This slack may be sufficient reason to optimize profits, if several structures are feasible. Similarly, with markets not always in equilibrium, certain attributes may be relatively cheaper in some securities than in others, which again implies a scope for constrained profit maximization or cost minimization.

3 Model domains

In this section we discuss some optimization models used in risk-management applications. For each model, we briefly describe its theoretical background and then present the basic components of a mathematical prototype. Also, when relevant, we discuss potential or actual extensions of the models. We establish first some common notation. Unless stated otherwise in subsequent sections the following are used:

$U=\{1,2,3,...,I\}$ denotes the universe of securities,
$i \in U$ indicates a security from the universe,
$T=\{1,2,3,...,T_{max}\}$ denotes a set of discrete points in time,
$t \in T$ indicates a point in time,
C_{it} indicates the cashflow from security $i \in U$ at time $t \in T$,
x_i is the nominal holdings of security $i \in U$.

3.1 A bond portfolio immunization model

Immunization is a portfolio strategy used to match interest-rate risk of an asset portfolio against a future stream of liabilities, in order to achieve net zero market exposure. There is a large literature on portfolio immunization, see for instance Bierwag (1987), Fabozzi and Pollack (1987), Granito (1984), and Platt (1986). We describe here the fundamentals.

3.1.1 Background

Portfolio immunization is in essence a hedging strategy based on the principles of section 2.1. As both assets and liabilities are interest-rate sensitive (i.e., sensitive to the same common factor) a hedging strategy can be set up which eliminates net sensitivity to that factor. It was shown in section 2.1 that immunization is achieved when a portfolio is selected with net zero present value sensitivity to the factor of interest.

To compute the interest rate sensitivity of a cashflow, consider the yield–price relation. Let:

r_i denote the cashflow yield, and
P denote the present value.

The present value of a security is given by:

$$P_i = \sum_{t \in T} C_{it}(1+r_i)^{-t} \tag{5}$$

Differentiating with respect to cashflow yield, gives the present value sensitivity k_i – or *dollar duration* – of the cashflow:

$$k_i = -\sum_{t \in T} tC_{it}(1+r_i)^{-(t+1)} \tag{6}$$

It is seen that dollar duration is additive. Thus, the dollar duration of a portfolio is given by:

$$k_p = \sum_{i \in U} k_i x_i \tag{7}$$

Given the present value, P_L, and dollar duration, k_L, of liabilities, an immunized portfolio must satisfy the two conditions that present values and dollar durations on assets and liabilities be equal:

$$\sum_{i \in U} P_i x_i = P_L \tag{8}$$

$$\sum_{i \in U} k_i x_i = k_L \tag{9}$$

3.1.2 The optimization model

Immunized portfolios can be established in many ways. It is therefore natural to examine whether they can be put together optimally. The most commonly used objective has been to maximize the asset portfolio yield. The idea is that since the portfolio return is "risk-free" we might as well maximize it. The portfolio yield is given implicitly by equation (5) which is a non-linear expression. It turns out, however, that a first-order approximation to the true portfolio yield is the dollar duration weighted average yield of the individual securities in the portfolio, i.e.:

$$r \approx \frac{\Sigma_{i \in U} k_i r_i x_i}{\Sigma_{i \in U} k_i x_i} \tag{10}$$

The denominator in (10) is given by (9) to be equal to k_L, hence maximizing approximate portfolio yield is equivalent to simply maximizing the numerator in (10). In this case, the core immunization model is a linear programming problem:

[IMMUNIZATION1]

$$\underset{x \in \Re^I}{\text{Maximize}} \sum_{i \in U} k_i r_i x_i$$

$$\text{subject to } \sum_{i \in U} P_i x_i = P_L$$

$$\sum_{i \in U} k_i x_i = k_L$$

$$x_i \geq 0$$

3.1.3 Variations and extensions

To see how the core model can be extended, first consider which risks have been controlled and which have not. Systematic interest-rate risk has been controlled. More specifically, the risk of a local parallel shift to all bond yields has been controlled. Other risks have not. Indeed, as pointed out in section (2.1), the exposure to all other risks has most likely been maximized. We will use our classification of risks to see how the core model should be augmented to limit exposure to each of them.

The result of the model [IMMUNIZATION1] will be a barbell portfolio (i.e., a portfolio consisting of a short maturity and a long maturity bond only). The reason is that long bonds are most efficient in maximizing dollar duration times yield, and that a short bond is most efficient in reducing overall dollar duration to its target value. Thus, the portfolio cashflow will be very dispersed. This property is desirable in two ways. Firstly, it ensures a high positive convexity of the net portfolio, so the immunized portfolio duration attains its minimum at the current yield levels. Secondly, it can be shown that more dispersed cashflows on the asset than on the liabilities is a sufficient condition for being able to retire liabilities as they occur, see Bierwag (1987).

At the same time, however, the dispersed cashflow implies that the asset portfolio is highly exposed to *shape risk*. Suppose, for example, that the liabilities consist of a single payment in ten years and the optimal asset portfolio consists of a ten-year zero coupon bond and cash. If the yield curve tilts so yields rise more than proportionally on long bonds than on short bonds, the asset portfolio will lose value more rapidly than the liability, thus causing negative net worth. We will show an economical solution to this problem in the factor immunization model of section 3.2. However, two other approaches are worth mentioning. The first one breaks up the liability cashflow in maturity buckets and matches present value and dollar duration on each bucket. This is equivalent to solving an immunization model [IMMUNIZATION1] for each maturity bucket in order to get a closer cashflow match. The second approach minimizes the portfolio cashflow dollar convexity.[1] This is the second-order derivative of the yield–price relation and is given by:

$$Q_i = \sum_{t \in T} t(t+1) C_{it} (1 + r_i)^{-(t+2)} \tag{11}$$

for a single security. Dollar convexity is additive, so the portfolio dollar convexity is given as:

$$Q_p = \sum_{i \in U} Q_i x_i \tag{12}$$

Convexity minimization must be constrained, since according to the aforementioned condition for liability retirement the net portfolio dollar convexity should be non-negative. Therefore, the immunization model becomes:

[IMMUNIZATION2]

$$\underset{x \in \Re^I}{\text{Minimize}} \sum_{i \in U} Q_i x_i$$

$$\text{subject to} \sum_{i \in U} P_i x_i = P_L$$

$$\sum_{i \in U} k_i x_i = k_L$$

$$\sum_{i \in U} Q_i x_i \geq Q_L$$

$$x_i \geq 0$$

which is again a linear programming problem.

Convexity constraints are also a remedy to *volatility risk*. Because of bond convexity, variations in bond yields will cause the immunization conditions to break. However, by forcing the net portfolio convexity to be non-negative, net worth has a minimum at current yields. If an upside due to volatility is not desired, but tracking performance is required, the above convexity minimization problem is appropriate. Convexity constraints become the more important when securities with embedded options are available in the selection universe. In this case, immunization is based on option adjusted dollar durations, which for such securities is only a local measure of interest-rate risk. By incorporating option adjusted dollar convexity constraints, the stability of net worth performance can be greatly improved. The global performance is not ensured in this approach, however, and another model is developed later – the global return shaping model of section 2.3 in chapter 2 – to improve immunization for such portfolios.

The common approach for the control of *sector risk* is to impose constraints to diversify across user defined sectors. These constraints state that nominal value or present value of bonds in a certain sector must lie within certain percentage limits of total nominal value or present value. This will clearly force more securities into the optimal portfolio and thus also help reduce *residual risk*. *Credit risk* can also be controlled to some extent using this approach, by choosing some sectors to represent rating classes.

As was mentioned above immunization is a dynamic strategy. Both sides of (8) and (9) depend on time, but not necessarily in the same way. The differential effect on asset and liability dollar duration from the passage of time is often denoted *duration drift*. When dollar durations drift apart, one side will be more exposed to interest-rate risk than the other. In order to maintain dollar duration equality, the portfolio must therefore be periodically rebalanced.

The need to rebalance implies that an immunized portfolio will be subject to *liquidity risk*. The most efficient approach for the control of liquidity risk is to demand that securities be bought in large lot sizes since these are more liquid than small lots. The constraint will be that either you buy at least l_i dollars of nominal value or nothing at all. To represent this requirement in an optimization model requires binary variables, turning the problem into a mixed-integer programming problem (MIP). Note, as an aside, that, since bonds are traded in certain denominations, the portfolio holdings should already be integer multiples of these denominations to be tradeable. Thus the immunization models [IMMUNIZATION1] and [IMMUNIZATION2] may already be a mixed-integer program.

Letting $y_i \in \{0,1\}$ denote the binary variable, l_i denote the minimum lot size to be purchased, and u_i the maximum permissible lot size (due, for example, to diversification requirements). The minimum lot size constraint is enforced by:

$$l_i y_i \leq x_i \leq u_i y_i$$

3.2 A factor immunization model

3.2.1 Background

Factor immunization is an enhanced immunization technique which explicitly attacks shape risk. This is achieved by relaxing the implicit assumption in the immunization model that the term structure of interest is flat and only shifts in parallel. To see the effect of relaxing this assumption, recall that a straight, default-free bond can be synthesized by a portfolio of zero coupon bonds with maturity dates $t \in T$. Each zero has its own yield, the collection of which is the term structure. To eliminate arbitrage, the bond price must equal the value of the portfolio of zeros.

Let r_t denote the yield of a zero coupon bond maturing at t. Then the bond price is given by:

$$P_i = \sum_{t \in T} C_{it}(1+r_t)^{-t} \tag{13}$$

To find the generalized interest sensitivity of the bond, differentiate (13) with respect to r_t:

$$dP_i = -\sum_{t \in T} C_{it} t (1+r_t)^{-(t+1)} dr_t \tag{14}$$

It can easily be verified that, if the term structure is flat (i.e., $r_t = r$) and only moves in parallel (i.e., $dr_t = d_r$), the derivative is dollar duration. If the term structure is not flat, but is only shifted in parallel, another dollar duration measure results, viz. the Fisher–Weil dollar duration, see Maloney and Yawitz (1985). However, one can postulate a model relating changes in the term structure to common factors. Let:

j denote an index of the set of factors J,

F_j denote the level of factor j, and

a_{jt} denote an array of coefficients.

Then a linear factor model is given by:

$$dr_t = \sum_{j \in J} a_{jt} dF_j \tag{15}$$

stating that, if factor j is increased by one unit, the term structure is affected as specified in a_{jt}. Substituting (15) into (14) and rearranging terms yields:

$$f_{ij} \doteq \frac{\partial P_i}{\partial F_j} = -\sum_{t \in T} t a_{jt} C_{it} (1+r_t)^{-(t+1)} \tag{16}$$

f_{ij} is known as the ith bond's *factor loading* for factor j. Note that by the specification in (15), the factors can be chosen to be independent. Hence, to immunize the effect of all factors on the shape of the term structure, one can simply immunize against each of the factors by matching dollar factor sensitivities on assets and liabilities. Since the model comprises all statistically significant shifts to the term structure, market risk is simultaneously immunized.

3.2.2 The optimization model

An optimally factor-immunized portfolio can be established by maximizing portfolio yield subject to present value equality and factor dollar duration matching on assets and liabilities for all factors. Thus, the core model becomes:

[FACTOR]

$$\underset{x \in \Re^I}{\text{Maximize}} \sum_{i \in U} k_i r_i x_i$$

$$\text{subject to} \sum_{i \in U} P_i x_i = P_L$$

$$\sum_{i \in U} f_{ij} x_i = f_{Lj}, \forall j \in J$$

$$x_i \geq 0$$

All terms are as defined in section (3.1), with the addition of f_{ij} that denotes factor loading of security i for factor j and f_{Lj} that denotes the factor loading of the liability. This model is a linear programming problem or, when tradeability considerations are added, a mixed integer program.

3.2.3 Variations and extensions

The addition of factor constraints has eliminated shape risk from the immunization models [IMMUNIZATION1] and [IMMUNIZATION2]. Otherwise, the same risks apply. However, the factor immunized portfolio cashflow matches the liability cashflow much better. Since the factor immunized portfolio is also more diversified than the immunized portfolio (it contains $|J|$ bonds), residual risk is also reduced. The cost of this increased protection, of course, is a loss in yield. To enhance further risk control – sector, residual, credit, and so on – the same measures as for the immunization model apply.

3.3 A bond dedication model

3.3.1 Background

It should be clear by now that one can add protection to the immunization model by imposing constraints that make asset cashflows look more like the liability cashflows. Why not take the full step and require cashflow matching instead of matching on average? Clearly, this would eliminate risk completely. There are two primary reasons why exact cashflow matching is not used very often. The first is that it is mostly infeasible to achieve exact cashflow match using traded instruments. The second is that cashflow matching appears to be more expensive than immunization. We have already argued that the high cost of dedication compared to immunization is only apparent when one considers risks, portfolio maintenance, and rebalancing costs.

To attack the problem of infeasibility, one can establish an almost cashflow matched portfolio. Dedication is a technique to do so. The basic idea is to establish a portfolio with cashflows which always suffice to cover liability payments, if necessary after reinvestment to the liability dates. To eliminate reinvestment risk, a conservative reinvestment rate (possibly 0%) is assumed. Should actual reinvestment rates turn out to be higher than the conservative rate, the strategy will gradually generate a surplus.

3.3.2 The optimization model

To formulate an optimal dedication model, use the following notation:

τ an index of liability payment dates from the set $T_l \subseteq T$,
$\Delta\tau$ length of time interval between liability payment dates $\tau - 1$ and $\tau \in T_l$,
L_τ liability payment at time τ,
s_τ cash holdings (surplus) at time τ,
ρ reinvestment rate, and
$D_{i\tau}$ reinvested value of bond cashflows between liability dates $\tau - 1$ and $\tau \in T_l$.

The total number of payment dates can potentially be very large. Instead of looking at every date, we can recast the model to consider only liability dates.[2] To do so we compute the reinvested value of bond cashflows occurring between liabilities:

$$D_{i\tau} = \sum_{t \in [\tau - 1, \tau]} C_{it}(1+\rho)^{\tau - t} \tag{17}$$

Next we balance cashflows at each liability date. On a liability date, reinvested bond cashflows and cash holdings are received, the liability is paid, and possibly a surplus is generated which is carried forward to the next liability date at the reinvestment rate:

$$\sum_i D_{i\tau} x_i + s_{\tau - 1}(1+\rho)^{\Delta\tau} = L_\tau + s_\tau \tag{18}$$

Imposing non-negativity constraints on the surplus enforces the precedence of asset cashflows to liability cashflows, and dedication is achieved. The appropriate objective is to minimize the cost of the portfolio, viz. the cost of bond purchases and initial cash holdings. Thus, letting $\tau = 0$ denote the settlement date, the optimal dedication model becomes:

[DEDICATION]

$$\underset{x \in \Re^I}{\text{Minimize}} \sum_{i \in U} P_i x_i + s_0$$

$$\text{subject to } \sum_{i \in U} D_{i\tau} x_i + s_{\tau-1}(1+\rho)^{\Delta\tau} = L_\tau + s_\tau, \forall \tau \in T_l$$

$$s_\tau \geq 0, \forall \tau \in T_l$$
$$x_i \geq 0$$

This is a linear program, or potentially a MIP if tradeability is desired.

3.3.3 Variations and extensions

The dedication strategy in principle eliminates all risk but default (credit) risk when choosing a sufficiently conservative reinvestment rate. Since the strategy is critically dependent on the timely flow of cash the selection universe is normally restricted to contain only highest credit straight bonds.

It should be noted that other constraints may be relevant as well. One type of risk inherent in the strategy is holding period risk, i.e., the risk that for currently unknown reasons the portfolio must be liquidated at some future point. Therefore, liquidity risk considerations may be important. Sector and minimum lot size constraints may be appropriate means for the control of liquidity risk.

In some cases, it is known before dedicating a liability stream that the investor's budget does not allow a full dedication. In these cases, a version of the model may be used to fund liabilities as far out in the future as possible. Typically, the funding is constrained to meet the first liabilities fully and only partially fund a liability if all previous ones have been completely funded. This problem requires the use of binary decision variables on the funding of each liability as well as sequencing constraints. The objective is to maximize the number of fully funded liabilities, i.e., the sum of the binary variables. The problem thus becomes a MIP.

One problem in the dedication strategy is that it assumes that liabilities are known with certainty or that good estimates are available for the full time span. For many investors this may not be true. Instead, they know the liabilities for a relatively short time period and have good estimates of the overall duration of liabilities. For these investors, a hybrid between immunization and dedication may be relevant. In the *horizon matching*, or *combination matching*, model the purpose is precisely to provide dedication for part of a liability stream and overall dollar duration matching. One version of this model is:

[COMBINATION]

$$\underset{x \in \Re^I}{\text{Minimize}} \sum_{i \in U} P_i x_i + s_0$$

$$\text{subject to } \sum_{i \in U} D_{i\tau} x_i + s_{\tau-1}(1+\rho)^{\Delta\tau} = L_\tau + s_\tau, \forall \tau \in T \subset T_l$$

$$\frac{\Sigma_{i \in U} P_i x_i + s_0}{\Sigma_{i \in U} k_i x_i} = \frac{P_L}{k_l}$$

$$s_\tau \geq 0, \forall \tau \in T \subset T_l$$
$$x_i \geq 0$$

3.4 Worst-case analysis of option positions

The above models implicitly assume that investors have symmetric preferences on risk in the sense that both positive and negative net worth deviations constitute risk. However, many investors are concerned with insuring against negative net worth while still keeping some upside potential on returns. Such investors may be more adequately described as having asymmetric preferences, i.e., assigning different penalties to surplus and deficit. Similarly, by the nature of liabilities, the target return distribution need not be symmetric.

To construct tailored return distributions derivative instruments are very effective. Especially, because of the asymmetric payoff, options are well-suited for shaping an overall risk exposure profile. In this and the next section, we describe two models for such portfolios including derivative instruments. In this section, we describe a model which analyzes a given option portfolio under worst-case scenario and in the next section we describe a model aimed at structuring portfolios with given return distributions.

3.4.1 Background

The need for analyzing option portfolios arises because risk/return tradeoffs grow increasingly more complex as more options are included in a portfolio. The literature is full of exotic names for different option positions, see for example Cox and Rubinstein (1985), Fabozzi (1989), and Sykes (1987). For example, a *bull spread* is obtained by buying a call with low strike price and selling a call with a high strike price (both options with the same expiration date and underlying instrument). This position uniformly gains when prices rise and loses when prices drop, thus constituting a directed bet. A *straddle* is established when buying a call and a put with the same strike price, underlying instrument, and time to expiration. This position loses when prices do not move much but gain when prices move sufficiently in either direction. Thus, this is a bet on volatility. Such simple strategies are easy to analyze. However, what happens when the two positions are combined? Obviously, this depends on the mixture of the two and the payoff pattern is bound to be more erratic and complicated than for the simple strategies.

For these reasons, it becomes increasingly more difficult to assess complex option positions. In some instances, it may be desirable to gain a quick overview of worst-case and best-case performance of a position, along with an identification of the types of market moves that cause these cases to occur. This is the task of the following optimization model. To simplify matters, we will only consider European options on interest-rate futures and analyze effects of instantaneous changes in the risk-generating factors which are the interest rate and market volatility.

To specify the model we need pricing equations for futures and options. Using the Black model, Black (1976), we have the price of a European futures call option:

$$C(F,S,\tau,r,\sigma)=e^{-r\tau}(FN(d_1)-SN(d_2)) \tag{19}$$

where:

C is the call premium,
F is the futures price,
S is the strike price,
τ is term to expiration,
r is the risk-free interest rate assumed to lie in the interval $(\underline{r},\bar{r})$,
σ is the futures price volatility assumed to lie in the interval $(\underline{\sigma},\bar{\sigma})$,
$N()$ denotes the cumulative standard normal distribution function,
d_1 is the function defined by:

$$d_1=\frac{\ln(F/S)+\sigma^2\tau/2}{\sigma\sqrt{\tau}} \tag{20}$$

d_2 is the function defined by:

$$d_2=d_1-\sigma\sqrt{\tau} \tag{21}$$

Similarly, a European futures put option is priced according to:

$$P(F,S,\tau,r,\sigma)=e^{-r\tau}(-FN(-d_1)+SN(-d_2)) \tag{22}$$

Finally, the futures' price is given by the cost of carry model:

$$F(t,r)=F_{0t}e^{rt} \tag{23}$$

where F_{0t} is the adjusted spot price of the underlying bond, i.e., the present value of cashflows occurring after the maturity date of the futures contract.

The portfolio in question may contain options and futures maturing on different dates, and the same level of interest rates and volatility may not apply to all these dates. A simple model is obtained by letting interest rates and volatilities move independently across maturities and instruments. However, in a more consistent framework, functional dependencies across instruments and maturities as well as possible relations between volatility and interest rates, may be specified. For simplicity, we stick with the independence assumption here.

3.4.2 The optimization model

We can now collect the above in an optimization model to identify worst-case scenarios for portfolios containing futures and options on these. The portfolio composition is known and consists of nominal x_{it} in futures on security i maturing at $t \in T$, y_{ijt} and z_{ijt} in call and put options respectively on futures on security i, maturing at $t \in T$, and with strike price indexed by $j \in J$.

Whenever interest rates or volatilities move, the value of the position is affected. The objective of the optimization model is now to identify those changes in interest rates and volatilities which minimize the present value of the portfolio. The allowable movements of rates and volatilities are bounded, and present value minimizing movements therefore constitute a bounded worst-case scenario. When changing the direction on the objective, the model isolates a best-case scenario. The model is:

[OPTIONS]

$$\underset{r_t, \sigma_t}{\text{Minimize}} \quad \sum_{i \in U} \sum_{t \in T} \left(F_{it} x_{it} + \sum_{j \in J} (C_{ijt} y_{ijt} + P_{ijt} z_{ijt}) \right)$$

$$\begin{aligned}
\text{subject to } & F_{it} = F(F_{i0}, t, r_t) \\
& C_{ijt} = C(F_{it}, S_j, t, r_t, \sigma_t) \\
& P_{ijt} = P(F_{it}, S_j, t, r_t, \sigma_t) \\
& \underline{r_t} \leq r_t \leq \overline{r_t} \\
& \underline{\sigma_t} \leq \sigma_t \leq \overline{\sigma_t}
\end{aligned}$$

The model is non-linear, having a linear objective and non-linear equality constraints.

3.4.3 Variations and extensions

The core model given here can be criticized on several grounds. While realistic extensions are possible they go beyond the scope of this chapter. At the same time the resulting mathematical models become increasingly more complex and hence very difficult to solve.

The first criticism of [OPTIONS] is that the pricing formulae may not be accurate. Thus, in principle, the Black futures option model cannot be used to price options on interest futures. However, it is often used in practice and may be thought of as being roughly right for the purposes of the above model. Also, if necessary, other option pricing formulae could be incorporated. Second, the model is only valid for European options. Approximate formulae for pricing American options could be included, however. Third, to enhance consistency, the model should probably incorporate functions covering autocorrelations of interest rates and volatilities. Fourth, the model is restricted to a narrow set of interest and volatility sensitive instruments. But, as long as closed form pricing formulae exist for other instruments, these can be incorporated as well, to increase security coverage.

One technical problem with the model is that it is not likely to be convex. Therefore, the solution is likely to be locally and not globally optimal. Furthermore, problems with non-linear constraints (especially equality) are very difficult to solve. See the discussion in sections 3.2.2 and 3.6 of chapter 2.

3.5 Mean-variance models

3.5.1 Background

All the models examined so far hedge against different forms of systematic risk. Residual risk has been dealt with using *ad hoc* constraints to ensure diversification. A systematic way for dealing with residual risk is possible if we assume that residual risk is accurately represented by a function of the mean and variance in the return of securities only. Assume that the investor's preference can be represented by some (derived) utility function over the mean and variance of the portfolio's returns, thus favoring portfolios with higher means and lower variances. The optimal portfolios for this investor are those that achieve the highest expected return for a given level of variance and the smallest possible variance for a given level of return. Such portfolios are called *mean-variance* portfolios. A broader class is that of minimum-variance portfolios that include the portfolio with the smallest possible variance at every level of expected return. A detailed analysis of mean-variance models is given in Ingersoll (1987). See also the seminal paper by Markowitz (1952), whose name is associated with the models in this section, and chapters 3 and 4.

3.5.2 The optimization model

To develop the minimum-variance model we define:

Q the variance/covariance matrix $\{q_{ij}\}$ between securities $i,j \in U$,
μ_i expected return of security $i \in U$,
μ_p target expected return for the portfolio,
x_i fraction of the portfolio that consists of security $i \in U$.

The minimum-variance portfolio when unlimited borrowing is possible is obtained by solving the model:

[VARIANCE1]

$$\underset{x \in \Re^I}{\text{Minimize}} \ \frac{1}{2} \sum_{i \in U} \sum_{j \in U} q_{ij}, x_i x_j$$

$$\text{subject to} \sum_{i \in U} \mu_i x_i = \mu_p$$

$$\sum_{i \in U} x_i = 1$$

The model can be solved analytically using the first-order optimality conditions to give an optimal portfolio:

$$x^* = \phi' Q^{-1} e + \omega' Q^{-1} \mu$$

where e is a vector of all 1's, μ is the vector of expected returns $\mu = (\mu_1, \mu_2, \ldots, \mu_I)'$, and ϕ and ω are Lagrange multipliers associated with the constraints of [VARIANCE1].

If a riskless asset is available then model [VARIANCE1] can be simplified. The budget constraint $\Sigma_{i \in U} x_i = 1$ can be eliminated since the riskless asset can absorb any residual or can be used to finance purchases. The target return constraint is expressed in excess form, and the modified model is:

[VARIANCE2]

$$\underset{x \in \Re^I}{\text{Minimize}} \ \frac{1}{2} \sum_{i \in U} \sum_{j \in U} q_{ij} x_i x_j$$

$$\text{subject to} \sum_{i \in U} (\mu_i - r) x_i = \mu_p - r$$

where r is the return of the riskless asset, and is thus known with certainty. The solution to this problem can be obtained once more analytically as:

$$x_i^* = \phi' Q^{-1} (\mu - re)$$

and the residual investment in the riskless asset is:

$$x_0^* = 1 - e' x^*$$

It can be shown, using elementary calculus arguments, that when holdings are not constrained the optimal portfolio consists of a combination of only two portfolios: the riskless asset and one more. Hence, it would appear that the scope of optimization techniques in mean-variance models is rather trivial. This is not the case as we will see in the next section. But first we point out two limitations of the core models [VARIANCE1] and [VARIANCE2]. As already pointed out, mean-variance models only diversify residual risk. Systematic risk considerations may be addressed by imposing a target beta level for the portfolio, thus diversifying its exposure to systematic market risk. If β_i is the beta factor of security i then a target β_p for the portfolio can be attained by adding the constraint:

$$\sum_{i \in U} \beta_i x_i = \beta_p$$

to either [VARIANCE1] or [VARIANCE2]. Furthermore mean-variance analysis is based on the assumption that returns are multivariate normally distributed. Hence, the portfolio is normally distributed as well – being a weighted combination of such distributions – and hence it can be fully characterized by its mean and variance only. Including skewed instruments, like options, in the universe would violate this assumption and render the models meaningless.

3.5.3 Variations and extensions

While mean-variance portfolios are efficient for investors with preferences only on mean and return, the models [VARIANCE1] and [VARIANCE2] achieve only one of the two requirements for efficiency: lowest variance for a given target of expected return. Obtaining the highest return for a given level of variance, say q_p, would result in a non-linear constraint of the form:

$$\sum_{i \in U} \sum_{j \in U} q_{ij} x_i x_j = q_p$$

This difficulty is avoided in the model [MEAN] we develop next. In addition to the budget constraint we add general linear constraints on the permissible combinations of x in the form $Ax = b$. These constraints are used to eliminate sector risk by grouping together investments in securities of a sector s and set a target b_s for the exposure to this sector. Liquidity risk can be eliminated once more by adding minimum lot size constraints. Finally, we add a portfolio turnover constraint in the form $h(x,x^0) \leq \bar{h}$. This restriction ensures that the change between the current holdings x^0 and the desired portfolio x should be bounded by $\bar{h}$. This constraint is essential in mean-variance models: the covariance matrix is in most practical applications almost singular, and hence the optimal decision changes significantly with small changes in the problem data. To avoid big changes when continuously reoptimizing the portfolio using [MEAN] the turnover constraints are imposed. Additional constraints are often added to eliminate small trades. These constraints are of the form "zero–or–range": either an instrument is not traded (i.e., it remains at the current holding level) or it should be traded by more than a specific amount. Bringing the discussion to a conclusion we formulate the mean-variance model most commonly used in practice, see for example Perold (1984), as:

[MEAN]

$$\underset{x\in\Re^I}{\text{Minimize}} \sum_{i\in U}\sum_{j\in\Re^J} q_{ij}x_ix_j - \lambda\sum_{i\in U}\mu_i x_i$$

subject to $Ax=b$

$$h(x,x^0)=\max\left\{0, \sum_{i\in U}(x_i-x_i^0)\right\}\leq\tau$$

$$\underline{u}_i y_i \leq x_i - x_i^0 \leq \bar{u}_i y_i$$
$$\underline{l}_i z_i \leq x_i - x_i^0 \leq \bar{l}_i z_i$$
$$y,z\in\{0,1\}$$

The parameter λ is used to judiciously tradeoff the return component against the variance component and hence to trace out a complete spectrum of efficient portfolios. This model is a mixed-integer non-linear program. Tradeability considerations can also be added using the mixed-integer non-linear programming formulation. Mulvey (1987) has shown that some instances of [MEAN] can be represented using network optimization and thus very large problems can be solved quite efficiently; see for example Dembo *et al.* (1989).

3.6 Utility models

3.6.1 Background

The mean-variance model of the previous section is a static, single-period model with an unspecified time horizon. Hence, it fails to consider the long-term impact of short-term objectives. The typical example of what could go wrong with the static, single-period model, is the retaining of investments with high transaction costs and consistently negative returns. Losses during a single time period may not offset the transaction costs, while repeated losses over a longer time horizon may warranty the sale of the investment. At the same time the mean-variance model is descriptive of investors with a quadratic utility function of the form:

$$u(r)=r-\frac{br^2}{2}$$

(see Ingersoll (1987)), where r denotes return. The expected value of $u(.)$ is:

$$V(\mu,\sigma^2)=\mu-\frac{b}{2}(\mu^2+\sigma^2)$$

where μ and σ^2 are the expected value and variance of return respectively. Hence, the mean-variance model is equivalent to one of maximizing expected utility. Quadratic utility functions pose technical difficulties. First, a concave quadratic function is decreasing beyond a point and precautions should be taken to remain in the lower range of the function. Second, absolute risk aversion increases with quadratic utility functions. This property is usually unrealistic.

The utility models introduced next extend the mean-variance models in two important directions: they are multiperiod models, and at the same time they permit the use of any applicable utility function that may be appropriate for each investor.

3.6.2 The optimization model

The multiperiod utility model is developed here based on the work of Grauer and Hakansson (1985). We develop the model for the isoelastic utility function:

$$u(1+r)=\frac{1}{\gamma}\{(1+r)^{\gamma}+(\gamma-1)\}$$

where γ is a constant indicating risk aversion. The model can be readily extended for more general functions. We define the following:

s index from the set of scenarios S,
x_{it} nominal holdings in security i at time period t,
x_{0t} nominal holdings in the risk-free instrument at period t,
x_{bt} amount borrowed in period t,
r_{it} anticipated return on instrument i at period t,
r_{0t}, r_{bt} return of risk-free asset and borrowing rate respectively,
π_{ts} probability of scenario $s \in S$ occurring at time period t.

Under scenario s the return on security i takes the value r_{its} which is known with certainty.

The model can now be defined as follows:

[UTILITY]

$$\underset{x\in\Re^I, x_{0t}, x_{bt}}{\text{Maximize}} \quad \sum_{s\in S} \pi_{ts}\frac{1}{\gamma}\left(1+\sum_{i\in U} x_{it}r_{its}+x_{0t}r_{0t}+x_{bt}r_{bt}\right)^{\gamma}$$

$$\text{subject to} \quad \sum_{i\in U} x_{it}+x_{0t}+x_{bt}=1$$

$$\Pr\left\{1+\sum_{i\in U} x_{it}r_{it}+x_{0t}r_{0t}+x_{bt}r_{bt}\geq 0\right\}=1$$

$$x_{it}, x_{0t}\geq 0$$
$$x_{bt}\leq 0$$

The first constraint is the budget constraint which we have also encountered in the mean-variance model [VARIANCE1]. The second constraint ensures solvency. It is a probabilistic constraint (Pr{} denotes probability of the random event in the brackets) and as such it would be difficult to include in the model. Nevertheless, the objective function imposes infinite penalty for non-solvency. Hence, the probabilistic constraint is redundant and can be ignored when solving the model.

3.6.3 Variations and extensions

The primary advantage of model [UTILITY] is that it can generate investment decisions for a wide range of risk-bearing attitudes: from risk neutrality ($\gamma=1$) to infinite risk aversion ($\gamma=-\infty$). The multiperiod model is also consistent with a model that maximizes expected terminal utility of wealth: if w_T is the wealth at the end of the time horizon, and $u(w_T)=\frac{1}{\gamma}w_T^{\gamma}$, with $\gamma\leq 1$ then maximizing $E[u(w_T)]$ generates the same sequence of optimal portfolios as the sequence of single-period models for every $t\leq T$. Finally, we point out that input data for [UTILITY] could be based on observable rather than expected returns: one may use the observed return for periods $0,1,2,\ldots,t-1$ to develop the probability distribution of returns at period t. Tradeability requirements and restrictions on the minimum allowable trade can also be added to this model, much in the same way as in [MEAN]. The resulting model then becomes a mixed-integer non-linear program.

3.7 Multiperiod, stochastic planning models

3.7.1 Background

The majority of the models examined thus far are static and single-period models. Even the multiperiod utility model of section 3.6 is myopic in the sense that investment decisions made at period t do not depend on expectations for any time period beyond $t+1$. In cases where uncertainty prevails at all the stages of the planning horizon – as is often the case – and furthermore some corrective action (i.e., recourse) would be possible between periods t and $t+1$, then stochastic programming models become more appropriate. Such models are not very common at present in financial planning due to their complexity and the significant requirement for input data. Nevertheless some very interesting models have appeared in the literature. Even more important, these models were developed for concrete problems in banking and asset allocation and were used with real data to demonstrate their superior performance over more elementary, deterministic, models. Some examples of stochastic planning models are: (1) the dynamic model for bank portfolio management of Bradley and Crane (1972), (2) the bank asset/liability model of Kusy and Ziemba (1986), (3) the active portfolio management model of Mulvey and Vladimirou (1992), (4) the stochastic programming model for mortgage-backed financing of Zenios (1991), and (5) the stochastic dedication models of Hiller and Schaack (1990). There is not a unified stochastic planning model that encompasses the details of all these models. Hence, we present in the next section a rather generic formulation. Interested readers should consult the references for more details.

3.7.2 The optimization model

Let S be the set of scenarios (for example, on instrument returns). Define:

x_0 vector of investment decisions for the first time period, to be made based on information already known at $t=0$,

x_1^s second-period decisions to be made for each scenario $s \in S$,

π_s probability of scenario $s \in S$,

$U(x_0,x_1^s)$ utility of return, as a function of the decision variables.

Of course $x_1^s, \forall s \in S$ has to be determined *a priori*, before any scenarios have been realized. Otherwise the model simplifies to a sequence of single-period myopic models solved after the fact $s \in S$ has been observed. The decisions x_0,x_1^s are made such that an expected utility function is maximized (expectations being calculated as a combination of utility values under all scenarios, each weighted by the probability π_s). At the same time all accounting, policy, diversification, and other constraints should be satisfied under every scenario. Using matrices A_0,A_1^s,B_1^s to represent general constraints the optimization model becomes:

[STOCHASTIC]

$$\underset{x_0,x_1^s \in \Re^I}{\text{Maximize}} \sum_{s \in S} \pi_s U(x_0,x_1^s)$$

$$\begin{aligned} \text{subject to } & A_0 x_0 = b_0 \\ & B_1^s x_1^s + A_1^s x_0 = b_s, \forall s \in S \\ & x_0 \geq 0 \\ & x_1^s \geq 0, \forall s \in S. \end{aligned}$$

Note that a set of constraints is imposed for each scenario and hence the resulting problem is a very large non-linear program. The first-stage decisions x_0 do not depend on knowledge about the future (of course!) and hence they are identical under all scenarios.

3.7.3 Variations and extensions

The most important extension of model [STOCHASTIC] is in multistage/multiperiod planning. Unfortunately, the problem suffers from a combinatorial explosion of scenarios as the number of time periods increases. Stochastic planning models remain a virtually unexplored field in financial modeling. Nevertheless the need to capture uncertainty in a systematic way is of great importance in the environment of our applications. This remains an open area for further research.

ACKNOWLEDGMENTS

The research of S.A. Zenios was partially supported by NSF grants ECS-8718971 and CCR-8811135, AFOSR grant 89-0145 and grants from Digital Equipment Corporation and Union Bank of Switzerland.

NOTES

1 Or it minimizes the cashflow dispersion, M^2, defined as the present value weighted square deviation of cashflow timing around the duration of the portfolio. This is based on the risk minimization strategy in Fong and Vasicek (1984). When including dispersion constraints, the problem becomes non-linear.
2 This is in fact an identical formulation to the much larger problem of full cashflow accounting, see Zipkin (1989).

REFERENCES

G. Bierwag (1987), *Duration Analysis*. Ballinger Publishing Company, New York.

F. Black (1976), "The pricing of commodity contracts," *Journal of Financial Economics*, September.

S. P. Bradley and D. B. Crane (1972), "A dynamic model for bond portfolio management," *Management Science*, 19(2): 139–51.

A. Brooke, D. Kendrick, and A. Meeraus (1988). *GAMS: A User's Guide*. The Scientific Press, California.

J. C. Cox and M. Rubinstein (1985), *Options Markets*. Prentice-Hall.

H. Dahl, A. Meeraus, and S. A. Zenios (1989), "Some financial optimization models: III. An algebraic modeling system library," Report 89-12-03, Decision Sciences Department, The Wharton School, University of Pennsylvania, Philadelphia.

R. S. Dembo, J. M. Mulvey, and S. A. Zenios (1989), "Large scale nonlinear network models and their application," *Operations Research*, 37: 353–72.

F. J. Fabozzi (ed.) (1989), *The Handbook of Fixed-Income Options: Pricing, Strategies & Applications*. Probus Publishing Company.

F. J. Fabozzi and I. M. Pollack (eds.) (1987), *The Handbook of Fixed Income Securities*. Dow-Jones, Irwin.

H. G. Fong and O. A. Vasicek (1984), "A risk minimizing strategy for portfolio immunization," *The Journal of Finance*, December.

K. Garbade (1986), "Models of fluctuation in bond yields – an analysis of principal components," Technical Report, Bankers Trust Company, Money Market Center, June.

M. R. Granito (1984), *Bond Portfolio Immunization*. Lexington Books, D. C. Heath and Company.

R. R. Grauer and N. H. Hakansson (1985), "Returns on levered actively managed long-run portfolios of stocks, bonds and bills," *Financial Analysts Journal*, September: 24–43.

R. S. Hiller and C. Schaack (1990), "A classification of structured bond portfolio modeling techniques," *Journal of Portfolio Management*, 17(1).

J. E. Ingersoll, Jr. (1987), *Theory of Financial Decision Making. Studies in Financial Economics*. Rowman & Littlefield.

M. I. Kusy and W. T. Ziemba (1986), "A bank asset and liability management model," *Operation Research*, 34(3): 356–76.

R. Litterman, J. Scheinkman, and L. Weiss (1988), "Common factors affecting bond returns," Technical Report, Goldman, Sachs & Co., Financial Strategies Group, September.

K. J. Maloney and D. E. Logue (1989), "Neglected complexities in structured bond portfolios," *The Journal of Portfolio Management*, 15(2).

K. J. Maloney and J. Yawitz (1985), "Interest rate immunization and duration," Working Paper, School of Business Administration, Dartmouth University.

H. Markowitz (1952), "Portfolio selection," *Journal of Finance*, 7: 77–91.

J. M. Mulvey (1987), "Nonlinear network models in finance," *Advances in Mathematical Programming and Financial Planning*, JAI Press.

J. M. Mulvey and H. Vladimirou (1988), "Solving multistage stochastic networks: an application of scenario aggregation," Report SOR 88-1, Princeton University.

(forthcoming), "Stochastic network optimization models for investment planning," *Management Science*.

A. F. Perold (1984), "Large scale portfolio optimization," *Management Science*, 30(10): 1143–60, October.

R. B. Platt (ed.) (1986), *Controlling Interest Rate Risk: New Techniques & Applications for Money Management*. John Wiley & Sons.

P. Ritchken (1987), *Options: Theory, Strategy, and Applications*. Scott, Foresman and Company.

W. F. Sharpe (1985), *Investments*. Prentice Hall.

D. Sykes (1987), "CMO concepts," Technical Report, Bear Stearns, Mortgage Related Products Research, July.

J. Walmsley (1988), *The New Financial Instruments*. John Wiley & Sons.

S. A. Zenios (1991), "Massively parallel computations for financial modeling under uncertainty," in J. Mesirov (ed.), *Very Large Scale Computing in the 21st Century*. SIAM, Philadelphia, PA, pp. 273–94.

P. Zipkin (1989), "The structure of structured bond portfolios," Technical Report, Columbia University, Graduate School of Business.

2 Some financial optimization models: II Financial engineering

HENRIK DAHL, ALEXANDER MEERAUS and
STAVROS A. ZENIOS

1 Introduction

In this chapter we address the use of optimization models in *financial engineering*. We give this term to the process of creating new packages of old risk attributes. By repackaging and stripping risk attributes from existing instruments the financial engineer improves the marketability of the products and fits the needs of individual investors. Financial engineering also takes a service role in financial operations. For example, in order to obtain an AAA rating on an issue certain properties should hold under both best- and worst-case scenarios. Optimization models could automate the process of analyzing the scenarios.

This chapter is organized as follows: section 2 discusses three models from financial engineering and section 3 provides a brief overview of existing solution methodologies. Emphasis is placed in this section on the availability and capabilities of software for the solution of the optimization models presented earlier. It aims to convey a first exposure of the techniques to financial analysts, and may have little to offer to an operations research expert. An appendix provides an optimization model for the estimation of the term structure of interest rates, and explains Monte Carlo simulation techniques for generating interest-rate scenarios. Models like the one described here are often used to generate key input data to several of the optimization models discussed in both chapter 1 and this one. Chapter 1 presents optimization models for risk management. A library of well-documented financial optimization models and data, developed in the general algebraic modeling system GAMS of Brooke, Kendrick and Meeraus (1988), is given in the technical report by Dahl, Meeraus and Zenios (1989).

2 Model domains

In this section we discuss some examples from financial engineering. For each model, we briefly describe its theoretical background and then present the basic components of a mathematical prototype. Also, when relevant, we discuss potential or actual extensions of the models. The notation established in chapter 1 is still employed, unless stated otherwise in a specific section.

2.1 Mortgage settlement

2.1.1 Background

The delivery and settlement of mortgage-backed securities (MBS) between the primary dealers – federal agencies and investment banks – is governed by an extensive set of rules and regulations to ensure uniform practices. An MBS may be traded for immediate settlement – within five business days of the trade date – or for forward delivery. Forward delivery settlements are employed by originators with known quantities of mortgages that have not yet been pooled, and hence pool information is on a *to-be-announced* (TBA) basis. The Public Securities Association has set forth the "Uniform Practises for the Clearance and Settlement of Mortgage-Backed Securities" (for a description, see Bartlett (1989)). These practises specify allowable variances in delivering TBAs, acceptable denominations, the size and number of pieces in a delivery.

The fact that some variance is allowed to the originators leaves room for structuring the TBA in a cost-effective manner. Even if the opportunity for profit is ignored, however, the need to adhere with the PSA practises creates a rather complex feasibility problem. In the next section we explain one by one the uniform practises and develop an optimization model.

2.1.2 The optimization model

Rule 1 The value of pools assigned to a TBA lot must be between the agreed lower and upper limit, unless the pool is failed, in which case the value of the lot must be zero. Let:

i index the set of TBAs,
l index the set of lots,
p index the set of pools,
σ_i be the agreed variance on delivery of TBA i,
L_{il} be the size of lot l for TBA i,
w_{il} be a binary decision variable, indicating either that lot l of TBA i is delivered ($w_{il}=0$) or is failed ($w_{il}=1$), and
V_{pil} denote the value assigned from pool p to lot l of TBA i.

Then the delivery condition may be written as:

$$(1-\sigma_i)L_{il}(1-w_{il})\leq \sum_p V_{pil}\leq(1+\sigma_i)L_{il}(1-w_{il}) \tag{1}$$

Here, if $w_{il}=1$, both ends of the inequality become zero, and hence the value allocated must be zero. Otherwise, it must stay between agreed bounds.

Rule 2 The adjusted face value of a pool can either be used for allocation, or it is "boxed," i.e., it is left over for future allocations or other business activities. Let:

B_p denote the amount of pool p boxed, and

A_p denote the adjusted face value of pool p.

Then we have:

$$\sum_i \sum_l V_{pil} + B_p = A_p \tag{2}$$

Rule 3 For a number of other constraints, the allocation pattern must be represented by binary decision variables, denoting whether a pool is allocated to a particular lot or not. To determine these decision variables, z_{pil}, we can use the following constraint:

$$V_{pil} \leq (1+\sigma_i) L_{il} z_{pil} \tag{3}$$

This constraint simply forces the decision variable z_{pil} to 1, if the allocated value is positive. If the allocated value is zero, z_{pil} may be 0 or 1.

Rule 4 Pools must be diversified in the allocation. In particular, it is illegal to use only "class 3" pools, i.e., pools with delayed delivery which are highly risky:

$$\sum_{p \notin C3} V_{pil} \geq 1 - w_{il} \tag{4}$$

where $C3$ denotes the set of "class 3" pools. In itself, this constraint only forces the allocated value to be strictly positive (≥ 1) if the lot is not failed. However, if the pool is failed, the value must be zero by equation (1). Other constraints will determine the actual sizing of the allocation.

Rule 5 Pools must be allocated in a given minimum amount ($25,000 for GNMAs) of original face value, and with increments in certain steps ($5,000 for GNMAs) of original face value. Exception is allowed for one piece, designated the tail piece.

To formalize this requirement, remember that the value allocated is denominated in adjusted face value. Therefore, to convert to original face value, define O_p as the original face value of the pool. Then the original face value allocated is $V_{pil}O_p/A_p$. Next define:

u^0_{pil} as the binary decision variable whether or not to allocate the initial chunk,

u_{pil} as the number of subsequent full pieces to allocate, i.e., an integer variable,

u^T_{pil} as the binary decision variable whether or not to allocate the tail piece, and

T_p as the size of the tail denominated in original face value.

Then, the denominated condition may be written (here for a GNMA) as:

$$V_{pil}O_p/A_p = 25u^0_{pil} + 5u_{pil} + T_p u^T_{pil} \tag{5}$$

This condition determines how pools could be split to allow a more efficient allocation.

Rule 6 The tail of a pool cannot be split under any circumstances. Therefore, the tail piece is either fully allocated or boxed:

$$\sum_i \sum_l u^T_{pil} + u^{BT}_p \leq 1 \tag{6}$$

where u^{BT}_p denotes the binary decision variable whether or not to box the tail of a given pool.

Rule 7 These splitting rules are not only valid for allocations to TBAs but also to what is boxed. We define:

w^{0B}_p, the binary decision variable whether or not to box the first chunk, and

w^B_p, an integer variable indicating the number of additional chunks boxed.

Then the splitting rule is enforced by:

$$B_p O_p/A_p = 25w^{0B}_p + 5w^B_p + T_p w^{BT}_p \tag{7}$$

Rule 8 In addition, the step increment on allocations and boxed pools only is allowed after the first chunk has been allocated or boxed:

$$u_{pil} \leq M u^0_{pil} \tag{8}$$

$$u^B_p \leq M u^{0B}_p \tag{9}$$

Here, M is a large number.

Rule 9 A pool must be split if it is to be allocated to more than a single lot or if it is not fully allocated to a single lot, in which case part of original face value remains in the box. To formalize this, it is convenient to use two auxiliary variables:

z^{B1}_p is a binary variable which is 1 if something or nothing – but not everything – of a pool is boxed. It is zero if everything is boxed.

z^{B2}_p is a binary variable which is 1 if something or everything of a pool is boxed. It is zero if nothing remains in the box.

These two variables are determined by:

$$z_p^{B1} \geq \frac{A_p - B_p}{A_p} \tag{10}$$

and:

$$z_p^{B2} \geq 1 - \frac{A_p - B_p}{A_p} \tag{11}$$

With these two binary variables, it is seen that $z_p^{B1} + z_p^{B2} = 1$ if everything or nothing remains in the box. Otherwise the sum is 2. This can be utilized to determine splitting. Let:

s_p be a binary variable stating whether a pool is split or not, and let
n be the total number of lots in the allocation.

Then, s_p is given by:

$$\sum_i \sum_l z_{pil} + z_p^{B1} + z_p^{B2} - 2 \leq n s_p \tag{12}$$

Rule 10 To allow for constraints on the allowed number of pools satisfying a particular lot allocation, we need to define when one pool or two pools are "big." One pool is "big" in a particular lot allocation if it is greater than or equal to the minimum value requirement:

$$2\sigma_i L_{il} F_{il}^1 \geq V_{pil} - (1 - \sigma_i) L_{il} \tag{13}$$

where F_{il}^1 is a binary variable stating whether or not a lot is filled by a single "big" pool. To understand the constraint, consider the case where every single pool allocated to the lot amounts to less than the lower bound on adjusted face value. In that case, the right-hand side becomes negative, and F^1 can be either zero or one. On the other hand, if a single pool has a value higher than the lower bound, the right-hand side becomes positive, and F^1 must be 1. The multiplication by 2 on the left-hand side is necessary for the case where a single pool fills the lot to the upper bound.

It becomes even messier to define what it means to allocate two large pools to a lot. Two pools are "big" – taken together – if their value is at least as large as the lower allocation bound. This may be formulated as:

$$2\sigma_i L_{il} F_{i,l}^2 \geq V_{pil} + V_{p'il,p' \neq p} - (1 - \sigma_i) L_{il} \tag{14}$$

with F_{il}^2 as the binary variable flagging a "big" two-pool allocation to the lot.

Rule 11 We are now at the crux of the allocation: No more than a specified number of pools may be allocated to a lot. For GNMAs only three pools may be allocated, so as a first condition we have that:

$$\sum_p z_{pil} \leq m_{il} \tag{15}$$

where m_{il} denotes the maximum number of pools allowed in a single lot. In addition, if a single pool is "big," only that pool may be allocated:

$$\sum_p z_{pil} \leq 1 + (1 - F_{il}^1) m_{il} \tag{16}$$

Furthermore, if two pools together are "big," only those two pools may be used in the given lot:

$$\sum_p z_{pil} \leq 2 + (1 - F_{il}^2) m_{i,l} \tag{17}$$

Rule 12 Finally, we need to specify the objective of the allocation. We choose to maximize profits which is given as the allocated amount multiplied by the spread between the agreed forward price and the market price at the time of delivery, minus the costs of splitting pools. Let:
P_i^F denote the forward price of TBA i,
P_i^M denote the market price of TBA i, and
P_p^S denote the cost of splitting pool p.
We can now specify the complete optimization model:

$$\text{Maximize} \sum_p \sum_i \sum_i (P_i^F - P_i^M) V_{pil} - \sum_p P_{psp}^S$$

$$\text{subject to } (1 - \sigma_i) L_{il} (1 - w_{il}) \leq \sum_p V_{pil} \leq (1 + \sigma_i) L_{il} (1 - w_{il})$$

$$\sum_i \sum_l V_{pil} + B_p = A_p$$

$$V_{pil} \leq (1 + \sigma_i) L_{il} z_{pil}$$

$$\sum_{p \notin C3} V_{pil} \geq 1 - w_{il}$$

$$V_{pil} O_p / A_p = 25 u_{pil}^0 + 5 u_{pil} + T_p u_{pil}^T$$

$$\sum_i \sum_l u_{pil}^T + u_p^{BT} \leq 1$$

$$B_p O_p / A_p = 25 u_p^{0B} + 5 u_p^B + T_p u_p^{BT}$$
$$u_{pil} \leq M u_{pil}^0$$
$$u_p^B \leq M u_p^{0B}$$

$$z_p^{B1} \geq \frac{A_p - B_p}{A_p}$$

$$z_p^{B2} \geq 1 - \frac{A_p - B_p}{A_p}$$

$$\sum_i \sum_l z_{pil} + z_p^{B1} + z_p^{B2} - 2 \leq ns_p$$

$$2\sigma_i L_{il} F_{il}^1 \geq V_{pil} - (1 - \sigma_i) L_{il}$$
$$2\sigma_i L_{il} F_{il}^2 \geq V_{pil} + V_{p'il, p' \neq p} - (1 - \sigma_i) L_{il}$$

$$\sum_p z_{pil} \leq m_{il}$$

$$\sum_p z_{pil} \leq 1 + (1 - F_{il}^1) m_{il}$$

$$\sum_p z_{pil} \leq 2 + (1 - F_{il}^2) m_{il}$$

2.1.3 Variations and extensions

The model just described is a very complex mixed-integer program, and cannot be solved by general purpose software for integer programming. It is currently considered one of the open problems in financial optimization. Most commonly it is addressed using heuristic procedures that guarantee to satisfy the constraints. Given the magnitude of the market segment that is influenced by the PSA practises we believe that this is an important problem that will receive attention from operations research experts in the forthcoming years, in search of effective solution procedures.

2.2 Structuring collateralized mortgage obligations

2.2.1 Background

The basic idea behind a Collateralized Mortgage Obligation (CMO) is to restructure the cashflows from an underlying mortgage collateral into a set of high quality bonds with different maturities. The bonds will be retired in a sequence given by the stated maturities, and possibly according to stated priorities. Thus, the bond with the lowest maturity is fully retired before any other bond is amortized. During the amortization period all interest-bearing bonds receive interest.

By repackaging the collateral cashflow in this manner, two things are achieved. First, the life and risk characteristics of the collateral are restructured. The mortgage collateral is subject to prepayment risk, which implies that the future cashflow is uncertain and the life of the mortgage may vary widely. When repackaged into CMO bonds – or tranches – these risks are better controlled. Thus, the short maturity bonds are guaranteed to be retired first, implying that their lives will be less uncertain, although not completely fixed, since all cashflows from the collateral (less a service fee) are passed through to the bondholders. Furthermore, even the long maturity bonds will have less cashflow uncertainty than the underlying collateral. Therefore, the CMO structure allows the issuer to target different investor groups more directly than when issuing straight mortgage-backed securities (MBS). The low maturity tranches may be appealing to investors with short horizons while the long maturity bonds may be attractive to pension funds and life insurance companies. Each group can find a bond which is better customized to their particular needs than straight MBS, so they may be willing to pay a premium for the added protection and customization. Thus, the issuer may gain profits by targeting the issue to different market segments, rather than simply selling the collateral.

The second objective of a CMO issuance is to utilize particular market conventions for arbitrage purposes. The convention in mortgage markets is to price bonds with respect to their weighted average life (WAL), viz. the principal payment weighted average time to the cashflow. A bond with a WAL of three years will be priced at the three year treasury rate plus a spread, while a bond with a WAL of seven years will be priced at the seven year treasury rate plus a spread. The WAL of the CMO collateral is typically high, implying a high rate for (normal) upward sloping yield curves. By splitting the collateral into several tranches, some with a low WAL and some with a high WAL, better rates are obtained on the short tranches while worse rates may result for long tranches. Overall, however, the issue may have a better rate than the collateral. The reason why this works out is that the WAL is not a good measure of riskiness or effective maturity. Other measures like effective duration, which embodies both the principal and interest payment components and the option components of the mortgage-backed issue, are theoretically superior to WAL, and would likely result in more efficient pricing, but such measures are difficult to compute, since they require options models, prepayment models, and the like. In addition, even though these measures are computed by investment banks, the results are hard to compare across institutions because the underlying models differ. In order to obtain some standardization, the convention is to use simple measures like the WAL.

When issuing a CMO, several restrictions apply. First, to obtain an AAA rating, it must be demonstrated that the collateral can service the payments on the issued CMO bonds under both a best-case and a worst-case scenario. These scenarios are well defined and standardized, and cover the two extreme cases of full immediate prepayment of the collateral and no prepayment at all. Second, the structure must satisfy the sequencing scheme imposed by stated maturities and/or priorities. Third, to price the bonds, the expected WAL of each tranche must lie within certain bounds. This is again caused by the market convention, which allows for a slack in pricing the bonds. Thus, a bond with a WAL between three years and 3.44 years is thought of as comparable to a three-year treasury, while a seven-year treasury may be compared to a CMO tranche with a WAL between seven and 7.94 years. Fourth, for particular tranches the performance may have to be analyzed under different scenarios and be guaranteed to lie within certain ranges. This, for instance, is the case for the so-called Planned Amortization Classes (PACs) and Targeted Amortization Classes (TACs), see Roberts *et al.* (1988), Bartlett (1989) or Walmsley (1988).

The CMO market has been one of the fastest growing fixed-income markets since the first issuance in 1983, and much effort has gone into developing new structures designed to take advantage of a particular market environment. Today, a typical CMO issuance contains "regular" bonds, tranches which guarantee a fixed cashflow for a wide range of prepayment rates (PACs and TACs), bonds which only receive interest or principal (IOs and POs), floating-rate tranches, accrual tranches, etc. (for an overview of such innovations, see Walmsley (1988)). The computational complexities involved in ensuring feasibility of a CMO structure and the need for rapid identification and exploitation of market opportunities, in connection with the possibility of finding several feasible CMO structures for the same collateral, makes CMO structuring a natural candidate for optimization.

2.2.2 The optimization model

To capture the essence of the optimization problem, we will describe a model for structuring CMOs with the following generic structure:

1 A number of "normal" bonds with different stated maturities, and coupons, which are retired sequentially according to maturity.
2 A single accrual tranche, i.e., a bond which receives no principal payment until all the above bonds are fully retired, and where interest is added to the outstanding principal at all times. Introducing an accrual tranche in the structure allows the issuer to allocate more cash payments to short maturity classes, hence issue a larger portion of the CMO in tranches with low yields.

3 A residual which receives any excess cashflows from the collateral, net of service fees.

The objective of the CMO structuring model is to maximize the proceeds of the issuance, i.e., the total present value of the tranches, by choosing the size of each tranche. The issue must satisfy the constraints mentioned above.

To ensure that the CMO can be retired under best- and worst-case scenarios, the expected cashflow of the collateral is modified to ensure overcollateralization. We shall not go into details here, only mention that the modification consists in computing the so-called Bond Value for every payment date. Disregarding effects of payment delays, etc. the Bond Value is the lesser of the outstanding collateral balance (to cover the case of immediate prepayment) and the present value of the collateral cashflow, assuming a zero prepayment rate, and discounting by the highest CMO bond coupon. The ratio of the Bond Value to the outstanding collateral balance at any point in time is next multiplied with the expected principal payments on the collateral (given a prepayment rate assumption), to determine the maximal rate of retirement that the collateral can support at any time under both best- and worst-case conditions. If the collateral net coupon is lower than the coupon on any tranche, further adjustments are needed. The best- and worst-case constraints can be treated parametrically and need not be part of the optimization model itself. Instead, they are covered by adjustments to the model input data.

To formalize the rest of the constraints, we first consider the cashflow accounting in the CMO. Let:

i index the set of tranches,

x_{it} be the outstanding principal for each tranche at date t,

p_{it} be the principal payment of tranche i at date t,

c_i be the periodic coupon rate of tranche i.

By definition, the principal payment (which may be negative for the accrual tranche) amortizes outstanding principal:

$$p_{it} = x_{it-1} - x_{it}$$

and the total cashflow consists of interest payments and principal payments:

$$C_{it} = c_i x_{it-1} + p_{it}$$

In addition, the CMO cashflow is related to the cashflow on the collateral. First, all adjusted principal payments from the collateral are allocated to the CMO normal tranches and accrual tranche. Let:

n index the subset of tranches which are normal, and

m denote the accrual tranche, and

$\mathcal{P}_t$ denote the adjusted principal payments from the collateral.

Then we have:

$$\sum_n p_{nt} = \mathscr{P}_t + c_m x_{mt-1} - C_{mt}$$

where the two last terms on the right-hand side cover the contribution from accrual on the accrual tranche.

We also need to make sure that principal payments are only made to a tranche if this is amortizing. To formalize this, let:

z_{it} be a binary variable which is 1 if tranche i is amortizing at time t and 0 otherwise, and

$\mathscr{C}_t$ be the expected cashflow (less service fee) from the collateral.

Then we have:

$$p_{nt} \leq \mathscr{C}_t z_{nt}$$

for normal tranches, and:

$$C_{mt} \leq \mathscr{C}_t z_{mt}$$
$$p_{mt} \leq \mathscr{P}_t z_{mt}$$

for the accrual tranche. The two conditions for the accrual tranche are needed because principal payments can be negative on that tranche (during the accrual period).

Finally, since the CMO is a pass-through, all cashflows from the collateral (less a service fee) are allocated to the CMO. Therefore, letting R_t be the payment to the residual at any point in time, we have that:

$$\sum_i C_{it} + R_t = \mathscr{C}_t$$

Given this cashflow accounting, we may specify the constraints on WAL. Let:

$\underline{W}_n$ be the lower bound on the WAL for tranche n, and

$\bar{W}_n$ be the upper bound.

Then we have:

$$\underline{W}_n \leq \frac{\Sigma_t t p_{nt}}{p_{n0}} \leq \bar{W}_n$$

Next we impose the sequencing requirements. We introduce a matrix of variables y_{it}. Then sequencing is enforced by:

$$y_{it} \geq y_{it+1}$$
$$y_{it} = y_{i-1t} + z_{it}$$
$$y_{it} \geq 0$$

In combination, these constraints imply that once a tranche stops amortizing, it cannot resume retirement later, and that lower maturity tranches are retired before higher maturity tranches (the tranche indexing is done with respect to stated maturity). An additional constraint:

$$\sum_i z_{it} = 1$$

ensures that only a single tranche retires at a time.

Further constraints are straightforward. First, normal tranches may not accrue interest, or, in other words, their principal payments must be non-negative. Second, by the end of the program, all tranches must be fully retired. Third, outstanding balances and cashflows must always be non-negative:

$$p_{nt} \geq 0, \ x_{iT_{max}} = 0, \ x_{it} \geq 0, \ C_{it} \geq 0, \ R_t \geq 0$$

Finally we specify the objective as maximizing the proceeds P:

$$P = \sum_i \sum_t C_{it}(1 + r_i)^{-t} + \sum_t R_t(1 + r_r)^{-t}$$

where r_i and r_r are the yields of the tranches and the residual, respectively.

Summarizing, the full mixed-integer programming model is:

[CMO]

$$\underset{x_{i0} \in \Re^I}{\text{Maximize}} \ \sum_i \sum_t C_{it}(1 + r_i)^{-t} + \sum_t R_t(1 + r_r)^{-t}$$

subject to

$$p_{it} = x_{it-1} - x_{it}$$
$$C_{it} = c_i x_{it-1} + p_{it}$$

$$\sum_n p_{nt} = \mathscr{P}_t + c_m x_{mt-1} - C_{mt}$$

$$p_{nt} \leq \mathscr{C}_t z_{nt}$$
$$C_{mt} \leq \mathscr{C}_t z_{mt}$$
$$p_{mt} \leq \mathscr{P}_t z_{mt}$$

$$\sum_i C_{it} + R_t = \mathscr{C}_t$$

$$\underline{W}_n \leq \frac{\Sigma_t t p_{nt}}{p_{no}} \leq \overline{W}_n$$

$$y_{it} \geq y_{it+1}$$
$$y_{it} = y_{i-1t} + z_{it}$$

$$\sum_i z_{it} = 1$$

$$p_{nt} \geq 0$$
$$x_{iT_{max}} = 0$$
$$x_{it} \geq 0$$
$$C_{it} \geq 0$$
$$R_t \geq 0$$
$$y_{it} \geq 0$$

2.2.3 Variations and extensions

This model is quite complex and difficult to solve. First, it contains a very large number of binary variables: for ten tranches and 360 (monthly) time periods it has 3,600 binary variables z_{it}. Second, empirical evidence with attempts to solve this problem with existing software for mixed-integer programming indicate that the problem is very hard: the relaxed problem gives very poor estimates of the solution, and changes in the sequencing variables seem to have a very small effect on the objective function. The problem is further complicated when one wishes to include other tranche types, like PACs, TACs, multiple accrual tranches, and the like. This is in general considered one of the open problems in financial optimization.

2.3 Global return shaping

2.3.1 Background

Several new instruments in the fixed-income area are characterized by uncertainty regarding future cashflows. In most cases, the cashflow depends on the development of interest rates. Thus, floating-rate securities typically have coupon payments set with reference to the realization of an interest-rate index some period before the coupon is paid. This mechanism carries over to derived products, like interest-rate swaps, inverse floaters, etc. Cashflows on interest-rate futures and options also depend on the development of rates. Interest-rate futures are marked to the market every day with margin payments determined by the change to the futures price, i.e., the change to interest rates. Similarly, the optimality of exercising a bond option depends on interest rates. Again, this dependency carries over to instruments with embedded options like callable bonds. In some cases, interest-rate movements are not the only determinant of payments. This is the case for mortgage-backed securities (MBS), where prepayments depend on general economic developments, geographical location, and other factors that influence the decision of the mortgage owner to relocate and thus prepay. However, even for these instruments, interest-rate moves are normally thought to be the dominant influence.

While the performance of regular straight bonds is generally well captured in simple summary measures like price, duration, and so on, this is not the case for most derivative securities. Thus, small changes to interest rates may have large impacts on the duration of an MBS. Indeed, contrary to regular bonds, if rates fall enough duration may go negative, even though at initial rates both duration and convexity were positive. Therefore, it is desirable to describe the performance of such new fixed-income instruments using the whole distribution of returns.

In this section, we shall present a framework for asset/liability management using general fixed-income derivatives. We address a question which is in essence opposite to that addressed in section 3.4 in chapter 1. There, the issue was to isolate best- and worst-case performance of a given portfolio. Here the question is: "How can an investor establish a portfolio of fixed-income securities – bonds and derivatives – such that it mimics the behavior of liabilities?"

To answer this question, we first briefly describe security valuation under uncertainty, second present an optimization model for global return shaping, and finally discuss some extensions.

2.3.2 Valuation under uncertainty

Consider first a deterministic world where the forward short-term interest rate is known in advance. In this world, the price of a payment, C_T, made at time T, can be determined recursively. Let r_t denote the one-period rate at time t. The value of the payment one period before maturity is then given by:

$$P_{T-1}=C_T(1+r_{T-1})^{-1}$$

Similarly, the value two periods before maturity is:

$$P_{T-2}=P_{T-1}(1+r_{T-2})^{-1}=C_T(1+r_{T-1})^{-1}(1+r_{T-2})^{-1}$$

Continuing the argument gives a present value of:

$$P_0=C_T\prod_{t<T}(1+r_t)^{-1} \tag{18}$$

We have just found the present value of a single payment along a particular interest-rate path. We could repeat the exercise along another interest-rate path. Also, it is seen that value additivity is respected: the present value of two payments is simply the sum of their present values. In addition, the payment could be path dependent. With all these enhancements, we can write the present value distribution of a security along a set of interest-rate paths as:

$$P_{0s}=\sum_{\tau}C_{\tau s}\prod_{t<\tau}(1+r_{ts})^{-1} \tag{19}$$

s indexes interest-rate scenarios from the set of plausible scenarios S,
P_{0s} is the present value of the security – or portfolio – given scenario s,
τ denotes payment dates,
$C_{\tau s}$ denotes the payment at date τ, given scenario s, and
r_{ts} denotes the one-period forward rate at time t under scenario s.

Equation (19) gives us a method to price securities along each interest path. To find the fair price of the security, we resort to the "local expectations hypothesis," see Cox *et al.* (1981), which states that in equilibrium, each security is priced to yield a locally expected rate of return equal to the risk-free interest rate. Phrased differently, the hypothesis says that the equilibrium price equals the expected present value over all interest-rate paths. Thus, we have that:

$$P = E_s\{P_{0s}\} \tag{20}$$

where E_s denotes expectation.

Clearly, if all possible interest-rate paths and their probabilities were known, it would be no problem to find security prices. However, normally interest rates are assumed to be stochastic. The difficulty in security pricing therefore occurs in finding the appropriate stochastic process for forward rates. Here, several different models could be postulated: diffusion processes, mean-reverting processes, or jump processes. Given a particular model of stochastic forward rates, we already have a method to calibrate that process in (19) and (20): select parameters of the stochastic process in a way that generates prices for a set of reference bonds (typically treasuries) which replicate market prices.

Having calibrated the stochastic process, we can use it to simulate a set of forward-rate paths r_{ts}, apply a cashflow model to generate $C_{\tau s}$, and determine prices and present-value distributions over simulated scenarios by (20) and (19), respectively.[1]

2.3.3 The optimization model

Once the set of security values along the different forward-rate paths has been established, this may be used to shape portfolio-return distributions. Here, we shall consider a simple setup which follows the reasoning behind the immunization model in section 3.1, chapter 1. The aim is to construct an asset portfolio which mimics liabilities with respect to interest-rate sensitivity. In particular, we wish to find the minimum-cost portfolio which preserves non-negative net worth over all interest-rate scenarios. Let:

P_{is} denote the present value of security i under scenario s,
$\mathcal{P}_i$ denote the price of security i,
L_s denote the value of liabilities in scenario s.

Then the optimization problem may be written as:

[SHAPE]

$$\underset{x \in \Re^I}{\text{Minimize}} \sum_{i \in U} \mathscr{P}_i x_i$$

$$\text{subject to} \sum_{i \in U} P_{is} x_i \geq L_s, \forall s$$

Note that we do not include any non-negativity constraints on holdings in this linear programming model. The reason is that most derivatives and offbalance instruments can be sold short without any restrictions (except possibly for varying margin requirements). On the other hand, it goes without mention that the model can be enhanced with diversification constraints and tradeability constraints as discussed in chapter 1.

2.3.4 Variations and extensions

The return shaping model above finds the least-cost immunized portfolio in the Reddington sense (see Schaefer (1984)): if we can find a portfolio which preserves solvency against instantaneous random interest-rate movements, then by self-financed rebalancing over time (and neglecting transaction costs), we can keep the portfolio net worth non-negative at any point in time.

The model also covers another view of immunization by which the aim is to lock in a minimum return over a target-holding period. In fact, if we think of liabilities as a zero coupon bond maturing on the horizon date, the model can be used to efficiently replicate this bond, and hence provide a portfolio with a return at least as large as that of the zero.[2]

The model could also be enhanced to provide multiperiod immunization. Using the model outlined above, the value distribution of any security across forward-rate paths is known at any point in time. Also, cashflows can be reinvested along these paths to give horizon value distribution. By forcing non-negative net present value at a sequence of dates, the model will provide a least-cost portfolio with good static as well as dynamic tracking properties.

This aspect makes the model useful for financial engineering applications. Since the liability side could be thought of as any given target return distribution, the model can be used to create or hedge new derivative products by existing ones. Thus, in principle, the model could be used to find ways to monetize the prepayment option of mortgage-backed securities, or the delivery option of bond futures, and so on.

3 Solution methodologies

In this section we survey the methodologies commonly employed for the solution of the optimization models introduced in chapter 1 and section 2. We briefly discuss all branches of optimization that are used in the solution of any of the models given earlier: (1) *linear programming* (LP), (2) *non-linear programming* with linear (NLP-L) or non-linear (NLP-NL) constraints, (3) *network programming* (NET), (4) *mixed-integer programming* (MIP), (5) mixed-integer non-linear programming (MINLP), (6) *global optimization*, and (7) *stochastic optimization* (SO). A final subsection looks into current activities with parallel and vector supercomputers for the solution of larger and more complex optimization models.

Several of the optimization techniques used in financial modeling are presently at a stage of maturity that allows their routine use in the applications. Those include linear and network programming. Some of the methodologies – for example, non-linearly constrained optimization – still belong to the state-of-the-art. Still others are in an early stage of development – for example, stochastic optimization and mixed-integer non-linear programming. For each class of techniques we give an overview of its current state. Primary emphasis is placed either on the availability of suitable software systems – for those areas that are considered mature – or on successful algorithmic approaches – for those areas that are still under development. The solution of large-scale problems receives special attention. In addition to the examples of software systems discussed below most scientific computing libraries provide routines for numerical optimization (see for example the IMSL and NAG libraries). In most cases such libraries are well tested, efficient, and reliable but lack the capability to solve large-scale problems.

A mapping between the applications of section 3 in chapter 1, section 2 in this chapter and the appendix, and the structure of the underlying optimization models is given in table 2.1. Most of the applications fit under more than one class of models. The exact representation depends both on the details that the user is trying to capture, and the complexity of the underlying optimization model that may be impossible to solve effectively. For example, cashflow matching for portfolio dedication can be represented using network models that can be solved very quickly. Adding policy and diversification constraints, together with convexity matching, results in a linear program that can be solved quite efficiently in most cases. Imposing restrictions on the purchase of even or odd lot sizes results in a mixed-integer program that can be difficult to solve.

A general introduction to mathematical programming – with emphasis both on modeling and algorithmic developments – is Bradley *et al.* (1977).

Table 2.1. *Classification of problem areas by model type*

Model (chapter: section)	LP	NLP-L	NLP-NL	NET	MIP	MINLP	SO
Dedication (1: 3.3)	1			0	2		
Immunization (1: 3.1)	1				2		
Factor immunization (1: 3.2)	1				2		
Worst (1: 3.4)		1					
Settlement (2: 2.1)					1		
Structuring CMOs (2: 2.2)			1		1		
Term structure (2: appendix)			1				
Mean-variance (1: 3.5)		1		0		2	
Utility models (1: 3.6)		1		0		2	
Stochastic models (1: 3.7)		0		0			1

Notes: 0: simplified model.
1: complete model.
2: generalized model.

A handbook on optimization techniques has recently been compiled by Nemhauser *et al.* (1989). Aspects of numerical optimization, with discussion on large-scale problems, is Gill *et al.* (1981). Extensive coverage of the field of integer programming is Nemhauser and Wosley (1988).

3.1 Linear programming

Linear programming (LP) is by far the most widely used optimization model. Originating in Dantzig's contribution of the simplex method in 1947, Dantzig (1963), LP has undergone significant development over the last four decades. Several variants and refinements of the simplex algorithm have been implemented in efficient and stable software. As a result it is today considered feasible to solve problems with hundreds of variables and constraints on personal computers. Large mainframes and supercomputers are employed to solve problems with thousands of variables and constraints. Some studies also report on the solution of linear programs with over half a million variables, Beasley (1987), but such reports are indicative of research in supercomputing and do not yet provide a routine procedure for the solution of problems of this size.

Recently Karmarkar (1984) proposed an interior point method for the solution of linear programs. The algorithm received widespread attention from the mathematical programming community and it appears to be a strong competitor for the simplex algorithm. A commercial system that integrates Karmarkar's algorithm on a parallel computing platform is developed and distributed by AT&T and it has been reported to solve linear programs for military airlift with 14,000 constraints and 321,000 variables in approximately one hour of computer time for the United States Air Force.

The most widely used and cited simplex based software is the MPSX system from International Business Machines Corp. (IBM (1987)), APEX-IV system from Control Data Corporation (CD (1982)), Scicon from Scicon Ltd (Scicon (1986)), and XMP, Marsten (1981). In conclusion, linear programming methods and software are reliable and very efficient. Practitioners feel comfortable that solving non-trivial linear programming problems will not tax their expertise – or lack thereof – in optimization.

3.2 Non-linear programming

Non-linear programming problems (NLP) are pervasive in financial applications. The need to accommodate measures of risk – either in a mean/variance framework or in a utility maximization framework – results in non-linearities in the objective function. Worst-case analysis of options and estimating the term structure of interet rates results in non-linear relations in the constraints. In either case the non-linearities capture the essence of the models and hence linear programming approximations will give only crude estimates.

We discuss in this section two sub-classes of non-linear optimization: problems with linear (NLP-L) and with non-linear (NLP-NL) constraints under the general assumption that all functions are continuously differentiable. The solution of non-differentiable problems is at this point the privilege of experts and none of the models introduced in section 2 falls in this category. The solution of unconstrained problems is discussed in most introductory textbooks of numerical analysis and is not given further attention here. A reference on practical aspects of optimization is Gill *et al.* (1981).

3.2.1 Non-linear programs with linear constraints

NLP could be arbitrarily difficult. First, the size of the problem is not the governing factor in determining *a priori* if the model at hand will be solved easily or not – in contrast to linear programs where the size of the problem is in general a good, but not the only, indicator. Second, the definition of the problem depends on user-supplied information on the non-linear function. This information is usually given in a subroutine that evaluates the non-linear function and its first- (and perhaps second-) order derivatives for use by the optimization algorithm. Due to these peculiarities solving NLP requires a certain degree of technical sophistication.

NLP-L optimization software, based primarily on *variable reduction* methods, has been the topic of research and development over the last decade. Several systems are designed with the sole purpose of testing a new algorithm or solving a specific problem. Some systems, however, were designed with the intent of being general purpose. It is also often the case that a software system is designed to perform better for some special structure of the constraints. MINOS of Murtagh, and Saunders (1977) is designed to handle large sparse problems both for efficiency of computation and memory usage. SOL/QPSOL of Gill *et al.* (1983) is intended for small dense problems. CONOPT of Drud (1985) takes advantage of the structure of large and sparse dynamic programming constraints. A relatively simple and robust algorithm, whose development was motivated by the quadratic form of mean-variance models, is simplicial decomposition of von Hohenbalken (1977). The algorithm has been specialized for non-linear problems with network constraints and has been used to solve asset allocation models (see Mulvey and Zenios (1987)). A survey paper on the status of non-linear programming and related software is Waren and Lasdon (1979).

3.2.2 Non-linear programs with non-linear constraints

The presence of non-linear constraints complicates NLP even further. Attaining feasibility of a set of non-linear equations may be possible only at the limit (i.e., an algorithm may take an infinite number of iterations to achieve a feasible point). Maintaining feasibility, especialy with non-linear equality constraints, is also very difficult. In general practitioners may not expect to get reliable software for non-linearly constrained problems even for medium-size problems. The most successful methods at present are successive quadratic programming (SQP) and successive linear programming, that attempt to solve the non-linearly constrained problem by solving a sequence of linearly constrained problems (see Gill *et al.* (1981)). SOL/NPSOL of Gill *et al.* (1983) was designed for the solution of small to medium and dense problems based on SQP methods. Non-linearly constrained optimization problems can also be solved by MINOS and CONOPT.

3.3 Network programming

Network models (NET) are usually characterized by their very large size and the sparsity of the constraints. The special structure of the constraint matrix allows the design of graph data structure for the efficient execution of the optimization algorithms. Algorithms and software for the solution of non-linear network problems are surveyed in Dembo *et al.* (1989). A recent discussion on the implementation of linear network algorithms is Brown and McBride (1984).

The general formulation is:

$$\underset{x \in \Re^n}{\text{Minimize}} \; F(x)$$

$$\text{subject to } A \cdot x = b$$
$$l \leq x \leq u$$

$x \in \Re^n$ is the vector of flows over the edges of the network, $b \in \Re^m$ is the vector of supplies (if $b_i > 0$) and demands (if $b_i < 0$) at the nodes of the network. $l, u \in \Re^n$ denote lower and upper bounds respectively on the flow over the edges. The $m \times n$ matrix A, known as the node–arc incidence matrix for the network, is characterized by the presence of at most two non-zero entries in every column: a positive entry indicates the origin node of an edge and a negative entry indicates the destination node of the same edge. The objective function $F(x)$ could be either linear or non-linear and possibly non-separable.

An important distinction is made between *pure* and *generalized* networks. In the former class the flow over an edge remains unchanged between the origin and the destination nodes. In the latter class the flow at the origin node is scaled by a constant factor at the destination node. The scaling factor – *multiplier* in the generalized network terminology – is used to indicate either change of units (for example, conversion between different currencies) or gains/losses (for example, appreciation in value). The presence of multipliers in the generalized network framework allows their use in modeling cash balance growth due to reinvestment, capital gain, and so on. Figure 2.1 illustrates a simple network (transportation) model for the estimation of least margins in covering positions in options. The linear programming model for estimating the margins is well known, see for example Rudd and Schroeder (1982).

The development of specialized algorithms for this problem class allows the solution of very large models in a fraction of the time required by general purpose optimizers. Software implementations for linear network problems are one to two orders of magnitude faster than general purpose LP codes, with generalized network codes being slower than the pure network counterparts. Algorithms for non-linear problems demonstrate again superior performance compared to general purpose NLP software, but the differences between pure and generalized network problems become insignificant. The GENOS system of Mulvey and Zenios (1987)

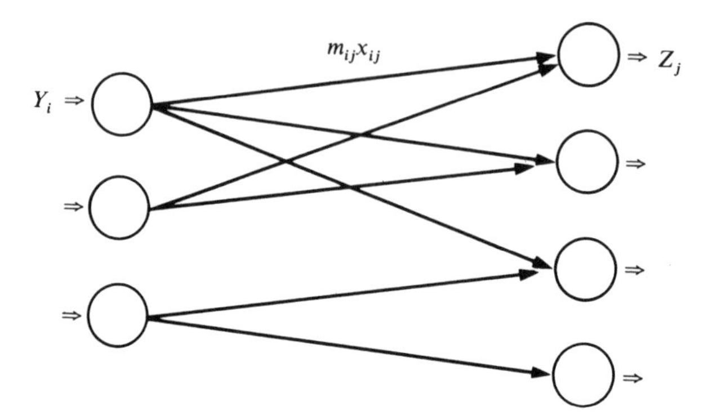

$i = \{1,2,3,\ldots,I-1\}$, index short stocks, and $j = \{1,2,3,\ldots,J-1\}$, index long stocks

m_{ij} , the margin required for a position involving one share or option of security i paired with one share or option of security j

I, J , are "null" securities to incorporate uncovered positions

x_{ij} , the number of contracts paired between securities i and j.

$$\underset{x_{ij}}{\text{Minimize}} \quad \sum_{i=1}^{I} \sum_{j=1}^{J} m_{ij} x_{ij}$$

$$\text{subject to} \quad \sum_{i=1}^{I} x_{ij} = Z_j, \forall j = 1,2,3, \ldots, J.$$

$$\sum_{j=1}^{J} x_{ij} = Y_i, \ \forall i = 1,2,3, \ldots, I.$$

$$x_{ij} \geq 0$$

Figure 2.1 Network model for the calculation of margins covering positions in options

can solve non-linear network models with or without gains and has been used in financial applications. It has also been interfaced with the algebraic modeling system GAMS, see Zenios (1990) for ease of representation of large-scale models. The linear network solver GENNET of Brown and McBride (1984) has been interfaced with Microsoft Windows for ease of representation of network models, including those arising in cashflow management.

3.4 Mixed-integer programming

Integer variables appear in several of the models of chapter 1 and section 2. Most financial instruments are purchased or sold in fixed batches with even-size lots being preferred over odd-size lots. The size of the batches may also vary between the first and all subsequent transactions. Integer variables also arise from modeling considerations, like the sequencing of tranches in structuring CMOs and the elimination of small transactions in mean-variance and utility models. The mixed-integer optimization model (MIP) takes the form:

$$\begin{aligned} &\underset{z\in\mathscr{Z}^n}{\text{Minimize}}\ c\cdot y \\ &\text{subject to } A\cdot z=b \\ &\qquad\qquad l\leq z\leq u \end{aligned}$$

where $z\in\mathscr{Z}^n$ is the vector of decision variables and $\mathscr{Z}^n$ is the set of integral n-dimensional vectors. The $m\times n$ matrix A represents general linear constraints together with the simple bounds $l,u\in\Re^n$ and the vector of coefficients $b\in\Re^m$. $c\in\Re^n$ is the vector of cost coefficients.

Quite often the set of integral vectors is replaced by the set of n-dimensional binary vectors $\mathscr{B}^n$ and the model can be rewritten in the form:

$$\begin{aligned} &\underset{y\in\mathscr{B}^n}{\text{Minimize}}\ c\cdot y \\ &\text{subject to } A\cdot y=b \\ &\qquad\qquad l\leq y\leq u \end{aligned}$$

Any optimization problem in integer variables can be transformed to a (much larger) optimization problem in binary variables using a straightforward transformation. Integer variable z is expressed as a combination of binary variables y:

$$z=2^0y_0+2^1y_1+2^2y_2+2^3y_3+\cdots$$

It is customary to refer to MIP as being the class of optimization problems with binary variables.

MIP problems are theoretically intractable and users cannot expect to solve very large problems efficiently, if at all. However, significant progress has been made in solving special cases of MIP models and general purpose software is available for solving MIP problems with up to 200 binary variables. One of the most widely used systems is ZOOM of Marsten (1987) which has been used, among other applications, in systems for financial planning: the BONDOPS system of BARRA (1984) and IFPS/OPTIMUM of EXECUCOM (1986). Other optimization systems that include integer programming capabilities include Sciconic of Scicon (1986) and LINDO of Schrage (1986).

3.5 Mixed-integer non-linear programming

Mixed-integer and non-linear programs (MINLP) appear primarily in the mean-variance and utility models. As one may expect, however, the solution of MINLP problems combines the difficulties of non-linear programs and MIP problems. Not much progress has been made in developing general purpose software systems for this area, and users that need to solve their models as MINLP could either resort to a heuristic approach or develop their own software based on published algorithms. For example, Perold (1984) developed a piecewise linearization scheme to handle the "zero-or-range" (i.e., minimum transaction) constraints in mean-variance models. The method was shown to perform very well on the test problems he solved.

An algorithm that has been used successfully in solving MINLP problems is generalized Benders' decomposition, that iterates between solving a sequence of non-linear subproblems in the continuous variables and linear master programs in the integer variables (MIP), Geoffrion (1972). Specializations of generalized Benders' algorithm for the case when the non-linear subproblems have a network structure have been developed by Nielsen and Zenios (1992) and have been used to solve mean-variance models with up to 200 binary variables. A methodology for solving MINLP problems based on the integration of existing software packages (i.e., MINOS for the NLP and ZOOM for the MIP) through a high-level modeling language (GAMS of Brook *et al.* (1988)) has been developed by Paules and Floudas (1989). The technique has been applied successfully in the solution of some small- to medium-size problems in chemical process synthesis with non-linearities both in the objective and the constraints.

Other algorithms that have appeared in the literature, and were applied successfully in solving small- to medium-size MINLP problems, are the outer approximation of Duran and Grossman (1986) and the specialization of MINOS by Mawengkang and Murtagh (1986).

3.6 Global optimization

Most of the non-linear programming techniques solve the optimization problem only for a local optimum. For minimization problems, for example, such a local optimum x^* has the property:

$$f(x^*) \leq f(x), \forall x \in X \bigcap B$$

where $f: \Re^n \rightarrow \Re$ is the continuous, real-valued function we optimize, $X \subset \Re^n$ is the feasible region of the optimization problem, and B is some neighborhood of x^*. It is possible that several local optima will exist. The problem of finding the one with the overall best value could be an important one, especially if the optimal function values differ substantially.

The codes for non-linear programming introduced in previous sections can identify only local solutions. (Precisely which local solution depends on the starting point.) No reliable codes exist to identify global solutions. Most of the financial optimization models introduced above can be proven to have only one local solution which is, hence, global. However, the model for the worst-case analysis of options (chapter 1, section 3.4) is likely to have multiple local solutions. Any efforts to solve that problem in practice need to consider methods for global optimization. A good starting point on the current status of global optimization is the chapter by Rinnooy Kan and Timmer in Nemhauser *et al.* (1989).

3.7 *Stochastic programming*

The dominating factor in all of the financial applications introduced above is uncertainty. It takes several forms: uncertainty in prices or the returns of each instrument, volatility of interest rates, uncertainty in the magnitude of liabilities, and so on. While several approaches for handling uncertainty have been introduced – variance reduction, duration matching, etc. – the ideal mathematical framework for dealing with all contingencies is that of stochastic optimization.

There are several commonly encountered variants of stochastic programming models: multistage or two-stage, with or without recourse. It is outside the scope of this section to give a detailed classification. A good introduction to the field is given by the collection of papers in Dantzig *et al.* (1981) and the chapter by Wets in Nemhauser *et al.* (1989).

We concentrate our discussion on the stochastic linear program with simple recourse (SLPSR). It is defined as:

$$\underset{x}{\text{Minimize}} \; cx + \mathrm{E}_{\xi}(min_{y^+,y^- \geqslant 0}(q^+y^+ + q^-y^-))$$

$$\text{subject to } Ax = b$$

$$Tx + Iy^+ - Iy^- = \xi$$

where $x \in \Re^n$ is the first-stage decision vector while (y^+, y^-) are the second-stage recourse decisions, and ξ is the random vector. This model is said to have simple recourse since the recourse variables are uniquely determined from the x and the realization of ξ.

A simplifying approach to solving (SLPSR) is to substitute the random variable ξ by its expected value $\bar{\xi}$ and then solve the resulting linear program. Bounds on the possible error from such simplifications indicate that this approach may produce solutions far from optimum. Hence, researchers in this area aim at the development of algorithms that will handle explicitly the inner optimization problem that involves the expectation of the recourse action.

Stochastic programming goes back to Dantzig (1955) and significant effort has been devoted to the design and implementation of algorithms for several variants of stochastic optimization. A major project effort under the auspices of the International Institute of Applied Systems Analysis (IIASA) resulted in a collection of programs and test problems that are in the public domain. While most of the programs were not developed to be of production level quality they are in general well documented and tested. Researchers now believe that parallel processing provides the mechanism for the solution of stochastic programs using decomposition methods such as those proposed in Rockafellar and Wets (1991). Mulvey and Vladimirou (1991) developed a software implementation of the Rockafellar and Wets algorithm for the special case of network recourse constraints (i.e., the recourse constraints are networks) and used it for the solution of multiperiod investment planning models. Nielsen and Zenios (forthcoming) developed parallel algorithms for stochastic network optimization problems, and used them to solve models arising from financial applications with several thousand scenarios. At the same time Karmarkar's projective algorithm has been specialized for the solution of block-angular problems that arise in the solution of SLPSR, see Birge and Qi (1988). It is in general considered feasible to solve SLPSR problems with up to 100 random variables and a few realizations (5–10) per variable. The computing resources required, even for such small models, could be substantial.

3.8 Parallel optimization and supercomputing

Parallel computer architectures have appeared as a competitive computing paradigm since the early eighties. It is believed that increased computing performance can be delivered only by coupling together multiple processors and coordinating them for the solution of a single problem. There are several competing philosophies on how a parallel computer should be organized: the major distinctions are between shared memory or distributed memory multiprocessors and massively parallel systems. In the first category we find systems with a small to medium number (8–1024) of relatively powerful processors. These processors may execute their own instruction stream on different segments of a problem, with periodic exchange of information. Synchronization is achieved either through access to a shared memory or by passing messages via a communication network. In the second category we find systems with thousands of simple processors, each with its own local memory. Such architectures tend to view problems as collections of large amounts of data on which relatively simple operations have to be performed.

Parallel computing has been receiving widespread attention among investment and brokerage houses. The motivations for the pursuit of bigger and faster machines in the general domain of financial operations is beyond the scope of this chapter. Large-scale optimization, however, is one of the motivating factors for the use of parallel and supercomputers. Several of the models can be large and intractable for von Neumann machines. At the same time financial planning takes place in a very volatile environment. The input data for a specific model (i.e., prices, interest rates, etc.) are constantly changing and a model that takes several hours to run will produce results that are suboptimal if not totally obsolete. At the same time analyzing uncertainty, via scenario planning or stochastic optimization, requires the repeated solution of a model under several contingencies. Parallel and vector supercomputers could provide fast and accurate solutions for several of the models.

The influence of optimization models in the introduction of parallel computing in financial operations should not be overemphasized. Many other facets of financial operations are well suited for multiprocessing. Data processing and financial simulations are two applications well suited for parallel computing – they can indeed be called embarrassingly parallel! A data retrieval system by Dow Jones relies on a massively parallel Connection Machine CM-2 with 65536 processors for handling large volumes of financial data. A study on the use of parallel computers for Monte Carlo simulation of mortgage-backed securities is documented in chapter 14. An alternative method for simulation of mortgage-backed securities on a massively parallel architecture is reported in Hutchinson and Zenios (1991).

Several of the software systems we referred to earlier have been streamlined for the architecture of vector supercomputers. Several new algorithms have been developed specifically for parallel computers. An indication of the size of the problems that can be solved with supercomputers is offered in the empirical study of Beasley (1987). A survey on the status of parallel optimization with an annotated bibliography is Zenios (1989) and the volume by Meyer and Zenios (1988) contains several papers on parallel optimization algorithms.

4 Conclusions

Large-scale optimization models have a critical role to play in several problems appearing in financial modeling. It is important to recognize their limitations: their value depends on the ability of the manager to recognize the financial risks of his operating environment, and model all factors that are considered relevant. Once that is achieved the power of optimization models is indispensable. With the increased complexity and competitiveness of the financial markets we expect to see the use of optimization models increase. At the same time operations research analysts face, in financial modeling, an excellent domain for the application of their analytic skills. Several interesting algorithmic and modeling issues are raised in financial optimization models, that present potentially fertile directions for research.

Appendix: Estimation of the term structure of interest rates

A1 Background

In this appendix we present a model for estimating the term structure of interest rates. Models on the evolution of the term structure of interest rates provide key input data to both the factor immunization and the return shaping models. In addition, the term structure is an important tool in bond pricing: if the equilibrium yields of single payment instruments maturing on the bond cashflow dates are known, the equilibrium bond price can be found as the value of the cashflow replicating portfolio of these zeros (see equation (13) in chapter 1, section 3.2.1).

The problem of estimating the term structure is that, although bond yields are observable, the yields of implied zeros may not be. Even in markets where zero coupon bonds (strips) are traded, the yields of these bonds may be heavily influenced by tax laws, liquidity premia, etc. Therefore, it is normally attempted to infer implied zero yields from actively traded bond prices (i.e., on-the-run treasuries), rather than using the strips curve. This is done by estimating the coefficients of a function relating zero coupon yields to term-to-maturity in a way which minimizes a measure of the pricing error. The error is defined as the difference between actual market prices and prices implied by the model.

Since the relationship between yields and prices is non-linear, the estimation procedure is also non-linear. Most statistics and econometrics packages contain routines for performing non-linear regression, but even so, there is scope for optimization in estimating the term structure. The primary reason is that certain inequality constraints typically appear in the estimation. For example:

1 the term structure should contain only positive zero coupon yields,
2 the forward rates implied by the estimated term structure should all be non-negative,

3 some model parameters may only make sense within certain bounds,
4 some shapes of the term structure may be *a priori* implausible, so the change of spot rates or forward rates from one point in time to another may be constrained to lie within certain limits.

In general, standard non-linear regression routines do not allow such constrained estimation, so casting the estimation in the form of a constrained non-linear programming problem provides added flexibility.

A2 The optimization model

Different models for the term structure are available in the literature. Typical approaches include representing zero yields or the discount function as a power series of terms to maturity, Chambers *et al.* (1984), using cubic splines, Shea (1984), or exponential splines, Vasicek and Fong (1982), to fit spot yields or the discount function, or models like that of Cox *et al.* (1985) that are based on theoretically postulated models on the evolution of the term structure. Here, we will show the estimation model in a rather generic form.

Let:
P_i be the market price of bond i,
P_i^* be the fitted price of bond i,
y_t be the fitted yield (i.e., spot rate) of a zero maturing at time t,
d_t be the price of a pure discount bond maturing at t,
f_t be the implied forward rate between time t and $t+1$, and let
α denote a vector of model parameters.

We also use underscores and overbars to indicate lower and upper bounds on the estimated variables α, d_t, y_t, f_t.

By definition, the discount function and spot yields are related by:

$$d_t = (1+y_t)^{-1} \tag{21}$$

and implied forward rates and yields are related by:

$$f_t = \frac{(1+y_{t+1})^{t+1}}{(1+y_t)^t} - 1 \tag{22}$$

Given a set of spot yields, the equilibrium price of a bond is given by:

$$P_i^* = \sum_{t \in T} C_{it} d_t \tag{23}$$

We now only need to specify the term structure model. This could take any of three forms, depending on whether yields, discount rates or forward rates are determined. We use the discount function model:

$$d_t = g(t,\alpha) \tag{24}$$

where $g(.,.)$ is the function that is assumed to describe the evolution of interest dates (e.g., exponential spline). This gives us the full estimation model. For simplicity, we assume that the objective is to minimize the sum of squares of pricing errors:

[TERMS]

$$\underset{\alpha}{\text{Minimize}} \sum_i (P_i - P_i^*)^2$$

$$\text{subject to } d_t = g(t,\alpha)$$

$$P_i^* = \sum_t C_{it} d_t$$

$$y_t = d_t^{-1/t} - 1$$

$$f_t = \frac{(1+y_{t+1})^{t+1}}{(1+y_t)^t} - 1$$

$$\underline{\alpha} \le \alpha \le \bar{\alpha}$$

$$\underline{d}_t \le d_t \le \bar{d}_t$$

$$\underline{y}_t \le y_t \le \bar{y}_t$$

$$\underline{f}_t \le f_t \le \bar{f}_t$$

A.3 Extensions: Monte Carlo simulation

Quite often the estimated term structure given above is perturbed through a simulation model to generate alternative plausible interest-rate scenarios. These scenarios are then used to price interest-rate contingent claims or complex instruments. See, for example, chapters 5 and 14.

Several techniques have appeared in the literature for developing plausible scenarios of interest rates that are consistent with the current observations of the term structure. These include the one-factor model of Black *et al.* (1990), the binomial lattice of Ho and Lee (1986) or the martingale model of Heath *et al.* (1988). We describe here an approach based on Monte Carlo simulation. While this is not necessarily the best approach, it is one widely used in practical applications. One severe criticism of Monte Carlo simulations is that the arbitrage-free property is not inherent in the model but instead a calibration factor (drift) is used to enforce it. The more recent models by Black *et al.* and Heath *et al.* cited above preserve the arbitrage-free property inherently, and in that sense they are more accurate.

To generate short-term interest-rate scenarios we use a lognormal distribution. The rate f_t used is generated as follows:

$$\log \frac{f_t}{f_{t-1}} = z*\sigma + R(f_{t-1}) + \mu$$

where:

z is a normally distributed random variable, $z \sim N(0,1)$,

σ is the volatility of interest rates, assumed to be constant across the time horizon.

$R(f_{t-1})$ is a mean-reversion term that forces the simulated rates f_t to stay within historically acceptable limits (e.g., in the range $l=6\%$, $u=16\%$). It is defined by:

$$R(f) = \begin{cases} 0 & \textit{if } l \le f \le u \\ -\gamma_o(f-u)^2 & \textit{if } f > u \\ \gamma_1[(l-f)/f]^2 & \textit{if } f < l \end{cases}$$

where γ_0 and γ_1 are constants estimated based on empirical observations.

μ is a drift factor estimated so that the model rates are consistent with the term structure. It is estimated by requiring that the present value of on-the-run treasuries, computed using the model rates, is in agreement with the current market prices.

ACKNOWLEDGMENTS

The research of S. A. Zenios was partially supported by NSF grants ECS-8718971 and CCR-8811135, AFOSR grant 89-0145 and grants from Digital Equipment Corporation and AT & T.

NOTES

1 The model is normally enhanced by adding a spread over forward rates, to capture sector and liquidity spreads and relative value, see Brazil (1988), Hayre and Lauterbach (1988), Sykes (1988) and chapter 14. These models go under the general terminology *Option Adjusted Spread* models, and they are currently the predominant models in valuating interest-rate contingent claims such as mortgage-backed securities.

2 The model not only replicates but also strictly dominates the target liability distribution, at least in the discrete world. The model could easily be relaxed to handle weaker stochastic dominance schemes, see Ingersoll (1987).

REFERENCES

W. W. Bartlett (1989), *Mortgage-Backed Securities: products, analysis and trading*. Institute of Finance, New York.

J. E. Beasley (1987), "Linear programming on CRAY supercomputers," Manuscript, Department of Management Science, Imperial College, London.

J. R. Birge and L. Oi (1988), "Computing block-angular karmarkar projections with applications to stochastic programming," *Management Science*, 34: 1472–9.

F. Black, E. Derman, and W. Toy (1990), "A one-factor model of interest rates and its application to Treasury bond options," Financial Analysts Journal, Jan.–Feb., 33–39.

S. P. Bradley, A. C. Hax, and T. L. Magnanti (1977), *Applied Mathematical Programming*. Addison-Wesley, Massachusetts.

A. J. Brazil (1988), "Citicorp's mortgage valuation model: option-adjusted spreads and option-based duration," *Journal of Real Estate Finance and Economics*, 1: 151–62.

A. Brooke, D. Kendrick, and A. Meeraus (1988), *GAMS: A User's Guide*. The Scientific Press, California.

G.G. Brown and R. D. McBride (1984), "Solving generalized networks," *Management Science*, 30(12), December.

D. R. Chambers, W. T. Carleton, and D. W. Waldman (1984), "A new approach to estimation of the term structure of interest rates," *Journal of Financial and Quantitative Analysis*, 3.

International Business Machines Corp. (1987), "Mathematical programming system extended and generalized under bounding," IBM Manual SH20-0968-1.

Control Data Corp. (1982), "APEX-IV Reference Manual," CDC Manual 84002550.

J. Cox, J. Ingersoll, and S. Ross (1981), "A re-examination of traditional hypotheses about the term structure of interest rates," *Journal of Finance*, 36.

(1985), "A theory of the term structure of interest rates," *Econometrica*, 53: 385–407.

H. Dahl, A. Meeraus, and S. A. Zenios, (1989). "Some financial optimization models: III. An algebraic modeling system library", Report 89-12-03, Decision Sciences Department, The Wharton School, University of Pennsylvania, Philadelphia.

G. B. Dantzig (1955), "Linear programming under uncertainty," *Management Science*, 1: 197–206.

(1963), *Linear Programming and Extensions*. Princeton University Press.

G. B. Dantzig, M. A. H. Dempster, and M. J. Kallio (eds) (1981), "Large-scale linear programming," IIASA Collaborative Proceedings Series, CP-81-51, Laxemburg, Austria.

R. S. Dembo, J. M. Mulvey, and S. A. Zenios (1989), "Large scale nonlinear network models and their application," *Operations Research*, 37: 353–72.

A. Drud (1985), "CONOPT: a GRG code for large sparse dynamic nonlinear optimization problems," *Mathematical Programming*, 31: 153–91.

A. Geoffrion (1972), "Generalized benders decomposition," *Journal of Oplimization Theory and Applications*, 10: 239–60.

P. E. Gill, W. Murray, M. A. Saunders, and M. H. Wright (1983), "User's guide for SOL/NPSOL: A FORTRAN package for nonlinear programming," Report SOL 83-12, Department of Operations Research, Stanford University.

P. E. Gill, W. Murray, and M. H. Wright (1981), *Practical Optimization*. Academic Press, London.

L. Hayre and K. Lauterbach (1988), "Stochastic valuation of debt securities," Working Paper, Financial Strategies Group, Prudential-Bache Capital Funding.

D. Heath, R. Jarrow, and A. Morton (1988), "Bond pricing and the term structure of interest rates," Technical Report, Cornell University.

T. S. Y. Ho and S-B. Lee (1986), "Term structure movements and pricing interest rate contingent claims," *The Journal of Finance*, XLI(5):1011–29.

H. von Hohenbalken, H. (1977), "Simplicial decomposition in non-linear programming algorithms," *Mathematical Programming*, 15: 49–68.

J. M. Hutchinson and S. A. Zenios (1991), "Financial simulations on a massively parallel Connection Machine," *International Journal of Supercomputer Applications*, 5(2): 27–45.

J. E. Ingersoll, Jr. (1987), *Theory of Financial Decision Making. Studies in Financial Economics*. Rowman & Littlefield.

N. Karmarkar (1984), "A new polynomial algorithm for linear programming," *Combinatorica*, 4: 373–95.

L. Lasdon, F. Palacios-Gomez, and M. Engquist (1982), "Nonlinear optimization by successive linear programming," *Management Science*, 28(10): 1106–20.

R. E. Marsten (1981), "The design of the XMP linear programming library," *ACM Transactions on Mathematical Software*, 7(4): 481–97.

H. Mawengkang and B. A. Murtagh (1986), "Solving nonlinear integer programs with large-scale optimization software," in C. L. Monma (ed.), *Annals of Operations Research, Volume 5*, Scientific Publishing Company, Switzerland. Special volume on *Algorithms and Software for Optimization*.

R. R. Meyer and S. A. Zenios (eds) (1988), *Parallel Optimization on Novel Computer Architectures*, vol. 14 of *Annals of Operations Research*, A. C. Baltzer Scientific Publishing Co., Switzerland.

J. H. Mulvey and H. Vlachimiron (1991), "Solving multistage stochastic networks: an application of scenario aggregation," *Networks*, 21: 619–43.

J. M. Mulvey and S. A. Zenios (1987), "GENOS 1.0: A generalized network optimization system. User's Guide," Report 87-12-03, Decision Sciences Department, The Wharton School, University of Pennsylvania, Philadelphia, PA 19104.

B. A. Murtagh and M. A. Saunders (1977), "MINOS user's guide," Report SOL 77-9, Department of Operations Research, Stanford University, California.

G. L. Nemhauser, A. H. G. Rinnooy Kan, and M. J. Todd (eds) (1989), *Optimization*. Volume 1 of *Handbooks in Operations Research and Management Science*, North-Holland, Amsterdam.

G. L. Nemhauser and L. A. Wolsey (1988), *Integer and Combinatorial Optimization*. John Wiley & Sons, New York.

S. S. Nielson and S. A. Zenios (1992), "Mixed-integer non-linear programming on generalized networks," in C. A. Floudas and P. H. Pardalos (eds.), *Recent Advances in Global Optimization*, Princeton University Press, pp. 513–42, (forthcoming) "Massively parallel algorithms for nonlinear stochastic network problems," Operations Research.

G. E. Paules and C. A. Floudas (1989), "APROS: algorithmic development methodology for discrete continuous optimization," *Operations Research*, 37: 902–15.

A. F. Perold (1984), "Large scale portfolio optimization," *Management Science*, 30: 1143–60.

B. Roberts, S. K. Wolfe, and N. Witt (1988), "Advances and innovations in the CMO market," Technical Report, Bear Stearns, Mortgage Related Products Research, September.

R. T. Rockafellar and R. J.-B. Wets (1991), "Scenarios and policy aggregation in optimization under uncertainty," *Mathematics of Operations Research*, 16: 119–47.

A. Rudd and M. Schroeder (1982), "The calculation of minimum margin," *Management Science*, 28: 1368–79, December.

S. M. Schaefer (1984), "Immunisation and duration: a review of theory, performance and applications," *Midland Corporate Finance Journal*, Fall.

G. S. Shea (1984), "Pitfalls in smoothing interest rate term structure data: equilibrium models and spline approximations," *Journal of Financial and Quantitative Analysis*, 3, September.

D. Sykes (1989), "Valuing debt securities under uncertainty: the option adjusted spread model," Mortgage Products Special Report, Bear Stearns, April.

O. A. Vasicek and H. Gifford Fong (1982), "Term structure modeling using exponential splines," *Journal of Finance*, 2, May.

J. Walmsley (1988), *The New Financial Instruments*. John Wiley & Sons.

A. D. Waren and L. S. Lasdon (1979), "The status of nonlinear programming software," *Operations Research*, 27(3): 431–56, May–June.

S. A. Zenios (1989), "Parallel numerical optimization: current status and an annotated bibliography," *ORSA Journal on Computing*, 1(1): 20–43.

(1990), "Incorporating network optimization capabilities into a high-level programming language," *ACM Transactions on Mathematical Software*, 16(2): 113–142.

3 New "financial market equilibrium" results: implications for practical financial optimization

HARRY M. MARKOWITZ

This chapter is divided into two parts. The first reviews some financial market equilibrium results presented in Markowitz (1987). The second considers the implications of these results for practical financial optimization.

1 Market equilibrium with restricted borrowing

The assumption that portfolios are chosen subject to the sole constraint $\Sigma X_i = 1$ is highly unrealistic. It would allow you to place \$1,000 with your broker, short a million dollars worth of stock A, and use the proceeds plus your \$1,000 to buy \$1,001,000 worth of stock B. If we make the other usual CAPM assumptions but assume that the constraint set is a bounded polyhedron then the following conclusions typically fail to hold: (A) the market portfolio is an efficient portfolio, and (B) expected returns are linearly related to betas. Conclusions (A) and (B) fail to hold even if *some* investors are allowed $\Sigma X_i = 1$ as their sole constraint while others are subject to bounded constraint sets. These results are presented in Markowitz (1987, chapters 11 and 12 to which chapter numbers refer in the following) The failure of (A) and (B) may be illustrated in terms of the standard three-security mean-variance analysis with constraints $\sum_{i=1}^{3} X_i = 1$, $X_i \geq 0$, $i = 1,2,3$.

Markowitz (1952) analyzed the standard three-security model graphically, plotting X_1 as the abscissa, X_2 as the ordinate, and leaving $X_3 = 1 - X_1 - X_2$ implicit. As shown in chapter 11, a frequently more revealing procedure is to transform this graph by:

$$\begin{bmatrix} Z_1 \\ Z_2 \end{bmatrix} = \begin{bmatrix} \alpha_1 \\ \alpha_2 \end{bmatrix} + \begin{bmatrix} k_{11} k_{12} \\ k_{21} k_{22} \end{bmatrix} \begin{bmatrix} X_1 \\ X_2 \end{bmatrix} = \alpha + KX \tag{1}$$

where α and K are chosen so that V is minimized at $Z=0$, and is of the form: V_{MIN} plus a sum of squares. Ignoring the trivial case in which $V \equiv 0$, but permitting a singular covariance matrix, the transformation results in either:

$$V = V_{MIN} + Z_1^2 + Z_2^2 \tag{2}$$

or, e.g.:

$$V = V_{MIN} + Z_1^2 \tag{3}$$

Here we will consider the case in which (2) results. In this case one can rotate the Z-coordinate system:

$$Y = HZ \tag{4}$$

with H orthogonal, so that portfolio expected return is:

$$E = a_o + a_1 Y_1 \tag{5}$$

$a_1 > 0$, as well as:

$$V = V_{MIN} + Y_1^2 + Y_2^2 \tag{6}$$

This is always possible when (2) applies and the security expected returns are not all equal. A coordinate system in which (5) and (6) hold is referred to as a *canonical form* of the portfolio selection model.

We do not alter the set of EV efficient portfolios if we seek E^*V rather than EV efficiency where:

$$E^* = \frac{E - a_0}{a_1} \tag{7}$$

Then:

$$E^* = Y_1 \tag{8}$$

Figure 3.1 presents standard three-security portfolio analyzes in canonical form, assuming that (2) applies. In the 1952 diagram the set of feasible portfolios is the area on and in the triangle:

$$T = \{X \in R^2 : X_1 \geq 0, X_2 \geq 0, X_1 + X_2 \leq 1\} \tag{9}$$

Transformations (1) and (4) send T into a triangle $\tilde{T} = \overline{\text{abc}}$. Any triangle $\tilde{T}$ may be such a transform.

The means, variances, and covariances which give rise to a particular $\tilde{T}$, such as $\overline{\text{abc}}$ in figure 3.1a, may be determined as follows. Arbitrarily choose a_0, $a_1 > 0$, $V_{MIN} \geq 0$. From the coordinates (Y_1^a, Y_2^a), etc., of the vertices a, b, c compute:

$$\mu_a = a_0 + a_1 Y_1^a$$
$$V_a = V_{MIN} + (Y_1^a)^2 + (Y_2^a)^2$$

and similarly for b and c. Compute:

$$\text{cov}(r_a, r_b) = V_{MIN} + Y_1^a Y_1^b + Y_2^a Y_2^b$$

and similarly for σ_{bc} and σ_{ac}.

Since the V of a point is V_{MIN} plus the square of the distance to the origin, among portfolios in a set S the one closest to the origin has minimum V. In particular, the Pythagorean theorem implies that, among feasible points with given E – i.e., on a given vertical line – the one closest to the Y_1-axis has minimum V.

In figure 3.1a portfolio d, where the perpendicular from the origin meets ab, has minimum feasible variance. The set of efficient portfolios consists of line segments $\overline{de}$, $\overline{ef}$, $\overline{fc}$. Suppose some investors buy portfolio g and the rest portfolio h. Then the market portfolio will be on the straight line connecting these, e.g., at i. This point is not an efficient portfolio.

In figure 3.1a the efficient set lies on segments of three critical lines. In practice, mean-variance efficient sets computed for large portfolios may contain segments from scores or hundreds of critical lines. This is not only true for the standard model but also for more complex models, e.g., models which allow short positions but require collateral.

Suppose we make the usual CAPM assumptions that all investors have the same beliefs and are subject to the same constraints, but assume a standard constraint set, or other constraints involving inequalities or non-negative variables. Typically the efficient set will consist of many segments and, in such cases, typically the market will not be an efficient portfolio.

I say "typically" since "accidents" like that in figure 3.1c are possible. As in figure 3.1a, $\tilde{T}$ = abc is the constraint set and d has smallest feasible V. In figure 3.1c, the efficient segments are $\overline{de}$, $\overline{ef}$, $\overline{fc}$. Suppose some investors choose portfolio g, the rest choose h. The market is on the line between g and h, possibly at i – which is efficient. Thus it is possible for the market to be an efficient portfolio even though investors select portfolios from more than one segment. But such a case is not typical; e.g., a slight shift in investors between g and h will move the market off the efficient set. It is possible for the market to have almost *maximum* V for given E, such as portfolio g in figure 3.1b.

Chapter 12 shows that the μ_i and σ_{ij} which give rise to figures 3.1a–c are consistent with market equilibrium. The following is an alternative, shorter demonstration. Suppose an economy of immortal investors has three securities, determines prices once a year, and for the last thousand years has

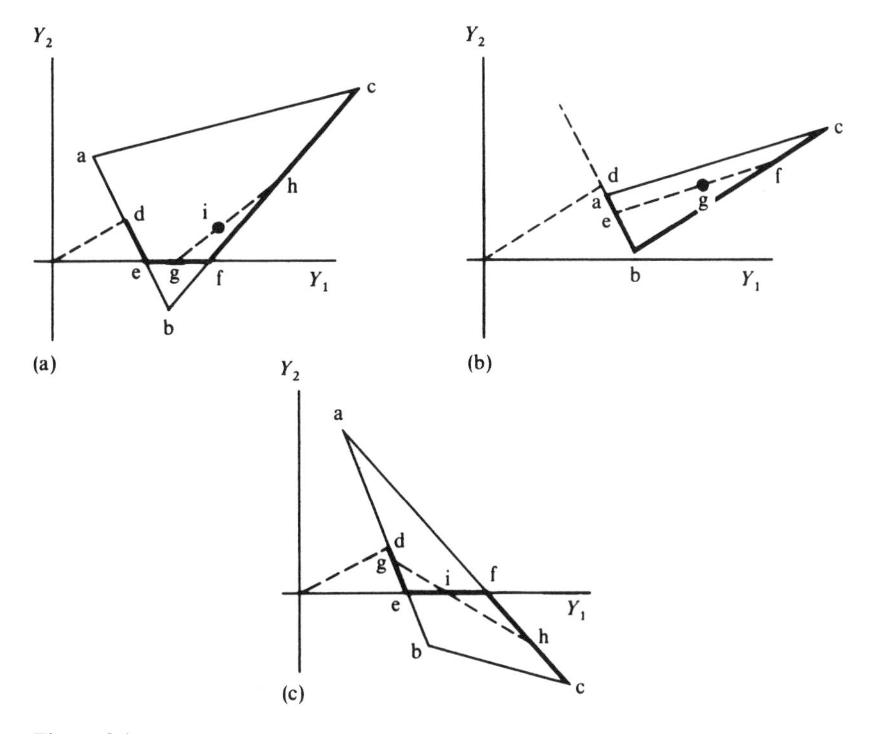

Figure 3.1

arrived at the same prices for each security (say \$1 per share for each of the three issues). The same prices are sure to happen next year, therefore the only uncertainty is in dividends. We assume means, variances, and covariances among dividends as shown in figure 3.1a (or 3.1b or 3.1c). Finally, we assume that the number of shares demanded with these parameters equal the fixed number of shares in existence. This provides a scenario by which any case in figure 3.1 is a long-run market equilibrium.

The reason why the Tobin (1958), Sharpe (1964), and Lintner (1965) model and the model with $\Sigma X_i = 1$ as its only constraint (analyzed by Roy (1952), Sharpe (1970), Merton (1972), and Black (1972)) imply that the market is an efficient portfolio is that both models imply that the set of efficient portfolios consists of at most one line segment. As long as the efficient set consists of one line segment then the market portfolio, which is a convex combination of points on this segment, must itself be on the segment, and therefore efficient. But real world efficient sets consist of many line segments.

Another conclusion of the TSL and RSMB standard CAPMs is that:

$$\mu_i = c_0 + c_1 \beta_i \tag{10}$$

where μ_i is expected return of the ith security and β_i is its regression against the return on the market portfolio. Chapter 12 shows that (10) holds if and only if the market portfolio lies on the Y_1-axis.[1] As our figures show, this is atypical if the efficient set consists of more than one segment.

2 Implications for practical portfolio optimization

We must distinguish between the assumptions of normative mean-variance portfolio theory and those of classical CAPMs.

Textbooks frequently state that normally distributed returns or quadratic utility functions are the underlying assumptions for the use of mean-variance analysis in practice. In contrast, Levy and Markowitz (1979) and others – see Markowitz (1987, chapter 3) for a summary of this literature – eschew these assumptions. Rather, they show that for a wide range of utility functions and historical distributions, if you know the mean and variance of the distribution, you almost know the distributions' expected utility. Thus, for investors with such utility functions, the selection of an appropriate portfolio from the mean-variance efficient frontier is a practical way of approximately maximizing expected utility.

Classical, positive CAPMs assume that all investors seek mean-variance efficiency. But one can use mean-variance analysis to select a portfolio in fact even if no one else in the world does so. Classical CAPMs – at least those which give sharp conclusions – assume that all investors have the same beliefs. But one can use mean-variance analysis to think through the implications for the portfolio as a whole of beliefs concerning individual securities even if no one else shares these beliefs.

Classical CAPMs, and the Arbitrage Pricing Theory of Ross (1976), make highly simplified assumptions concerning the constraint sets to which investors are subject. The Sharpe, Lintner CAPM assumes that investors can borrow all they want at the risk-free rate. The Roy, Sharpe, Merton, Black CAPM, and Ross's APT, assume that the investor can short and use the proceeds to purchase other securities. In contrast, typically when consultants or quantitative analysts perform mean-variance analyses to be used to allocate funds in fact, they are careful to model constraints reflecting law and fund policy. In particular, commonly used algorithms allow any number of linear equality and/or (weak) inequality constraints on the choice of portfolio in variables which may or may not be subject to upper- and lower-bound constraints.

Classical CAPM assumptions imply that the market portfolio is mean-variance efficient. In such a world you can obtain mean-variance efficiency without estimating means, variances, and covariances by investing in a capitalization-weighted index fund. This was part of the philosophical (and marketing) principles which underlay the development of today's $250 billion world-wide index fund industry.

We saw above that if we simply drop the unrealistic assumption concerning the constraint set then the market is no longer an efficient portfolio. In fact, figure 3.1b shows that the market can have almost maximum rather than minimum variance for given expected return.

This analysis considers what might be. Haugen and Baker (1990) consider what is. They show that one could in fact choose an index – other than a value weighted one – which, on an *ex post* (out of sample) basis, substantially dominates both the market and a large sample of random portfolios. They propose that the world would be better off if it tried to replicate, or use as a benchmark, their efficient index rather than an inefficient, capitalization-weighted market index. The argument in the preceding section supports this view in that it shows that an inefficient market portfolio is not an aberration which would disappear if investors were rational or had the same beliefs.

The rise of capitalization-weighted index funds is, in part at least, one of the practical consequences of classical CAPMs. Another is risk-adjustment procedures often used in the search for classes of stocks with superior performance. According to equation (7), in a classical CAPM world the expected return on any stock or portfolio is linearly related to its beta. The standard procedure, then, for deciding whether stocks with certain characteristics provide superior returns is to "risk adjust" their average returns, that is, to alter their average return by a multiple of their beta. This is the procedure used, for example, in determining that small capitalization stocks have "abnormally" high average return. Their returns are abnormal as compared to what they would be in a CAPM world.

But we saw above that if we drop the unrealistic assumptions about investor constraints, expected returns are typically not linearly related to betas. In the case of the "small cap" anomaly in particular, Levy (1990) shows that the small cap phenomenon is a prediction of, rather than contradiction to, the Levy (1978) "generalized CAPM"; and the Markowitz (1990) discussion of the Levy paper shows that the small cap effect does not contradict the CAPM in Markowitz (1987), described in the preceding section of this chapter.

Markowitz (1990) shows that the appropriate risk-adjustment formula for the CAPM described in the first section of this chapter differs from that of equation (7). I do not, however, recommend that the risk-adjustment formula of Markowitz (1990) be universally used in the place of (10), since the former too is based on simplifying assumptions and poses measurement problems, as described in the Tobin (1990) discussion of my chapter. The point, rather, is that Markowitz (1990) provides further evidence that (10) is not a correct calculation when we add even the small bit of realism described in the first section.

The normative use of mean-variance analysis requires estimates of expected returns, as well as variance and covariances, as inputs. Thus it is a potential consumer of studies of the expected returns of classes of securities, or securities under certain circumstances. But the information that some class of securities has expected returns which are significantly greater or significantly less than that which it would have in a classical CAPM world, is exactly as useful as the information that under certain circumstances average returns are greater or less than that which would be predicted by a linear relationship with sunspots.

The central question is what level of expected return should be predicted for given securities under given circumstances. The betas of stocks are a conceivable determinant of their expected returns. But once we reject classical CAPMs, and realize that more realistic CAPMs offer no fixed relationship between expected return and beta, then beta plays no more special a role, *a priori*, in predicting expected return than does, say, total variance or P/E or P/B, etc.

If nothing else, the results presented in the first section of this chapter raise questions concerning passive management and research toward active management. Concerning passive management they reinforce the questions raised by Haugen and Baker: since the market is not an efficient portfolio, why do we try to replicate it or use it as a benchmark? Concerning research to support active management, they raise questions about every study which seeks to explain "risk adjusted" returns, or concludes that a contribution has been made when it is shown that returns to a given class of stocks or under a given set of circumstances contradict classical CAPM.

NOTES

1 The "only if" part assumes that market portfolio has positive variance, else β_i is undefined.

REFERENCES

F. Black (1972), "Capital market equilibrium with restricted borrowing," *Journal of Business*, July.

R. A. Haugen and N.L. Baker (1990), "The efficient market inefficiency of value-weighted stock portfolios," National Investment Services of America, Inc. (NISA). Dedicated Stock Management Research Series, No. 9.

H. Levy (1978), "Equilibrium in an imperfect market: a constraint on the number of securities in the portfolio," *American Economic Review*, 68(4): 643–58.

(1990), "Small firm effect: are there abnormal returns in the market," *Journal of Accounting, Auditing and Finance*, January, 5(1), Special Issue devoted to Proceedings of the Productivity Conference, 15–17 December 1988, University of Florida.

H. Levy, and H. M. Markowitz (1979), "Approximating expected utility by a function of mean and variance," *American Economic Review*, June.

J. Lintner (1965), "The valuation of risk assets and the selection of risky investments in stock portfolios and capital budgets," *Review of Economics and Statistics*, February: 13–37.

H. M. Markowitz (1952), "Portfolio selection," *The Journal of Finance*, 7(1): 77–91.

(1987), *Mean-Variance Analysis in Portfolio Choice and Capital Markets*. Basil Blackwell, New York.

(1990a), "Risk adjustment," *Journal of Accounting, Auditing and Finance*, January, 5(1), Special Issue devoted to Proceedings of the Productivity Conference, 15–17 December 1988, University of Florida.

(1990b), "Comment on Haim Levy's small firm effect: are there abnormal returns in the market," *Journal of Accounting, Auditing and Finance*, January, 5(1), Special Issue devoted to Proceedings of the Productivity Conference, 15–17 December 1988, University of Florida.

R. C. Merton (1972), "An analytic derivation of the efficient portfolio frontier," *Journal of Financial and Quantitative Analysis*, September: 1851–72.

S. A. Ross, (1976), "The arbitrage theory of capital asset pricing," *Journal of Economic Theory*, 13: 341–60.

A. D. Roy, (1952), "Safety first and the holding of assets," *Econometrica*, 20: 431–49.

W. F. Sharpe (1964), "Capital asset prices: a theory of market equilibrium under conditions of risk," *The Journal of Finance*, 19(3): September.

(1970), *Portfolio theory and capital markets*, McGraw-Hill, New York.

J. Tobin (1958), "Liquidity preference as behavior towards risk," *Review of Economic Studies*, February: 65–86.

(1990), "Comment on Harry Markowitz's risk adjustment," *Journal of Accounting, Auditing and Finance*, January, 5(1), Special Issue devoted to Proceedings of the Productivity Conference, 15–17 December 1988, University of Florida.

4 Empirical tests of biases in equity portfolio optimization

What do *ex-ante* mean-variance optimal portfolios look like *ex-post*?

PETER MULLER

1 Introduction

Quadratic optimization has been successfully used as an equity portfolio management tool for many years. The typical problem involves maximizing a linear term (expected return) while minimizing a quadratic term (variance of *active return*, the difference between a portfolio's return and a benchmark's return). Additionally, piecewise linear transaction costs can be part of the objective function, as can various constraints and quadratic penalties. A typical objective function is shown below:

Find h to maximize:

$$U = \alpha^{T} \cdot h - \lambda \cdot ((h - h_m)^{T} \cdot V \cdot (h - h_m)) - p^{T} \cdot (h - h_0)^{+} - q^{T} \cdot (h - h_0)^{-}$$

such that

$$|h| = 1,$$

where:

α: vector of expected returns,
h_m: vector of benchmark portfolio holdings,
h_0: vector of current portfolio holdings,
p: vector of purchase costs,
q: vector of sales costs,
λ: risk-aversion parameter,
V: covariance matrix of asset returns.

$(h - h_0)^{+}$ is defined as $h - h_0$ if $h - h_0 > 0$, and 0 otherwise;
$(h - h_0)^{-}$ is defined as $h - h_0$ if $h - h_0 < 0$, and 0 otherwise.

Forecasting expected returns is a difficult business. A skilled portfolio manager will do well to achieve a correlation of 0.10 (or an R-square of 1%)

between his forecasts and actual outcomes.[1] Forecasting risk (variance and covariance of returns) is an easier business, but it is of course impossible to do precisely. Recently there has been some debate in the literature questioning the validity of quadratic optimization as a portfolio construction tool.[2] The argument is that errors made in risk forecasts will be sought out by a quadratic program. The realized risk of an optimal portfolio will tend to be larger than the forecast risk. This will be true even if the variance forecasts are on average unbiased.

Here is an example: let's assume each National League hitter has a true batting average (percentage of time getting a hit). We sample the batting averages of a collection of baseball players two months into the season. While each player's batting average is an unbiased estimate of his true average, the batting average of the league's leading hitter will tend to be an overestimate of that player's true batting average. This is because it is more likely that our estimation error (estimated average minus true average) will be positive for players with higher estimated averages.

If our goal is to pick the best hitter in the league, we should still select the player with the highest average. Even though we are overestimating his true average, he remains the best choice.

If we make our selection two months into the season, we will tend to have a large bias in our forecast. Baseball fans are well aware of the possibility that the leading hitter in May will not be batting 0.300 come September. However, if we make our selection in August, we will do much better. Not only will our estimate of the leader's batting average have much less bias, but chances are our selection will hold on to the batting crown for the rest of the year.

So what does this have to do with portfolio optimization? The two main points of this chapter come from the above example. First, the magnitude of bias in portfolio optimization is extremely dependent on the strength of the underlying risk model. The better the risk model, the lower the bias. Second, although optimized portfolios will tend to be more risky *ex-post* than is predicted *ex-ante*, there is no better method for constructing a portfolio.

This chapter is divided into three sections. The first section provides a basic example of how errors in variance forecasts influence the portfolio selection process. The second examines the predictive accuracy of the BARRA E2 Multiple Factor-Risk Model[3] in forecasting the risk of both actual and optimized portfolios. The third and final section contrasts the *ex-post* performance of portfolios constructed using quadratic optimization with the performance of portfolios built using alternative methods.

Example of how errors in variance forecasts affect the efficiency of optimized portfolios

Let's start with a two-asset example. Assume that we want to create a portfolio combining equities and long-term bonds. We forecast the excess return (over the risk-free asset) for equities to be 6% and for bonds to be 3%. Let's suppose the "true" annual volatility of stock returns is 20%, of bonds 15%, and the correlation of returns of the two assets is 0.35.

Our objective is to divide a portfolio into these two asset classes so as to maximize expected return divided by the standard deviation of return. This quotient is commonly referred to as an Information Ratio (or Sharpe ratio). This function is shown below:

Expected excess returns:
  Stocks: $E[r_s]=6\%$
  Bonds: $E[r_b]=3\%$
"True" risk structure:
  Stocks: $\sigma_s=20\%$
  Bonds: $\sigma_b=15\%$
  Correlation: $\rho_{sb}=0.35$
Optimization:

$$\text{Maximize}\left\{\frac{\text{Expected return}}{\text{Risk}}=\text{Information ratio}\right\}$$

Find h_s, h_b which maximizes:

$$\frac{E[h_s \cdot r_s + h_b \cdot r_b]}{\text{std}(h_s \cdot r_s + h_b \cdot r_b)} = \frac{h_s \cdot E[r_s] + h_b \cdot E[r_b]}{\sqrt{h_s^2 \cdot \sigma_s^2 + h_b^2 \cdot \sigma_b^2 + 2 \cdot h_s \cdot h_b \cdot \rho_{sb} \cdot \sigma_s \cdot \sigma_b}}$$

such that $h_s + h_b = 1$.

A plot of the Information Ratio as a function of the percentage of stocks in the portfolio is shown in figure 4.1. A simple calculation shows the optimal portfolio has 64% in stocks and 36% in bonds, giving an Information Ratio of 0.317.

What if the forecasts of return volatility and correlation are made with error?[4] Figure 4.2 shows how the "optimal" portfolio changes using incorrect risk forecasts. Notice how there can be a large discrepancy between what we believe to be our Information Ratio and the true Information Ratio. However, there is less of a discrepancy between the true Information Ratio of optimized portfolios using the incorrect risk forecasts and the best achievable Information Ratio.

If a risk model is used for quadratic optimization, the strength of the risk

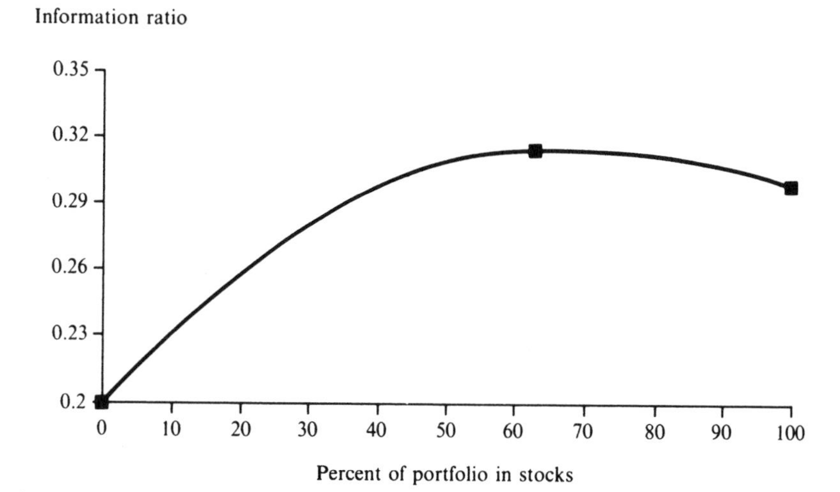

Figure 4.1 Two-asset example

Assumptions			Optimal portfolio		Expected return	Forecast		True	
σ_S	σ_B	ρ_{SB}	h_S	h_B	E(P)	σ_P	IR_P	σ_P	IR_P
20	15	0.35	64	36	4.9	15.6	0.317	15.6	0.317
22	13	0.30	45	55	4.3	13.8	0.314	14.2	0.306
20	15	0	53	47	4.6	12.7	0.361	14.6	0.314
20	10	0.35	33	67	4.0	11.0	0.365	13.8	0.289
15	15	0.35	85	15	5.5	13.7	0.405	17.9	0.310
			100	0	6.0			20	0.30
			0	100	3.0			15	0.20

Figure 4.2 How the optimal portfolio changes with a different risk model

model becomes much more important. This is because there is a direct relationship between errors made in variance/covariance forecasts and the magnitude of bias in the optimized portfolio. Two models may be fairly similar in their volatility forecasts for non-optimized portfolios, yet produce very different "optimal" portfolios. The process of optimization magnifies small differences.

A good example of this would be to consider what happens for a 500 asset optimization if historical variances and covariances are used to predict risk. If 200 previous monthly returns are used to build the covariance matrix, it will be possible for any set of 200 assets to find a portfolio which tracked the market perfectly each of the 200 previous months.[5] This portfolio will be considered to be "riskless" by the covariance matrix, but it is obvious that in the future the portfolio will not track the market perfectly.

2 Empirical tests of variance forecasts for optimized and non-optimized portfolios

Methodology

In this section we test risk predictions for various non-optimized and optimized portfolios. We set the following guidelines: We measure forecasts and realizations of *active risk*, or the standard deviation of a portfolio's return minus a benchmark's return. Institutional investors are typically more concerned with minimizing volatility relative to a benchmark, than with minimizing total volatility.

Our benchmark choice is the S&P500, a capitalization-weighted index of 500 US securities. Returns are calculated on a monthly basis, and realized standard deviations are calculated using twenty-four monthly returns. We performed analyses using several starting dates: January 1984, January 1985, January 1986, and May 1987.[6]

We evaluated the model's active risk predictions for three groups of portfolios. The first group consisted of fifty-six actual institutional portfolios held by BARRA clients at various dates. The second group consisted of portfolios optimized to have minimum active risk relative to the S&P500 (and no more than a specified number of assets[7]). The third group consisted of portfolios optimized to have minimum risk relative to the S&P500 and maximum exposure to a randomly generated return forecast.[8]

We evaluated the performance of each portfolio for the twenty-four months following the initial active risk prediction. For each portfolio we report the forecast active standard deviation and the realized active standard deviation of a buy-and-hold strategy. Portfolio characteristics change with mergers, bankruptcies, and large asset returns in general. This is especially true for smaller portfolios. For this reason, we also look at a second number, the *rolling* active risk prediction for each portfolio. This number is defined as the average active risk prediction (in variance space) made by the E2 model at the start of each of the next twenty-four months. Since this number takes into account the changing characteristics of our portfolios relative to the S&P500, we expect it to do a better job of forecasting the buy-and-hold risk of the portfolio.

Results

Tables 4.1, 4.2, and 4.3 report the forecast and realized active standard deviations for each group of portfolios. Both original and rolling risk forecasts are shown. We also report mean, minimum, and maximum monthly active returns, and a t-statistic measuring the likelihood that the mean active return is different from zero.[9] The ratio of realized standard deviation to forecast standard deviation is calculated as a measure of the accuracy of the E2 risk prediction for each portfolio. Figures 4.3, 4.4, and 4.5 display the relationship between forecast risk and realized risk.

When we measure the realized variance of a portfolio's active return we do so with error. Even a *perfect* prediction of risk will exhibit error.[10] The ratio of realized risk to forecast risk for the fifty-six actual portfolios ranged from a low of 0.62 to a high of 1.36 (using rolling forecasts). The best regression line (with zero intercept) for the rolling forecasts has a slope of 0.97 ($\pm$0.03), statistically indistinguishable from a perfect slope of 1.0. The slope using the original forecasts is 0.94 ($\pm$0.03).

For the seventeen optimized passive portfolios the slopes for original and rolling forecasts are 1.12 and 1.04 respectively ($\pm$0.04). For the twelve optimized active portfolios the corresponding slopes are 1.14 and 1.06 ($\pm$0.04). The ratio of realized to forecast risk ranges from 0.65 to 1.42 for the optimized portfolios.

The above results support our hypothesis that optimized portfolios exhibit higher realized risk than is forecast, even if the underlying risk model is on average unbiased. We *roughly* observe the bias to be approximately 20%, arrived at by dividing the average ratio for the optimized portfolios, 1.13, by the ratio for the non-optimized portfolios, 0.94.[11]

Table 4.1. *Managed portfolios*

Portfolio number	Number of assets	Date	Forecase active std. dev.	Realized active std. dev.	Ratio (real./for.)	Monthly active return Min.	Max.	Mean	(T-stat.)	Rolling forecast active std. dev.	Ratio (real./for.)
1	11	Jan-84	13.78	14.68	1.07	−5.57	11.01	−0.03	(−0.04)	13.76	1.07
2	43	Jan-84	15.79	14.61	0.93	−9.19	8.96	−1.80	(−2.09)	17.17	0.85
3	27	Jan-84	15.79	14.07	0.89	−7.59	7.51	−1.79	(−2.16)	14.37	0.98
4	39	Jan-84	12.32	12.66	1.03	−8.26	7.47	−0.66	(−0.88)	12.60	1.00
5	72	Jan-84	11.24	10.02	0.89	−4.51	6.92	−0.31	(−0.52)	11.71	0.86
6	44	Jan-84	10.45	9.80	0.94	−8.77	5.48	−0.43	(−0.74)	9.31	1.05
7	25	Jan-84	6.74	9.69	1.44	−3.81	7.01	0.76	(1.33)	7.12	1.36
8	27	Jan-84	7.12	8.80	1.24	−2.96	7.94	0.88	(1.70)	6.58	1.34
9	69	Jan-84	9.11	8.71	0.96	−7.23	4.78	−0.29	(−0.56)	8.57	1.02
10	42	Jan-84	8.40	7.40	0.88	−3.45	5.17	−0.62	(−1.43)	8.49	0.87
11	24	Jan-84	8.24	6.73	0.82	−4.54	3.33	−0.48	(−1.21)	7.50	0.90
12	36	Jan-84	5.95	6.09	1.02	−3.36	3.05	0.14	(0.38)	5.52	1.10
13	18	Jan-84	5.50	5.59	1.02	−2.57	3.49	0.33	(0.99)	5.26	1.06
14	90	Jan-84	5.39	5.13	0.95	−3.22	2.02	−0.47	(−1.54)	5.71	0.90
15	48	Jan-84	8.20	5.08	0.62	−2.86	2.34	−0.11	(−0.36)	8.79	0.58
16	36	Jan-84	5.48	4.84	0.88	−2.00	4.05	0.21	(0.74)	4.96	0.98
17	29	Jan-84	6.13	4.26	0.70	−2.39	2.14	0.22	(0.86)	5.95	0.72
18	59	Jan-84	4.95	4.11	0.83	−2.99	2.45	0.42	(1.72)	5.00	0.82
19	41	Jan-84	5.54	4.05	0.73	−2.17	2.06	0.24	(1.02)	5.48	0.74
20	79	Jan-84	4.87	4.04	0.83	−2.35	3.12	0.42	(1.75)	4.87	0.83
21	41	Jan-84	6.21	4.02	0.65	−1.45	3.33	0.44	(1.84)	6.03	0.67
22	1012	Jan-84	5.24	3.90	0.74	−2.83	2.42	−0.10	(−0.41)	5.00	0.78
23	52	Jan-85	17.29	16.82	0.97	−7.10	10.45	−1.25	(−1.27)	17.09	0.98
24	33	Jan-85	11.82	10.77	0.91	−4.73	7.92	0.40	(0.63)	12.87	0.84

25	69	Jan-85	10.04	9.85	0.98	−5.65	6.53	−0.60	(−1.04)	9.97	0.99
26	41	Jan-85	8.17	8.65	1.06	−4.63	5.29	−0.63	(−1.23)	7.57	1.14
27	57	Jan-85	9.26	7.73	0.84	−4.39	5.75	−0.07	(−0.15)	9.24	0.84
28	29	Jan-85	9.29	7.70	0.83	−5.26	4.15	−0.35	(−0.78)	8.62	0.89
29	34	Jan-85	5.88	7.54	1.28	−2.42	6.07	0.77	(1.73)	6.26	1.20
30	20	Jan-85	7.22	7.53	1.04	−3.80	6.40	0.31	(0.69)	7.72	0.98
31	40	Jan-85	6.34	7.50	1.18	−3.25	5.28	0.86	(1.94)	6.52	1.15
32	32	Jan-85	9.07	6.76	0.75	−4.08	2.83	−0.16	(−0.40)	8.12	0.83
33	37	Jan-85	7.90	6.55	0.83	−4.03	2.66	−0.01	(−0.04)	8.10	0.81
34	27	Jan-85	5.62	5.77	1.03	−2.75	4.31	0.69	(2.04)	5.59	1.03
35	39	Jan-85	5.43	5.33	0.98	−4.21	2.57	0.27	(0.85)	4.97	1.07
36	53	Jan-85	5.36	4.98	0.93	−1.90	3.61	0.58	(1.96)	4.84	1.03
37	42	Jan-85	5.97	4.70	0.79	−2.42	2.01	−0.26	(−0.94)	5.70	0.82
38	48	Jan-85	5.99	4.53	0.76	−3.11	3.88	−0.03	(−0.13)	5.68	0.80
39	80	Jan-85	6.50	4.42	0.68	−3.13	1.97	0.03	(0.12)	6.57	0.67
40	98	Jan-85	5.86	4.07	0.69	−2.18	1.98	−0.16	(−0.67)	5.00	0.81
41	40	Jan-85	6.61	4.01	0.61	−1.14	2.75	0.50	(2.10)	6.47	0.62
42	37	Jan-85	3.77	3.64	0.97	−1.36	2.00	0.44	(2.07)	3.94	0.92
43	60	Jan-85	3.14	3.18	1.01	−2.27	1.28	−0.21	(−0.63)	3.21	0.99
44	1110	Jan-85	4.72	2.82	0.60	−2.00	2.31	0.01	(0.07)	4.47	0.63
45	24	Jan-86	15.86	14.92	0.94	−5.75	12.43	0.75	(0.86)	14.36	1.04
46	42	Jan-86	11.98	12.17	1.02	−7.79	7.22	−0.78	(−1.09)	11.15	1.09
47	19	Jan-86	7.32	10.14	1.28	−4.76	6.75	0.84	(1.41)	8.31	1.22
48	152	Jan-86	7.50	8.30	1.11	−4.47	4.45	−0.12	(−0.24)	7.62	1.09
49	47	Jan-86	6.92	7.97	1.15	−3.79	5.63	−0.20	(−0.44)	6.35	1.25
50	59	Jan-86	3.49	3.88	1.11	−1.75	2.49	0.07	(0.29)	3.23	1.20
51	25	May-87	17.79	15.53	0.87	−10.65	7.58	−0.79	(−0.86)	15.20	1.02
52	68	May-87	13.64	14.04	1.03	−10.17	9.64	−0.24	(−0.29)	13.49	1.04
53	116	May-87	10.52	11.06	1.05	−5.00	8.41	0.12	(0.19)	10.15	1.09
54	113	May-87	10.52	11.06	1.05	−5.00	8.41	0.12	(0.19)	10.15	1.09
55	42	May-87	10.24	8.55	0.83	−3.77	5.58	0.15	(0.29)	8.32	1.03
56	206	May-87	7.17	5.39	0.75	−1.90	3.75	0.05	(0.16)	5.86	0.92

Table 4.2. *Optimized passive portfolios*

Universe	Number of assets	Date	Forecast active std. dev.	Realized active std. dev.	Ratio (real./for.)	Monthly active return				Rolling forecast active std. dev.	Ratio (real./for.)
						Min.	Max.	Mean	(T-stat.)		
SMALLCAP	536	Jan-84	6.40	4.26	0.67	−1.92	2.84	0.21	(0.83)	6.55	0.65
S&P500	100	Jan-84	1.06	1.41	1.33	−0.79	0.74	0.05	(0.62)	1.21	1.17
S&P500	200	Jan-84	0.51	0.73	1.42	−0.33	0.36	0.01	(0.34)	0.64	1.13
S&P500	50	Jan-84	1.80	2.71	1.51	−1.60	1.28	−0.13	(−0.80)	2.18	1.25
SMALLCAP	521	Jan-85	6.05	6.25	1.03	−3.61	2.51	−0.17	(−0.46)	6.42	0.97
S&P500	100	Jan-85	1.04	1.27	1.22	−0.52	1.11	0.12	(1.55)	1.28	0.99
S&P500	200	Jan-85	0.49	0.72	1.47	−0.41	0.43	0.06	(1.46)	0.66	1.09
S&P500	50	Jan-85	1.79	2.38	1.33	−1.90	1.01	0.16	(1.14)	2.05	1.16
SMALLCAP	505	Jan-86	6.35	8.35	1.31	−3.47	7.78	−0.17	(−0.35)	6.81	1.23
S&P500	100	Jan-86	1.04	1.12	1.08	−0.69	0.76	0.18	(2.77)	1.20	0.94
S&P500	200	Jan-86	0.48	0.72	1.49	−0.40	0.47	0.05	(1.18)	0.58	1.24
S&P500	50	Jan-86	1.85	2.61	1.41	−1.21	1.69	0.37	(2.42)	2.22	1.18
S&P500	99	May-87	1.34	1.25	0.93	−0.67	0.60	−0.04	(−0.57)	1.42	0.88
SMALLCAP	129	May-87	6.23	9.35	1.50	−4.13	6.62	0.19	(0.35)	7.21	1.30
S&P500	100	May-87	1.01	1.35	1.33	−0.96	0.74	−0.04	(−0.48)	1.24	1.08
S&P500	200	May-87	0.46	0.55	1.20	−0.31	0.22	−0.03	(−0.80)	0.61	0.90
S&P500	50	May-87	1.78	2.03	1.14	−1.70	0.57	−0.16	(−1.30)	2.00	1.02

Table 4.3. *Optimized active portfolios*

Risk aversion	Number of assets	Date	Forecast active std. dev.	Realized active std. dev.	Ratio (real./for.)	Monthly active return				Rolling forecast active std. dev.	Ratio (real./for.)
						Min.	Max.	Mean	(T-stat.)		
0.5	107	Jan-84	1.81	1.87	1.04	−0.76	1.60	0.15	(1.34)	2.12	0.88
0.05	45	Jan-84	3.55	3.63	1.02	−1.61	2.27	0.19	(0.87)	4.12	0.88
0.01	22	Jan-84	5.19	4.67	0.90	−1.61	3.16	0.20	(0.71)	5.59	0.83
0.5	113	Jan-85	1.68	1.78	1.06	−0.99	1.08	0.06	(0.55)	1.82	0.98
0.05	40	Jan-85	3.37	3.08	0.91	−2.43	1.50	−0.07	(−0.36)	3.54	0.87
0.01	20	Jan-85	6.02	5.76	0.96	−3.19	3.90	−0.35	(−1.04)	6.00	0.96
0.5	118	Jan-86	1.99	1.72	0.86	−1.25	1.25	0.20	(1.98)	2.12	0.81
0.05	37	Jan-86	3.99	4.36	1.09	−3.62	2.45	0.31	(1.19)	4.27	1.02
0.01	18	Jan-86	5.98	7.15	1.20	−6.41	3.29	0.63	(1.50)	6.82	1.05
0.5	115	May-87	1.70	2.27	1.33	−0.95	1.52	0.04	(0.30)	1.92	1.18
0.05	37	May-87	3.96	6.05	1.53	−2.78	3.55	−0.02	(−0.05)	4.49	1.35
0.01	20	May-87	5.78	8.66	1.50	−3.55	4.48	−0.01	(−0.01)	6.11	1.42

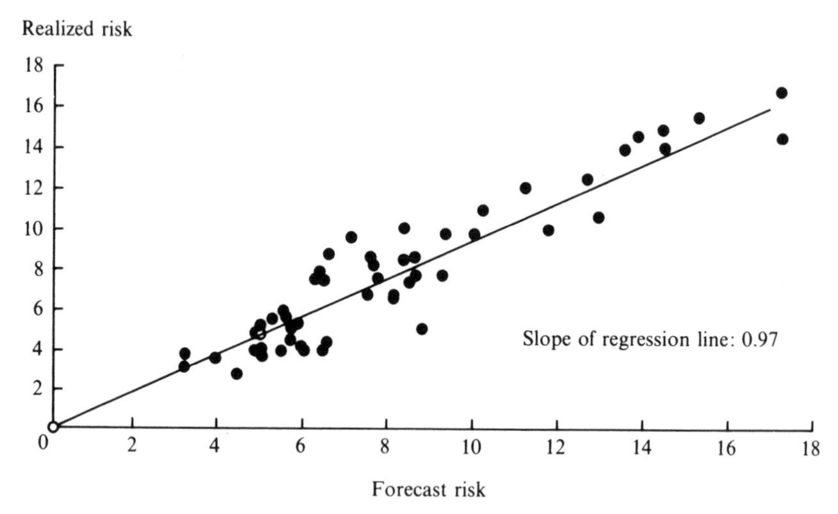

Figure 4.3 Managed portfolios

Note: Realized risk: actual active standard deviation of "buy and hold" portfolio over 24 months; forecast risk: "rolling" active risk prediction – average of 24 monthly risk predictions.

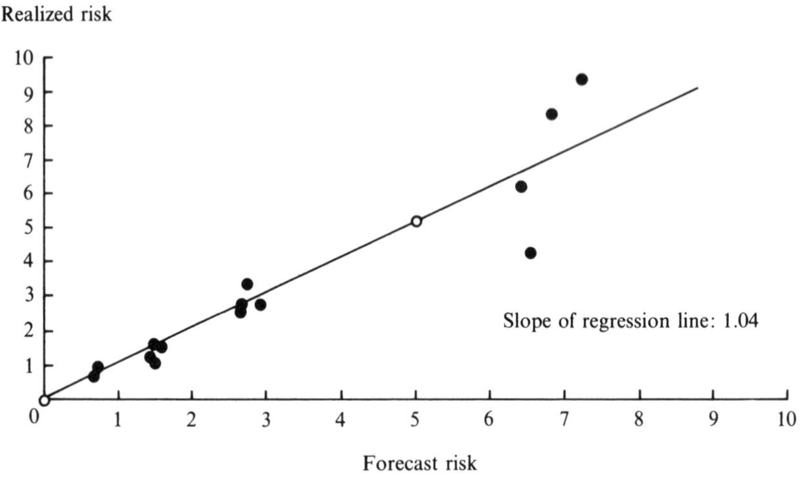

Figure 4.4 Optimized passive portfolios

Note: Realized risk: actual active standard deviation of "buy and hold" portfolio over 24 months; forecast risk: "rolling" active risk prediction – average of 24 monthly risk predictions.

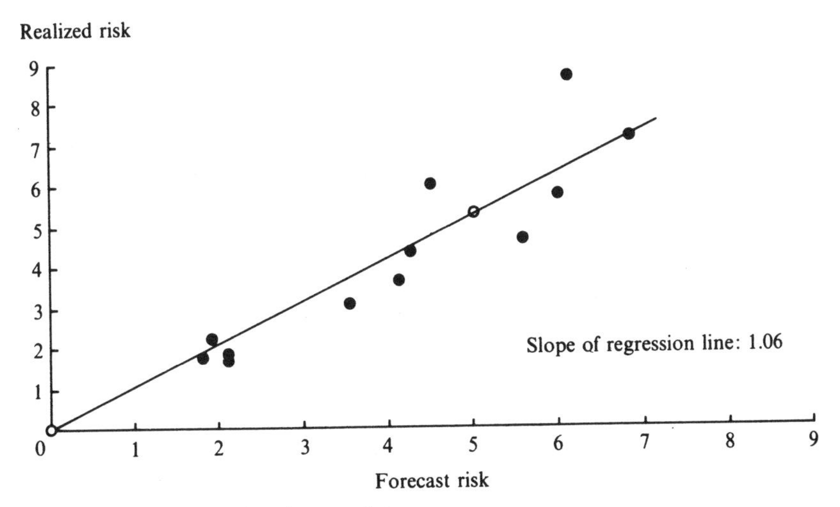

Figure 4.5 Optimized active portfolios

Note: Realized risk: actual active standard deviation of "buy and hold" portfolio over 24 months; forecast risk: "rolling" active risk prediction – average of 24 monthly risk predictions.

3 Comparison of optimization and alternative portfolio construction techniques: how do you build the best active portfolio?

Methodology

Optimization is not perfect, but is it possible to do better? The results in section 2 suggest that the bias induced by optimization is approximately 20%. Intuition suggests that an alternative portfolio construction technique will perform *as well or better* than optimization if that technique can build a portfolio with active risk no more than 20% higher than an optimized portfolio.

The following research compares the *ex-post* results of different portfolio construction techniques. We focus on a typical active manager's application. The manager has information, or return forecasts, for a group of stocks. Typically, even an excellent manager will have a correlation of no more than 0.1 between his forecasts and actual returns. The manager's goal is to create a portfolio which maximizes return while minimizing active risk.

For our research, we simulated a manger's possible information set by adding random noise to actual asset returns over the two years subsequent to each of our portfolio construction dates. This gave us return forecasts, or *alphas*. We added enough noise so the correlation between the alphas and

Table 4.4. *Realized information ratios*

Portfolio		Method			
Date	Risk aversion	Equal-weighting	Cap-weighting	Stratified sampling	Quadratic optimization
	LOW	1.10	1.30	0.63	2.16
January 1984	MED	0.95	2.24	0.64	1.89
	HIGH	0.73	1.31	0.69	1.75
	LOW	0.78	1.47	1.98	0.98
January 1985	MED	0.74	−0.53	1.29	1.68
	HIGH	0.50	−0.15	0.83	1.49
	LOW	1.17	0.91	0.69	2.08
January 1986	MED	0.69	0.98	0.33	2.29
	HIGH	0.60	0.99	0.51	2.51
	LOW	1.43	2.04	2.82	2.14
May 1987	MED	1.01	1.48	2.60	1.76
	HIGH	0.66	1.17	2.17	1.82
	Average:	0.86	1.10	1.27	1.88
	Standard Deviation:	0.27	0.79	0.89	0.40
	Maximum:	1.43	2.24	2.82	2.51
	Minimum:	0.50	−0.53	0.33	0.98

the actual returns was 0.1. We then built four sets of portfolios using quadratic optimization, three each in January 1984, January 1985, January 1986, and May 1987. Three portfolios were created at each date to correspond to varying levels of risk aversion. We did not allow the optimizer to sell short any asset, or allow the weight of any asset in a portfolio to exceed 10%.

We then looked at three common portfolio construction techniques as alternatives to quadratic optimization:

1 *Equal-weighing* top alpha stocks,
2 *Market-weighting* (*cap-weighting*) top alpha stocks,
3 *Stratified sampling* of top alpha stocks in each industry.

To simulate different levels of risk aversion we equal-weighted and cap-weighted the top 50, 100, and 150 stocks. For the stratified sampling approach we selected 1, 2, or 3 stocks in each industry. Thus we generated three portfolios at each date for each alternative method.

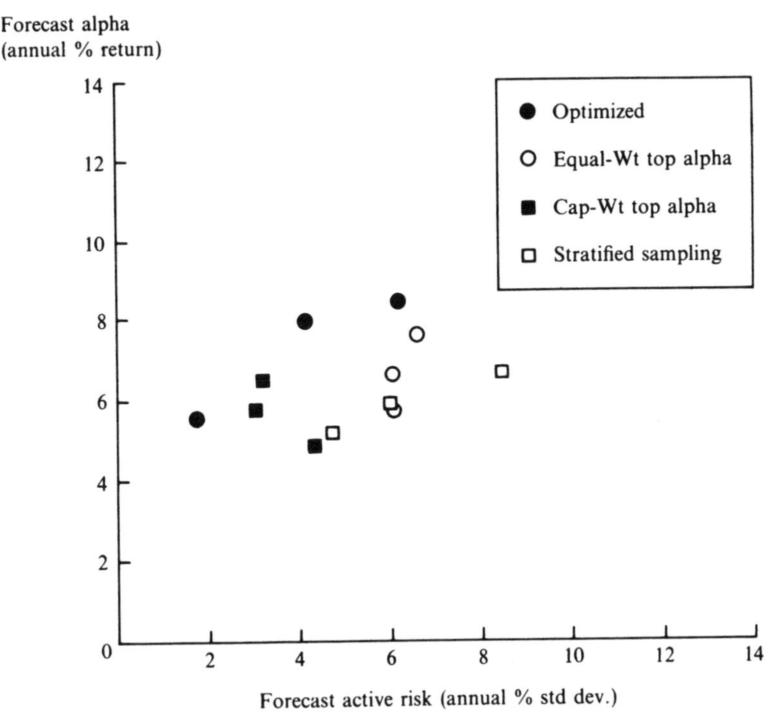

Figure 4.6 *Ex-ante* view of the world (January 1986 portfolios)

Results

We looked at the realized active risk and return of each of the portfolios following a two year buy-and-hold strategy. The results are displayed in table 4.4. Graphs of the *ex-ante* and *ex-post* Information Ratios are shown in figures 4.6, 4.7, 4.8, 4.9, and 4.10. Clearly, optimized portfolios will lie along the efficient frontier *ex-ante*. This can be seen in figures 4.6 and 4.8. For the January 1986 run (figure 4.7), the optimized portfolios were also the most efficient *ex-post*. For the May 1987 run (figure 4.9), the optimized portfolios were not the most efficient *ex-post*, but were reasonably close.

Table 4.4 shows that optimization produced the best *ex-post* portfolio seven out of twelve times, and the second-best portfolio four times. Stratified sampling was a distant second, giving the best portfolio four times and the second best twice. Notice that the cap-weighting method twice produced portfolios which underperformed the market!

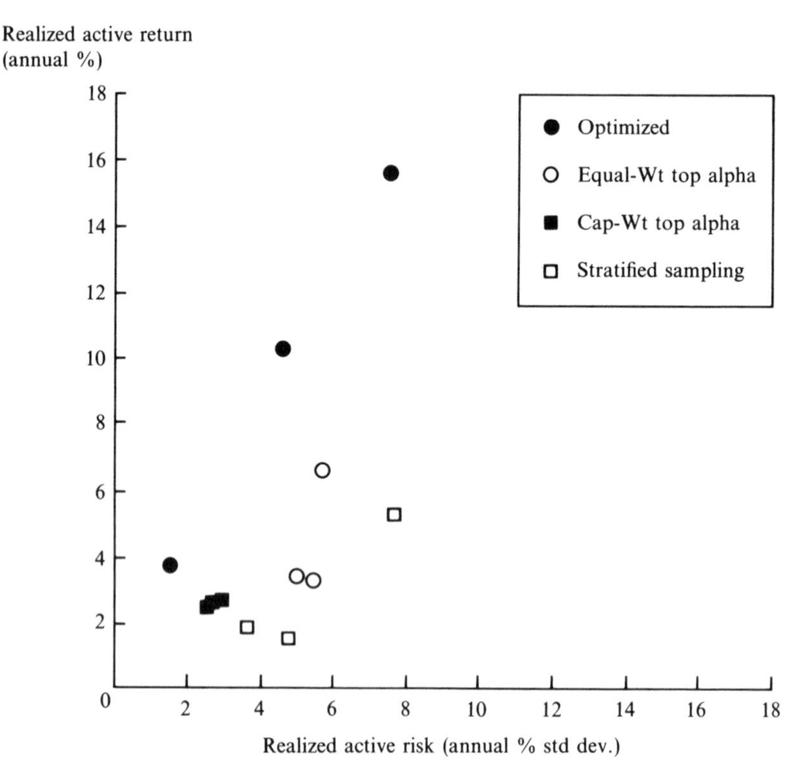

Figure 4.7 *Ex-post* view of the world (January 1986 portfolios)

Figure 4.10 shows the realized active risk versus active return for all forty-eight portfolios. Assuming perfect implementation of our alpha forecasts, we would expect a best possible return/risk ratio (Information Ratio) near 2.23.[12] On average, the optimized portfolios exhibited Information Ratios of 1.88. This is almost exactly what we would expect if there were a 20% bias in the portfolio construction process. The other techniques we used did not perform nearly as well. The average Information Ratio was 1.27 for stratified sampling, 1.10 for cap-weighting, and 0.86 for equal-weighting.

4 Summary

Our results suggest that even with the bias inherent in the process, quadratic optimization does a better job constructing portfolios than equal-weighting, cap-weighting, or stratified sampling. A few important

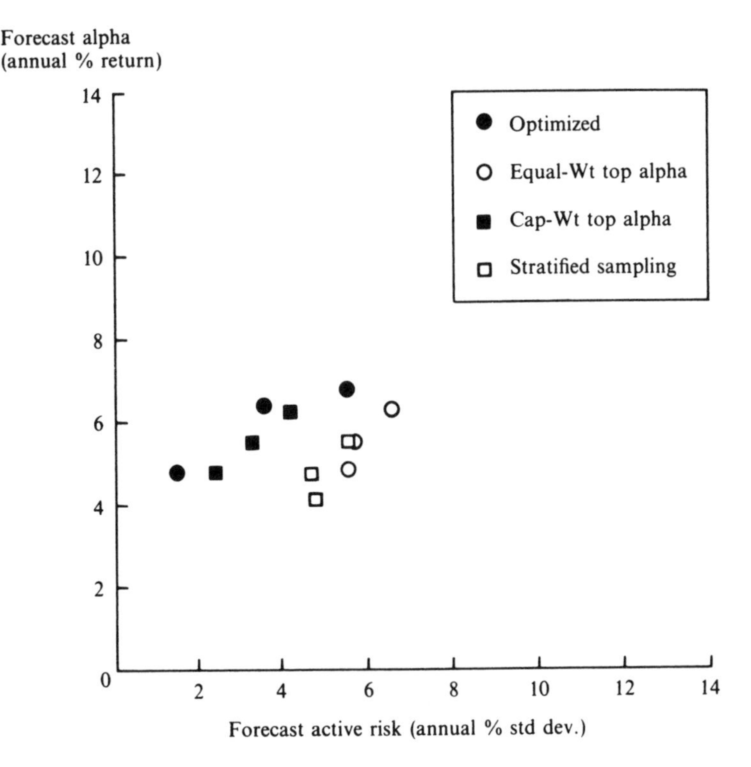

Figure 4.8 *Ex-ante* view of the world (May 1987 portfolios)

points should be made, however. First, although our results do point strongly in favor of optimization, our sample size is small.

Second, the choice of a risk model used in quadratic optimization is extremely important. If an inferior risk model is used in optimizing (for example, a covariance matrix constructed solely from historical stock return variances and covariances), optimization may well prove inferior to some of the other techniques.

Third, risk predictions for optimized portfolios should be scaled upwards to reflect the bias inherent in optimization. Our results suggest that multiplying risk forecasts by 1.2 is a good rule of thumb for optimized portfolios bought and held for two years. If optimization is used to rebalance portfolios more frequently, a larger multiplier should probably be used.

Note finally that we did not do anything special to "help" the optimizer along, beyond placing an upper bound of 10% on the holding of any

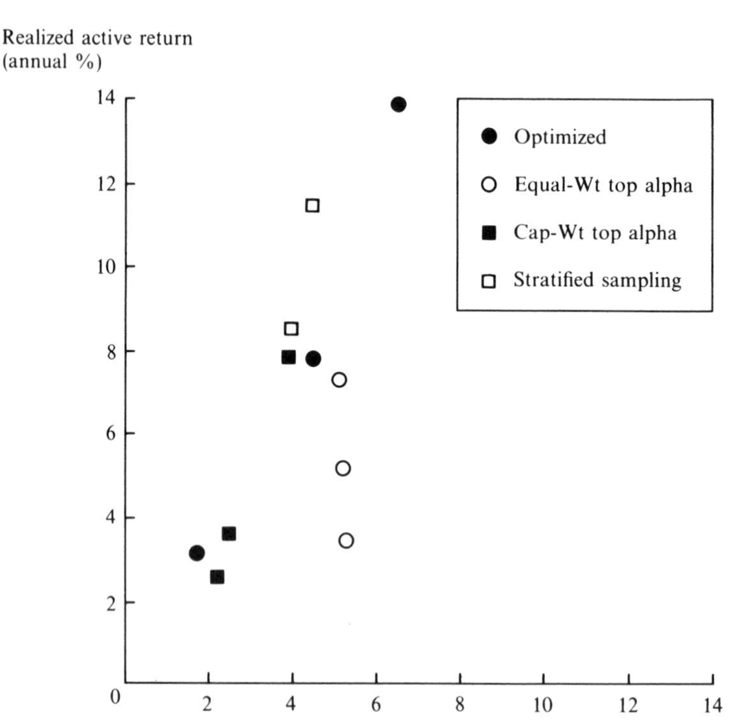

Figure 4.9 *Ex-post* view of the world (May 1987 portfolios)

security (which was seldom reached). Practitioners who use optimization typically go through a number of iterations in building an optimal portfolio. Large positions in any one asset are examined to see why they are chosen. Was it for risk reduction or for return maximization? Controversial assets[13] are often held at precisely market weight to neutralize possible large returns (in either direction). This type of fine-tuning of an optimized portfolio will many times improve performance (by reducing the bias in optimization). Of course, it is important not to fine-tune away real information!

Quadratic optimization is used today by many large money managers with great success. Our research has shown that optimization using a good risk model is still the best way to build a portfolio, even considering the method's inherent bias.

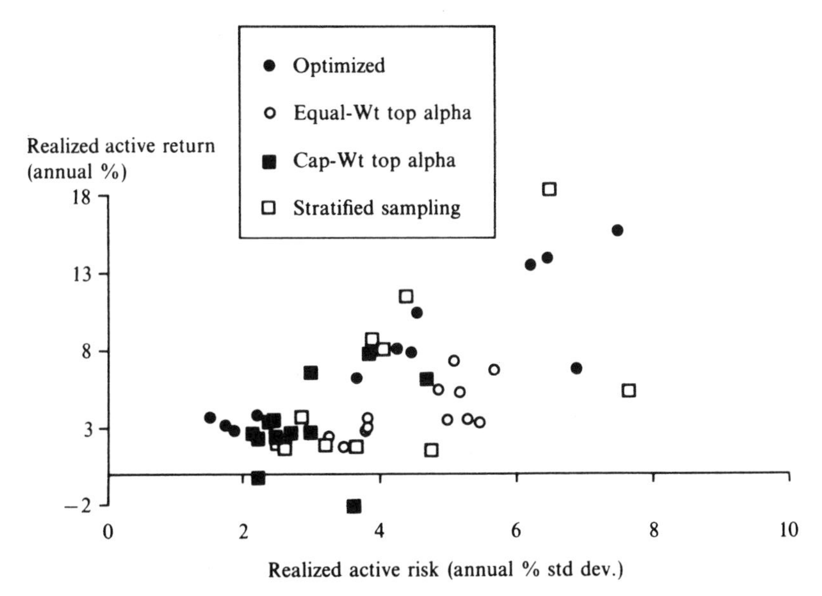

Figure 4.10 *Ex-post* view of the world

NOTES

I would like to thank Richard Grinold for his helpful suggestions and guidance and Rafi Zaman for his help in undertaking this research.

1 For further discussion, see Richard Grinold (1989).
2 For example, see Richard O. Michaud (1989).
3 BARRA is an investment consulting firm headquartered in Berkeley, CA. The BARRA E2 risk model for US equities is the most commonly used tool to forecast the risk of institutional portfolios.
4 For the purposes of demonstration we are assuming that our risk forecasts are incorrect but our return forecasts are unbiased.
5 To do this you solve 200 equations for 200 unknowns. Define $R_n(t)$ as the return to stock n over month t, and $R_M(t)$ as the return to the market over month t. Suppose we have N stocks ($n=1, \ldots, N$) and T months of data ($t=1, \ldots, T$). As long as $T \geq N$, we can solve for a "riskless" portfolio $\{h_n\}$ in the T equations:

$$1)\quad \sum_{n=1}^{N} h_n \cdot R_n(1) = R_M(1)$$

$$2)\quad \sum_{n=1}^{N} h_n \cdot R_n(2) = R_M(2)$$

$$\vdots$$

$$T)\quad \sum_{n=1}^{N} h_n \cdot R_n(T) = R_M(T)$$

For the case discussed in the text, $N = T = 200$.

6 One of the goals of this research was to test risk forecasts of the E2 model out of sample. The model was finalized and released before 1984. Our selection of starting dates was made so as to choose four out-of-sample periods with minimum overlap.
7 Without a restriction on the number of assets held, the optimizer will simply hold all the assets in the S&P500 at market weight. An iterative paring algorithm is used to select which assets in the S&P500 are used for optimization.
8 The relative importance of these conflicting objectives (minimizing risk and maximizing return) is determined by the risk-aversion parameter. Figure 1 shows how the risk-aversion parameter affects the optimization objective function.
9 Four of the fifty-six managed portfolios exhibited mean active returns greater than zero that were significant at the 95% level.
10 The standard error in measuring realized *variance* σ^2 is approximately equal to $\sqrt{2/(n-1)}\cdot\sigma^2$, where n is the number of months. In our case, $n=24$. If this error is normally distributed, a perfect risk prediction would exhibit a ratio of realized/forecast *standard deviation* greater than 1.14 or less than 0.86 33% of the time, and a ratio greater than 1.27 or less than 0.73 5% of the time.
11 This suggests that the risk forecast for a portfolio optimized using the E2 model should be multiplied by approximately 1.2 to get an unbiased forecast.
12 Using the Fundamental Law of Active Management.
13 Companies which may be taken over or companies involved in potentially large lawsuits. Recent examples are UAL Corporation (takeover) and Manville Corporation (asbestos litigation).

REFERENCES

R. Grinold (1989), "The fundamental law of active management," *Journal of Portfolio Management*, Spring.

J. D. Jobson and B. Korkie (1980), "Estimation for Markowitz efficient portfolios," *Journal of the American Statistical Association*, September.

R. Michaud (1989), "The Markowitz optimization enigma: is 'optimized' optimal?" *Financial Analysts Journal*, January–February.

Part II

Models

5 An economic approach to valuation of single premium deferred annuities

MICHAEL R. ASAY, PETER J. BOUYOUCOS and ANTHONY M. MARCIANO

1 Introduction

The 1980s have been a trying decade for life insurers. Volatile interest rates, new competition from other sectors of the financial services industry, and tightening investment spreads are some of the major new forces that have challenged the industry. The response has been to offer new products (the so-called "interest-sensitive" products), to promise higher rates, and to assume investment risks associated with higher book yields. The combination of these three actions has placed the industry in a precarious position as it faces the new decade.

Figure 5.1 shows interest rates during the 1980s for both six-month Treasury bills and thirty-year Treasury bonds. To help recap the developments that have shaped the industry, we have included three milestones (labeled A, B, and C). The first milestone (A) occurred during the early part of the decade when short-term rates shot to record heights. This caused disintermediation as policyholders fled to higher yielding alternatives offered in the capital markets (e.g., money market mutual funds). Associated with this flight to yield was the need for insurers to liquidate assets at a loss to meet the outflow. Policyholders' exercise of the options to surrender their policies, or to take out policy loans at substantially below-market interest rates, caused insurers both economic and accounting losses.

The insurers' response to this disintermediation was the creation of new so-called interest-sensitive products. Universal life and annuities were created which unbundled the insurance from the savings components in the policy, and had interest crediting rates tied much more closely to market interest rates. These policies have sold well but profit margins were much

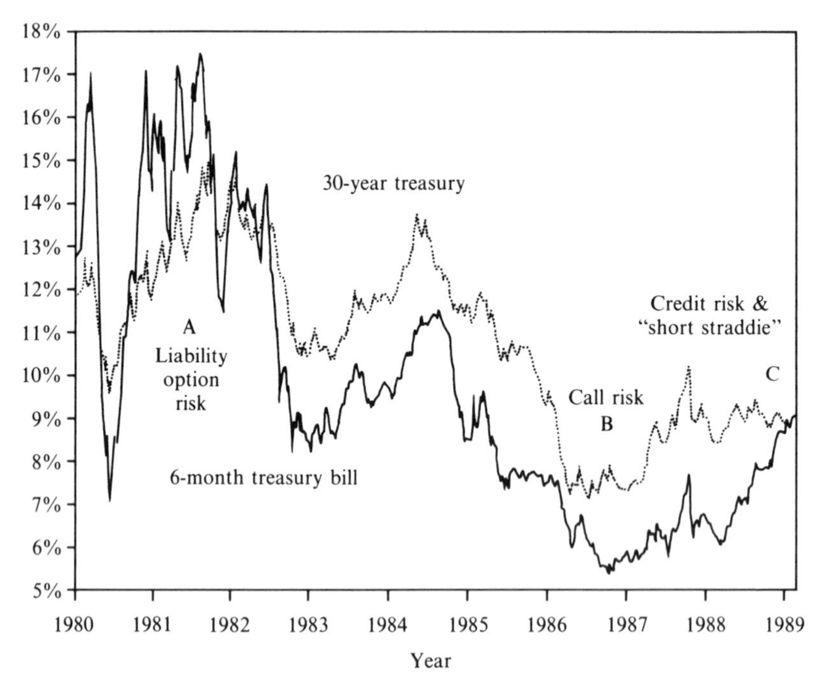

Figure 5.1 Interest rates during the 1980s (6-month bills versus 30-year bond)

narrower as ready non-insurance capital-market alternatives existed for the savings components.

While the first milestone meant trouble on the liability side of the balance sheet, the second milestone (B) affected insurer assets. From mid 1984 to early 1986, interest-rate levels fell by 500–600 basis points (bp). Normally this would have been a positive development, because insurers typically have invested long and should have reaped significant capital gains. However, many of the assets used to fund the new products included embedded call options that were owned by corporate issuers of debt and by homeowners backing securitized mortgages. The options were exercised against insurers when rates fell, reducing or eliminating capital gains and obliging insurers to reinvest funds at substantially lower rates. This was especially painful for insurers that had offered guaranteed rates to policyholders such as in long dated GICs when interest rates were higher. These losses were now locked in.

The third milestone (C) is occurring in 1988–9. Because insurers have realized that tight profit margins exist for many of their new products, they have taken on incremental credit risk with high yield securities. These asset portfolios contain embedded options, and the interest-sensitive liabilities – especially SPDAs and universal life insurance – contain many options that are held by policyholders. This combination of embedded options in assets and liabilities can be thought of as a "short straddle."[1] The insurer is subject to option risk for both rising and falling interest-rate scenarios.

The lesson of the 1980s is that these embedded options in both assets and liabilities can harm insurers if they are not properly managed (and priced). Insurers are generally aware of the existence of these options and are developing quantitative tools to help them understand their exposure and develop rational investment strategies.

This chapter will focus on assessing the costs and risks associated with one particular class of liabilities – SPDAs. As noted, the SPDA is one of the "new" generation interest-sensitive products. It is an accumulation savings vehicle with essentially *no* life insurance content, (upon death, the policyholder's estate gets the invested money plus interest back). Investors essentially purchase a bond with an interest rate set periodically, and interest earned is automatically reinvested in the contract. It is tax advantaged in that tax on interest is deferred until withdrawn from the policy. The tax feature plus insurer imposed penalties reduce the policyholder's incentive to move funds frequently, but withdrawals remain a problem if interest rates rise significantly and/or the crediting rate becomes uncompetitive with other savings products offered. To assess the cost and risks of the SPDA, we will introduce and explain an *insurance company liability valuation model.* This model is a part of a comprehensive asset/liability valuation model that treats both sides of the balance sheet in a consistent option-pricing framework. While this chapter focuses solely on SPDA liabilities, the model is capable of valuing any life insurance or annuity product, and a full range of publicly and privately traded assets.

In the pages that follow, we present both the rationale for valuation of insurance company liabilities and the specifics of the SPDA model. Section 2 describes the basic financial concepts useful in characterizing SPDAs. Section 3 discusses SPDA contract features, the policyholder lapse function, insurance company crediting strategies, and the competitive environment. Section 4 provides the results of the model for a specific policy and the sensitivity of those results to economic forces. In section 5, we discuss sensitivity to policy terms and crediting strategies. Section 6 focuses on implications for asset allocation, and section 7 presents our conclusions.

2 Economic valuation of SPDAs: an overview

The purpose of valuing SPDA liabilities is to develop asset/liability strategies to better manage profitability and the variability of surplus.

Financial valuation allows the insurer to quantify the value of options granted to policyholders and the expected cost of the SPDA given those options. It is a tool that provides the insurer insight into the economic impact of product management decisions (product design and investment and crediting strategies) and external variables to which the insurer must react (interest-rate levels, volatility, competitor and policyholder behavior).

Tools

What tools are required to value SPDAs? The short answer is (1) a theory of the term structure of interest rates and (2) an option-pricing model. While the term structure theory allows for the generation of plausible future interest rates, option pricing allows for the generation and valuation of SPDA cashflows. Option pricing is appropriate because the SPDA contains embedded options. For simplicity, we can characterize an SPDA as follows:

SPDA = Fixed-rate bond
− Insurer option to credit a lower (higher) rate
+ Policyholder option to lapse

where the fixed-rate bond is an accrual bond that compounds at the initial crediting rate, subject to an insurer option to change this rate. These first two components, taken together, define the insurer's crediting strategy. The policyholder lapse option will depend upon the crediting strategy as it relates to competition and transaction costs (surrender charges), among other things.

For the insurance model, we bring these two tools together in a Monte Carlo simulation.[2] This is the preferred method of valuation, because the SPDA cashflows are path dependent. That is, the cashflow in any given period depends not only upon the current levels of interest rates, but also upon the entire history of rates for that security (see, also, chapter 14).

How many replications in the simulation are necessary? Because Monte-Carlo simulation is a method of sampling a much larger universe of possible future interest-rate paths, sampling error is important. We have found that marketable assets, such as mortgage-backed securities, require over 1,000 interest-rate paths to approximate observable market prices (this incorporates variance reduction techniques that reduce the required number of paths). Similarly, to achieve precise valuation of SPDAs, we need a like number of replications.

The next question is: Given these tools what information does financial valuation provide? The short answer is (1) a measure of fair value (or

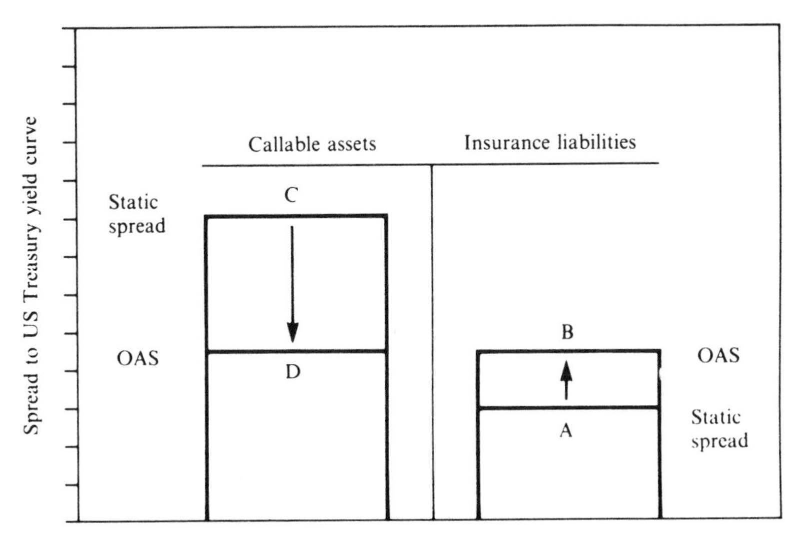

Figure 5.2 Impact of options on expected profitability (capital market assets versus insurance liabilities)

relative value) and (2) the sensitivity of this measure to changes in economic forces. This information allows us to compare diverse assets and liabilities in terms of their economic risk and return characteristics.

Profitability measures

Our model generates measures of profitability. We calculate these measures relative to the Treasury yield curve.[3] In this manner, we can calculate a static spread. This can be interpreted as the incremental return (or cost) over Treasuries in a world of constant interest rates. We may also calculate an option-adjusted spread (OAS). This is the net incremental return (or cost) over Treasuries in a world of volatile interest rates. The difference between the OAS and the static spread is the option cost, in basis points.

We summarize these profitability measures in figure 5.2. The static spread of a block of SPDAs is represented by point A. Because the insurer has written valuable policyholder lapse options, fair market valuation of these options suggests that the expected cost (OAS) is above the static cost. Simplistically, this can be thought of in terms of the impact of option exercise. If policyholders lapse when interest rates are high, these funds can be refinanced only at these higher rates. This is reflected by point B.

On the other side of the balance sheet, a callable asset may have a static spread represented by point C. In this case, the fair market valuation of these options suggests that the expected return (i.e., OAS) is below the static return. Simplistically, again, if the issuer calls the bond because interest rates have fallen, the insurer will have to reinvest the proceeds at lower rates. This is reflected by point D. Putting the two sides together, we find that the large static profitability (point C minus point A) shrinks (point D minus point B) when we account for options on both sides of the balance sheet.

Surplus variability measures

We calculate the present value of assets and liabilities by averaging results across thousands of replications. The present value of surplus is simply the difference. To assist in understanding the variability of the present value of surplus, we can generate asset and liability present values for various interest-rate levels. We show an example of this in figure 5.3. This chart shows the present values of assets and liabilities for instantaneous yield curve shifts of up to 200 bp. Notice that, in this example, the value of the liabilities exceeds that of the assets for both rising and falling interest rates. Consistent with the profitability analysis, the insurer is economically worse off when rates fall, owing to the increasing value of call options written in asset portfolios. When rates rise the insurer is also worse off, but in this case it is due to the increasing value of lapse options written in SPDAs.[4]

3 SPDA characteristics

This section describes the unique characteristics of SPDAs that are important to the valuation. These include policy features, insurer expenses, and behavioral assumptions (regarding the insurer, the aggregate policyholder, and the aggregate competitor). While these behavioral assumptions could easily be the subject of a separate paper (or papers), our objective here is to summarize major features while developing general functions that capture their essence.

Policy features

An SPDA is a tax-advantaged savings product offered by life insurers. Interest income on the single premium is accumulated on a tax-deferred basis, subject to a 10% penalty for withdrawal (to a non-SPDA vehicle) prior to age $59\frac{1}{2}$. This interest income can be characterized by an initial crediting rate, guarantee period, guaranteed minimum rate, and insurer crediting interest reset strategy (see below). This reset strategy is not a part

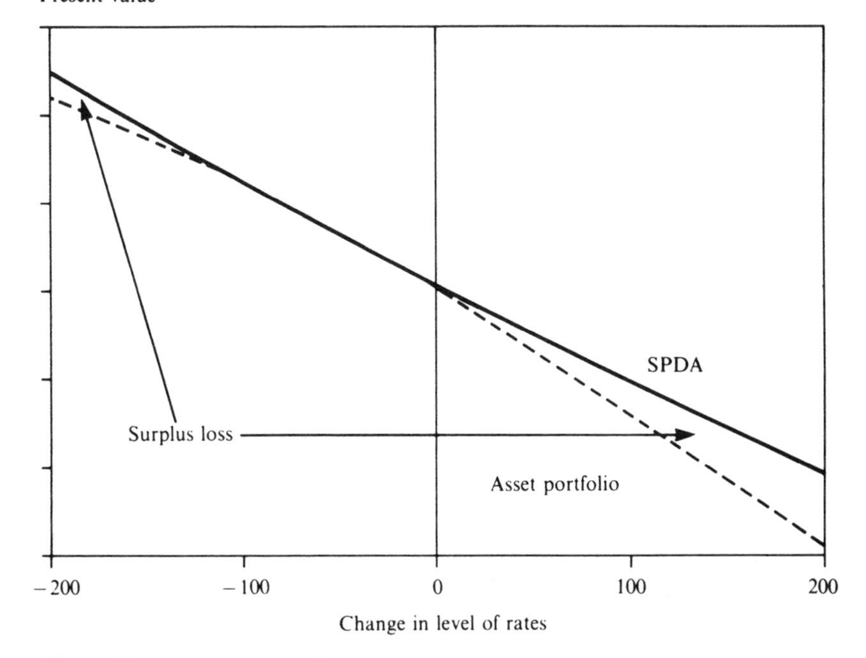

Figure 5.3 Present value of assets and insurance liabilities (plus and minus 200 basis points)

of the policy and hence, until realized, may only be perceived by the policyholder.

To encourage policyholders to keep their policies with the insurer, policies contain back-end load or surrender charges. However, the policies typically offer the policyholder the option of removing a portion of the accumulated cash value free of surrender charges. The accumulated value of the SPDA, adjusted for these surrender charges, is called the cash surrender value.

Insurer expenses

The surrender charges are imposed, in part, to offset acquisition costs. Sales commissions and other issuance expenses reduce the upfront proceeds to the insurer. For example, if acquisition costs are 5.15%, a $100 premium would result in upfront proceeds of $94.85 for the insurer. The insurer does this hoping to earn a higher rate on the proceeds, over time, than it credits to the SPDA.

Finally, the insurer incurs various ongoing expenses in both managing SPDA funds and performing administrative duties. These expenses can be thus categorized as investment expenses and administrative expenses. These expenses are important in assessing the overall cost of an SPDA block of business. We summarize these policy features and insurer expenses for the base case SPDA in table 5.1 at the end of this section.

Lapse behavior

SPDAs contain provisions that allow the policyholder to lapse (surrender) the policy at his discretion, subject to penalties. This means that, if the alternative ways in which the investor may deploy the funds are more valuable, he will withdraw them. While a detailed empirical study of lapse behavior is beyond the scope of this chapter, we can summarize here variables of likely importance. These can be categorized as either economic or non-economic.

Economic forces involve the investment characteristics of the policy relative to other alternative investments. Of key importance here are (1) the likely crediting rate on the policy as compared with alternative rates on other SPDAs available in the market, and (2) rates on other kinds of investment media in the capital markets. Since complete withdrawal from the SPDA market will trigger tax penalties, capital-market rates are likely to be less important than rates on other competitive SPDAs to which funds can be transferred without tax consequences. Consequently, the policyholder's perception of the insurance company's crediting policy relative to competitors is of key importance to the lapse decision.

Tempering the easy movement from one SPDA policy to another SPDA policy, however, is the lack of ready information on the universe of current rates available on policies, although aggressive companies do advertise heavily and sell through brokers. Also, most policies have stiff surrender charges (5–10% of the policy balance) in the early years of a policy, which generally make it expensive to lapse. This means the insurance company's crediting rate would have to be significantly lower than those of the competition to overcome the inertia created by these transaction costs.

There are, however, policyholder personal circumstances (including death) that lead to lapses independent of financial incentive. These non-economic or background lapses, while rational from the policyholder standpoint, may appear irrational from the insurance company's and financial market's vantage points. Just as non-economic prepayments on low-coupon residential mortgages occur, some base-line level of non-economic lapses will probably take place on insurance products (in this case when the crediting rate is high relative to the market).

There also seem to be cases of irrational non-lapse behavior. That is, the existing policy could become so financially unattractive relative to others available that all policyholders would be expected to lapse, without exception. Competition consistently paying 500–600 bp more on identical contracts would be an example of such circumstances. Again, like the borrowers who continue to pay 17% home mortgages when 10% is available, a fraction of SPDA holders will never lapse, no matter what the financial incentive.

Other important forces in consumer-lapse behavior include seasonality, policyholder age, policy seasoning, burnout, (how long superior rates from competitors have been available), and how the policy was sold – agent or broker. For the purposes of this chapter, however, we use a simplified model that captures a minimum lapse rate, a maximum lapse rate, financial incentive, inertia, and policy age. A convenient general functional form used extensively in the analysis of residential mortgage prepayments is the arctangent function.

The specific form used here is:

$$\text{annual lapse rate} = a + b\,\text{ARCTAN}\,[m(r - i - y) - n] \tag{1}$$

where a, b, m, and n are parameters chosen to give an interest insensitive annual lapse rate of 4%, a maximum lapse rate of 45%, and an accelerating lapse behavior as the differential between the competitor's and insurer's crediting rate grows more positive. The variables are: r, the competitor's rate, i, the company's crediting rate, and y, the inertial threshold. The latter depends on the applicable surrender charge in a particular period. We amortize this charge over an assumed three-year horizon to arrive at the value for y.

Figure 5.4 shows the lapse function with our base assumptions at three distinct ages in the life of an SPDA policy. Notice that the lapse function shifts to the left as surrender charges fall through time. That is, seasoned policies will have a greater propensity to lapse than newly issued policies. While better lapse functions can be developed, the arctangent provides a simple yet rich method of testing the importance of lapse behavior on the financial properties of the SPDA contract.

Crediting strategies

Insurance company

As we noted in the last section, policyholder lapse behavior is critically dependent on the relationship between the rate credited on the policy and those that other SPDA issuers are paying on similar instruments. The insurance company can therefore directly influence lapse

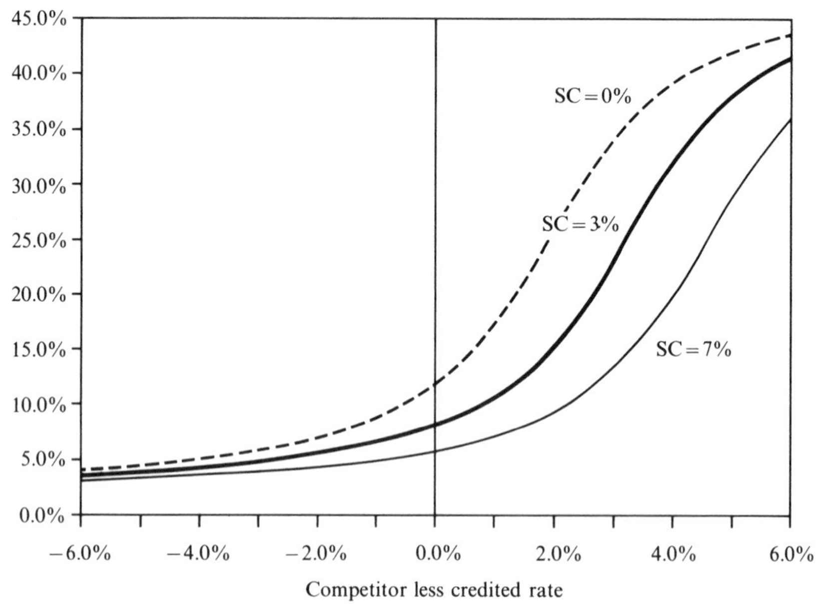

Figure 5.4 Policyholder lapse function (effect of surrender charge)

behavior, as one critical variable determining its magnitude is directly under the company's control. For example, the company can largely avoid lapses by paying a sufficiently high rate on the policies.[5] As we discuss below, meeting the competition will not always be the most economic strategy to follow. While higher crediting rates will prevent lapses, they are expensive. Economic theory suggests that, given the particular lapse behavior of policyholders, there will exist an optimal crediting rate level and a corresponding (non-zero) number of policies that lapse for each interest-rate environment. The "best" crediting strategy is one set sufficiently low that the savings in interest credited to the policyholders who don't lapse is equal to the earnings lost from those who do lapse.

While most insurers do not have an explicit crediting strategy, we can make some reasonable generalizations. First, when interest rates rise, book yield on the asset portfolio is an important constraint. Second, when market rates fall, the competitive nature of the SPDA business seems to dictate that "old money rates, do not stray too far below existing book yields." Hence, for both rising and falling interest rates, old money crediting rates, generally lag market interest rates.

To represent this behavior, we use a simple, yet flexible, autoregressive model. This model assumes that the insurer adjusts the crediting rate, on reset (and subject to a minimum rate), by a function of the then current differential between market rates and the current crediting rate. We can describe the crediting rate next reset as:

$$\text{Crediting rate } (t+1) = CR(t) + a[R(t+1) - CR(t)] \quad (2)$$

where $CR(t)$ is the current crediting rate, $R(t+1)$ is the market rate next reset period and "a" is a variable that defines the "speed of adjustment" to market rates.[6] We assume the market interest rate, $R(t)$, is the yield of the five-year Treasury. For our base-case strategy, we assume that "a" is 20%.[7] Hence, if market interest rates rise (fall) by 100 bp from one reset to the next, the new crediting rate will rise (fall) by 20 bp.

Competitors

It is plausible that competitors in the SPDA market face constraints similar to those of our company, so we can assume that they display similar crediting behavior. In the aggregate, however, there is one major difference. When interest rates rise the firm cannot credit current interest rates. But new entrants to the industry are not burdened by a low yielding portfolio, and thus they can afford to pay current prevailing rates available in the capital markets.[8] The existence of new profitable investment opportunities created by rising interest rates – and an industry weakened by portfolio losses – will encourage the formation of these new competitors. Also, the existing industry can raise "new money" by paying higher rates and investing the proceeds in instruments with higher yields. New and existing competitors drive a wedge between the company's crediting rate on old business and competitive rates in the market.

The composite competitors' new money rate therefore reflects current markets when interest rates rise and portfolio yield or "old money rates" when interest rates fall. We show this behavior in figure 5.5. We superimpose our insurance company's crediting rate (equation 2) for the same interest-rate environments on this function to show how crediting and competitive rates can diverge when rates rise.[9]

4 SPDA valuation: sensitivity to economic forces

The last two sections provided the necessary analytic tools to value SPDA contracts. This section uses these tools on the base-case SPDA contract (table 5.1) and shows how valuation is affected by the various economic

Table 5.1. *Base-case SPDA characteristics*

Policy characteristics	Level
Premium (upfront)	$100
Crediting rate (initial)	9.06%
Guarantee period	one year
Crediting rate resets (after guarantee)	annual
Minimum rate	3.00%
Surrender charges	7% year 1, declining one percentage point per year
Free withdrawal (% policy balance)	10%
Insurance company expenses	
Sales commissions (one time)	5%
Issuance expenses (one time)	15 bp
(Net proceeds from $100 premium)	$94.85
Investment expense	25 bp per year
Administrative expense	15 bp per year
Behavioral assumptions	
Crediting strategy (figure 5.5)	Equation 2
Competitors' strategy (figure 5.5)	max (five-year UST, or Equation 2)
Lapse relationship (figure 5.4)	arctangent

considerations that are not under the control of the insurance company. The economic influences include:

1 the level of interest rates and the shape of the yield curve,
2 the volatility of interest rates,
3 lapse behavior of policyholders, and
4 crediting behaviour of competitors.

The initial economic environment is the Treasury yield curve of 9 December 1988, which puts the five-year Treasury (the driver rate for the competitor and crediting rates) at 9.06% and the thirty-year Treasury at 8.97%. We assume one-month interest-rate volatility to be 16%. This means that if the current one-month interest rate is 9.06%, there is a 2/3 chance that at the end of one year, one-month interest rates will be within the range of 7.61%–10.51%. We have summarized lapse behavior in figure 5.4 and crediting behavior in figure 5.5.

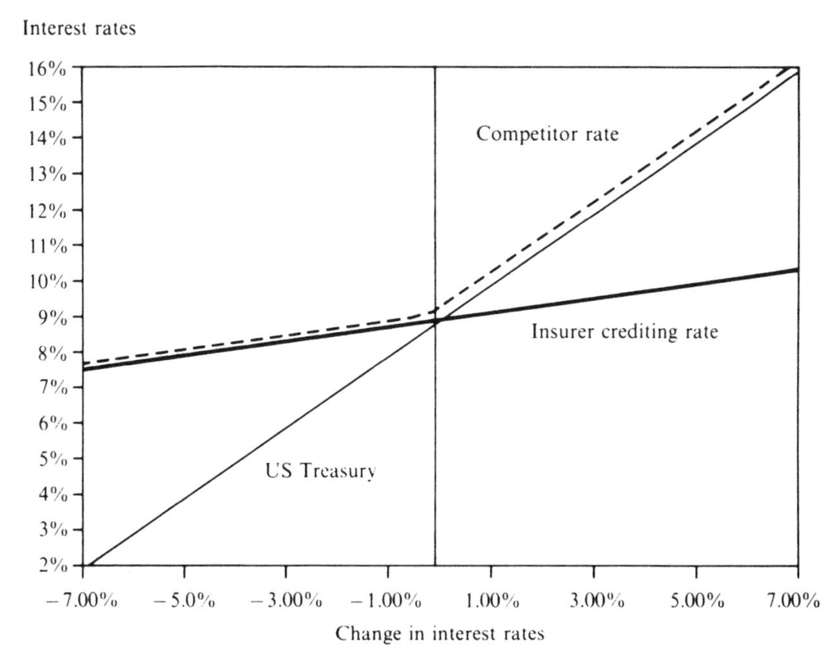

Figure 5.5 Crediting strategy

We will discuss forces that are under the control of the company – the insurance company's crediting strategy and contract provisions – in section 5.

Estimated SPDA costs

Table 5.2 shows the results of the valuation analysis. The option-adjusted spread of this SPDA is 126 bp over Treasuries. This means the all-in cost, including expenses and the value of the options granted to the policyholder, is 126 bp more than the return associated with equivalent Treasury cashflows. Viewed differently, the investment portfolio would have to yield at least 126 bp over Treasuries (on a risk- and option-adjusted basis) to break even on these policies.

The static spread to the Treasury curve is 94 bp. Because the initial SPDA rate of 9.06% is equal to the five-year Treasury, the spread is solely due to expenses, primarily amortization of the upfront commission. The remaining 32 bp represents option costs related to the policyholder's right to lapse.

Table 5.2. *SPDA cost analysis from Goldman Sachs valuation model*

Line	OAS	Static spread (bp)	Option cost (bp)	Duration (bp)	Convexity
1 Base case (table 5.1)	126	94	32	2.8	0.17
2 No expenses	13	−8	21	3.3	0.16
3 No lapse sensitivity	91	91	0	3.3	0.06
4 Low lapse sensitivity	107	93	14	3.2	0.15
5 Steep lapse sensitivity	140	94	46	2.7	0.19

Interestingly, if all expenses are removed from the analysis, the option cost falls to 21 bp (line 2 in table 5.2). This means that insurance company expenses by themselves have an option component. Commission costs are, in effect, amortized over the life of the policy in the OAS calculation. Lapses cause policies to shorten in effective maturity, and therefore commissions are amortized over a shorter period at a higher cost per year when rates rise. Of course, when rates fall, the amortization cost per year declines as the policy's maturity lengthens on slower lapse rates. On average, however, across all interest paths, the cost per period is higher than would be the case with a simple amortization in an unchanged interest-rate environment. This is because lapses increase more when rates rise than they slow when rates fall. From the analysis, we see that this makes a difference on the order of magnitude of 10 bp in option cost.

Does an OAS of 126 bp make this SPDA an "expensive" liability? To gain perspective, we compare an SPDA with an alternative form of funding – borrowing in the capital markets. Bondholders would stand behind policyholders in the event of a liquidation, and therefore bonds should have wider spreads to the Treasury yield curve than SPDA. Currently, highly rated insurance companies could borrow at around 50 bp over five-year maturity Treasuries, so SPDAs look expensive by this measure. This suggests that highly rated companies could replace their SPDA policies with cheaper capital market borrowing. But rating agencies, regulators, and investors do not seem to view insurance policies and borrowing in the same light. Issuing SPDAs is the business of the company, while borrowing is not an end in itself.[10]

The above cost analysis is, of course, critically dependent on the assumptions made. We have discussed the issue of the company's expenses, but policyholders' lapse behavior is still an unknown quantity. To convey

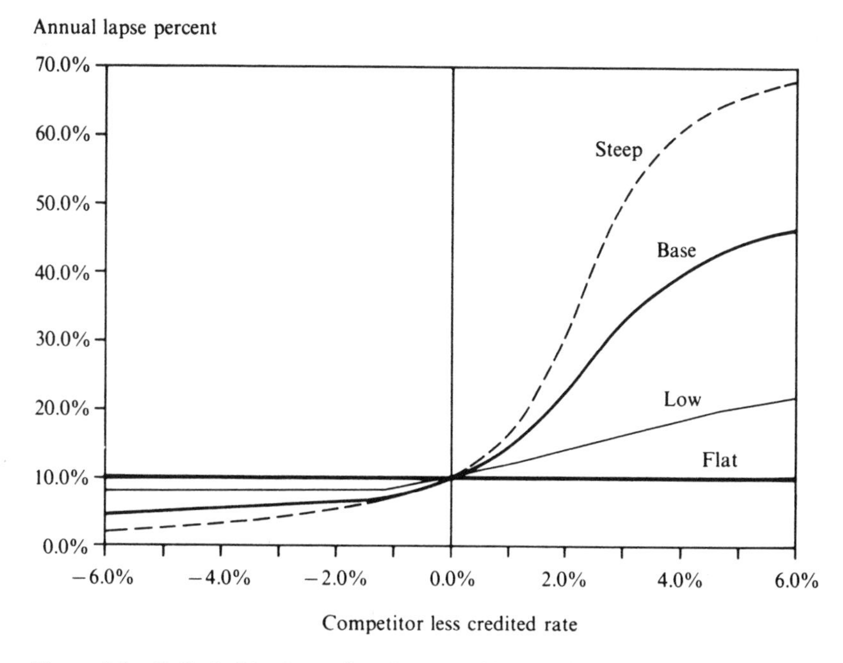

Figure 5.6 Policyholder lapse function (sensitivity analysis)

a sense of how important the lapse relationship is, table 5.2 also provides OASs on the several alternative lapse specifications shown in figure 5.6.

The first alternative (line 3 in table 5.2) is a book of business with no interest sensitivity in lapses ("Flat" in figure 5.6). Here we assume only the background lapse rate of 10% for all interest-rate paths. Note how the OAS falls to 91 bp. Actually, this policy has no option component at all, and the 91 bp reflects the amortization of the sales commissions and ongoing investment and overhead expenses. Line 4 shows a low interest sensitivity lapse function ("Low" in figure 5.6), which might be associated with policies sold through a captive sales force. The OAS here is 107 bp with a 14 bp option cost. Line 5 corresponds to "Steep" in figure 5.6, which has substantial interest sensitivity. Such policies typically could have been sold by brokers and represent "hot" money. The OAS here is 140 bp with an option cost of 46 bp.

Knowledge of how different sales methods affect lapses in an existing book of business can give clues as to the value of a captive sales force. In our example, if the lapse functions "Low" and "Steep" can be viewed as typical for captive and broker agents, captive agents add 33 bp per year in value to the SPDA business.

SPDA interest sensitivity

To characterize the interest-rate risk associated with the SPDA, this section focuses on three measures: option-adjusted duration (OAD), option-adjusted convexity (OAC), and an SPDA "price/interest-rate curve" for plus and minus 200 bp.

Option-adjusted duration

The OAD is an approximation of the sensitivity of the SPDA's present value to a parallel shift (up or down) in the initial yield curve. This can be interpreted as the percentage change in present value for a 100 bp shift in interest rates.[11] Hence, an OAD of 4 would reflect a 4% change in present value for a 100 bp shift up or down.

The OAD of the base-case SPDA policy is 2.8 years (table 5.2). Two elements contribute most directly to this number: first is the nature of the crediting strategy. Absent policyholder lapses, the duration of the SPDA will be approximately the duration of the crediting strategy. If the crediting strategy is representative of an underlying portfolio, the longer (shorter) this portfolio, the longer (shorter) the duration of the SPDA. This is apparent from line BAC in figure 5.7, where point A represents the base case. If the crediting stategy is made more insensitive to changes in interest rates (i.e., "*a*" is reduced from 20% to 10%, typical of a longer-duration portfolio), the OAD of the SPDA rises to 4.2 years (point B). If the crediting strategy is made more sensitive ("*a*" = 33.3%, typical of a shorter-maturity portfolio) the SPDA duration falls to 2.1 years (point C).

An important implication of this analysis is that, when the crediting strategy is directly tied to the asset portfolio, we cannot solve for the economic characteristics of the SPDA without prior knowledge of this portfolio. As we later show, dependency on the actual asset portfolio may make it difficult to define a defeasing asset portfolio (as actuaries often do), as well as minimizing the cost of the liability.

The second element determining SPDA duration is policyholders' lapse behavior. The greater the level of background lapse, the shorter the average life and duration of the SPDA. In figure 5.7, lowering background lapses to 4% per year from the base of 10% lengthens duration to 3.2 years (point D); increasing background lapses to 15% shortens it to 2.6 years (point E).

Option-adjusted convexity

The OAC represents, for a 100 bp parallel shift in the yield curve, the difference between the actual change in present value and the change predicted by duration. While OAD is a first-order approximation to the

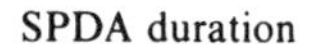

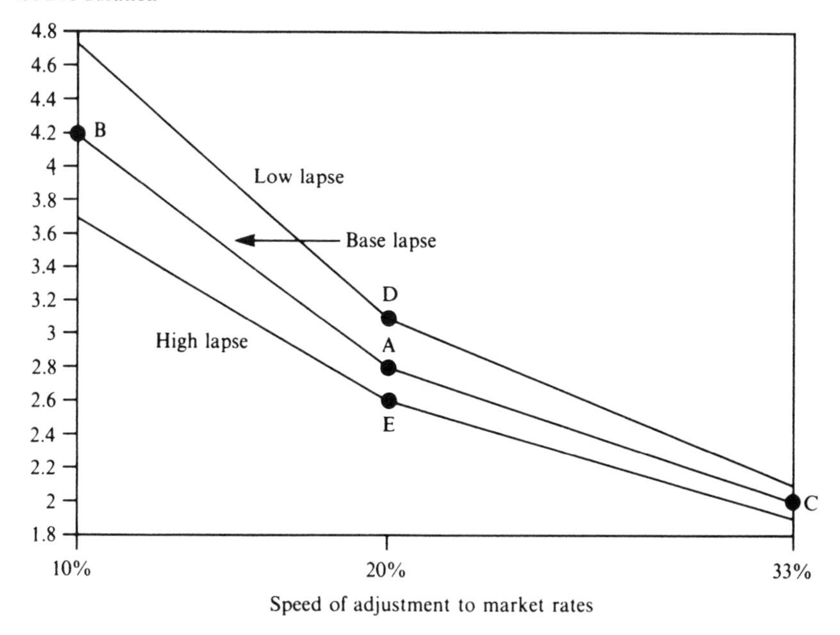

Figure 5.7 Sensitivity of duration of crediting strategy and lapse rate

change in present value, OAC reflects the curvature of the present value/interest-rate relationship. The greater the OAC, the greater the curvature and the poorer the price forecast provided by the OAD.

To interpret the OAC, consider a security with an OAD of 4, an OAC of 0.5, and a present value of 100. The OAD suggests a 4% change in present value to 104 if rates fall by 100 bp, and to 96 if rates rise by 100 bp. Taking the curvature, or OAC, into account, the present value change would be reflected as a 4.5% increase (to 104.50) for a 100 bp fall and a 3.5% decrease (to 96.50) for a 100 bp rise.

Long-duration assets and securities with embedded long options have significant positive convexity.[12] Securities whose options have effectively been written have small or negative convexity. Since an SPDA is an asset to the policyholder and since the policyholder has the option to lapse, we would expect the SPDA to be more convex than an equal duration Treasury. A 2.8-year Treasury has a convexity of 0.03, while – from table 5.2 – the SPDA has a convexity of 0.17, which confirms our intuition. Note also from table 5.2 that the greater the option component (the steeper the lapse functions), the greater the convexity.

While the OAD and OAC allow us to approximate the present value sensitivity to small changes in interest-rate levels, we are also concerned with the impact of larger interest-rate shocks. We can measure the sensitivity by calculating the present value at different interest-rate levels. In doing so, we generate a price/interest-rate curve.

SPDA price/interest-rate curve

Figure 5.8 shows the present value of the SPDA and an equal-duration Treasury for parallel and instantaneous yield curve shifts, upward and downward. We assume that at the point of zero change in interest rates, the value of the Treasury and SPDA liability are equal at $94.85. The durations are also equal at 2.8 years. As interest rates rise 400 bp, the SPDA liability falls in value to $87.08 while the asset portfolio declines to $84.71, for a present value loss in surplus of $2.37.

It is interesting to note that the SPDA value does decline significantly as interest rates rise. For, if policyholders were perfectly rational in an economic sense, the SPDA value would never fall below $93.70, the cash surrender value (CSV). However, because there is some non-economic lapse behavior, the present value of the SPDA can fall below the CSV line. For plus 400 bp, this policyholder inefficiency is worth $6.62 ($93.70–$87.08) to the insurer. Even so, the fact that the policyholder lapse option has value means that the value of the SPDA declines at a slower rate than that of the Treasury. The lapse option provides incremental value of $2.37 ($87.08–$84.71) to the SPDA policyholder over the Treasury.

Another way of interpreting this present value difference of $2.37 is by marking-to-market both sides of the insurer's balance sheet. The $2.37 represents the difference between the $10.14 capital loss on the assets ($94.85–$84.71) and the $7.77 capital gain on the SPDA liability. If the insurer were to offset the policyholder lapse option with capital-market instruments, the mark-to-market capital loss on the assets could be reduced to the capital gain on the SPDA.

5 SPDA valuation: policy terms and crediting strategies

Because policyholder lapse behavior is not purely rational in economic terms, the insurer can attempt to use the resulting inefficiencies to reduce the policy cost. Policy terms and crediting strategies are two variables over which the insurer has some control. These variables may be used to reduce the ultimate cost of the SPDA and to alter its interest-rate sensitivity characteristics. In this section, we will consider surrender charges, initial crediting rate, guarantee period, and crediting strategy.

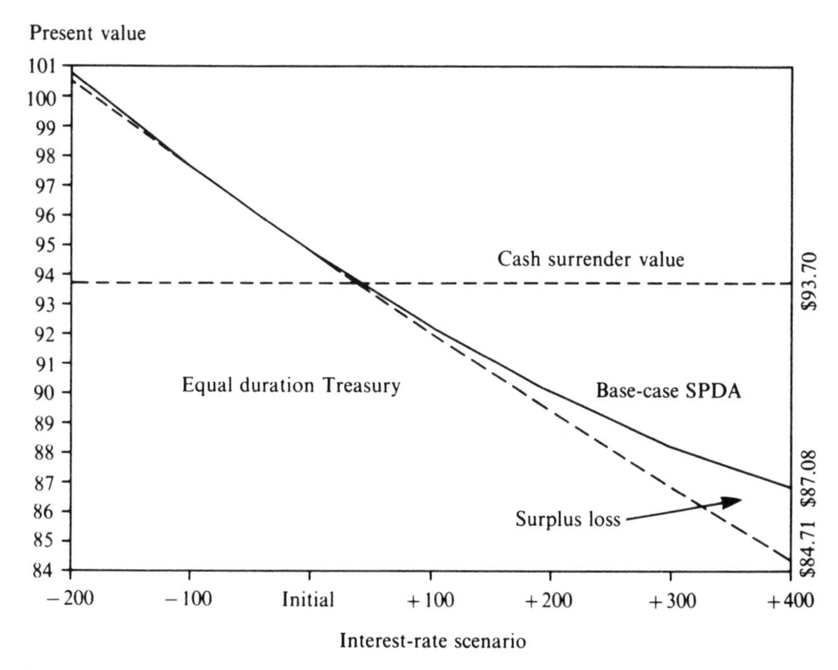

Figure 5.8 Present value of SPDA versus equal-duration Treasury

Contact provisions

Table 5.3 shows the effect of independently altering three major contract provisions: initial rate, guarantee period, and surrender charges on the SPDA's OAS. Lowering the initial crediting rate 100 bp and holding all other contract provisions the same reduces the OAS from 126 bp to 78 bp. The bulk of this reduction comes in the static spread, which, ignoring expenses, would put costs well below Treasuries. The option cost component, however, is little altered. Similarly, increasing the initial rate by 100 bp increases the OAS by 46 bp, again mostly by increasing the static spread. These results by themselves suggest a policy of crediting low rates. But of course, this ignores the marketing side of the coin: How many SPDA sales will be made when your initial crediting rate is 100 bp below the competition?

Another, more fruitful, way to view the analysis is to note that increasing the initial crediting rate 100 bp increases SPDA costs only by 46 bp (recall that, after the one-year guarantee period, the crediting rate reverts to equation 2). There may be some higher initial rate that brings in just enough

Table 5.3. *Effect of contract provisions on SPDA costs*

	OAS (bp)	Static spread (bp)	Option cost (bp)	Duration	Convexity
Base	126	94	32	2.8	0.17
Initial rate					
8.06%	78	49	28	2.9	0.19
10.06%	172	136	36	2.8	0.15
Guarantee period					
6 months	114	94	21	1.8	0.05
18 months	136	94	42	3.5	0.27
Surrender charges					
7% for all years	78	63	15	3.2	0.20
0% for all years	177	124	53	1.6	0.27

additional business that the additional profit, at lower profit margins, offsets the additional costs to all existing policyholders from the higher crediting rate. With some knowledge of how sales respond to the initial crediting rate, we could assess this tradeoff. The important point of the current analysis is that the model defines the marginal cost function for alternative initial crediting rates.

We can perform a similar analysis on the guarantee period. Reducing the guarantee period by six months lowers the option cost by 12 bp and shortens the duration by a year. Increasing the guarantee period by six months increases the cost by 10 bp and increases the duration by over half a year. Guarantees have the effect of making the policy more like a fixed-rate bond, as the crediting strategy is less sensitive to market interest-rate changes. This makes the policyholder's ability to lapse more important and the option cost higher.

Surrender charges are one of the most powerful tools in altering the financial characteristics of SPDAs. Surrender charges have two effects. First, they are a source of revenue. To the extent lapses occur no matter what the interest-rate environment, they reduce the cost of the SPDA portfolio in much the same way as does a lowering of the crediting rate. Second, they can be a powerful deterrent in reducing the incentive to lapse rationally. In effect, surrender charges raise the exercise price of the option and increase the financial benefit required by the policyholder before lapses can occur.

Table 5.3 also shows the impact of changing the surrender charge scale.

Raising surrender charges to 7% for all years through the seventh, instead of reducing them by one percentage point per year, lowers the OAS to 78 bp. The option cost is cut by more than half, while the static spread falls by a third to 63 bp. The duration of the policy is increased to 3.2 years, reflecting the reduced propensity to lapse. Thus, not only are SPDAs less expensive and less risky, but the asset portfolio can be longer.

Allowing greater movement out of policies by eliminating surrender charges for all years dramatically increases the option cost (to 53 bp) and reduces surrender charge collections, so that the static spread increases to 124 bp. Again we have ignored the marketing ramifications, but the tradeoffs represented in the model are dramatic.

Crediting strategy

As we described in section 3, the insurer has the option of changing the crediting interest rate. While we have valued an SPDA where this rate was allowed to change only symmetrically with changing interest rates, this need not be the case. This section will describe another crediting strategy, which attempts to reduce the cost of the SPDA by taking advantage of inefficient policyholder lapse behavior.

Figure 5.9 shows an alternative crediting strategy, in which crediting rates move asymmetrically. They fall with market interest rates but rise only 20% as fast as competition (and the market).[13] Compared with the base-case strategy, crediting rates decline much more in falling interest-rate environments. Of course, this asymmetrical strategy should lead to increased policyholder lapses. However, because policyholders do not exhibit perfectly economic lapse behavior, the insurer can tradeoff interest cost saved versus increased policyholder lapses. In our model, we can value this tradeoff by comparing option-adjusted spreads. These results appear in table 5.4.

The table shows that the new crediting strategy results in a reduction in OAS of 28 bp, all accounted for by reduced option cost. The crediting strategy contains option-like components arising primarily out of the ability to change crediting rates in an asymmetrical fashion. The increased value to the insurer of this reset option, beyond the higher cost resulting from the greater incentive to lapse, is 28 bp. This implies that the crediting strategy can be a powerful product management tool.

6 Implications for asset allocation

The preceding analysis provides a tool to examine the cost and risk of a particular SPDA policy. Once the contract design has been chosen, including this crediting strategy, we are in a position to examine asset

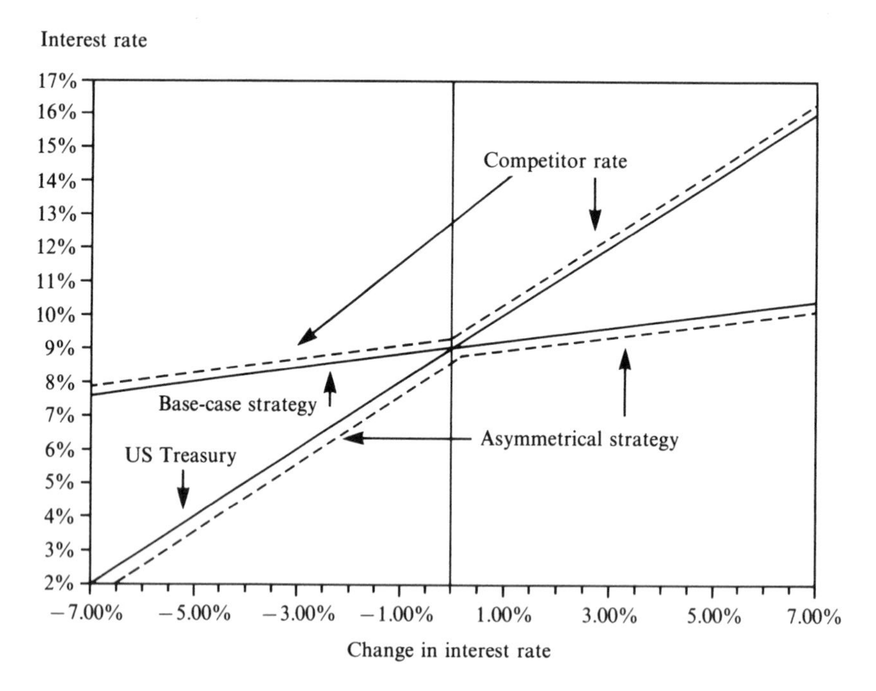

Figure 5.9 Interest-rate crediting strategy (base-case – versus asymmetric)

allocation. While some companies may wish to bear interest-rate and option risk, the following analysis attempts to use current capital market instruments to remove these risks while keeping the business profitable. For our base case SPDA, the OAS is 126 bp, and at least this much must be earned on the asset side for the business to remain profitable.

Table 5.5 shows a seemingly matched (core) portfolio deployed against the SPDA. This portfolio has no option risk, but it does contain credit risk consistent with such securities as perpetual floaters and BBB corporates. It has a static spread of 124 bp, which appears to cover the static spread on the SPDA of 94 bp. After we adjust for the options in the SPDA, however, the business is losing 2 bp (126 OAS on liabilities versus 124 OAS on assets), even without an adjustment for credit risk. While the portfolio is duration-matched, its convexity is insufficient to cover the liabilities. As we showed in figure 5.8, this business sustains surplus losses for rising interest rates. We can improve the performance of this portfolio, however, by including currently available capital-market instruments. Table 5.5 highlights valuation characteristics for two of these asset classes; principal-only mortgage-backed strips (POs) and interest-rate caps.

Table 5.4. *Impact of crediting strategy*

	OAS (bp)	Static spread (bp)	Option cost (bp)	Duration	Convexity
Base case	126	94	32	2.8	0.17
Asymmetric strategy	98	94	4	2.6	0.05

Table 5.5. *Representative securities*

	Price	OAS	Static spread	Option cost	Duration	Convexity
SPDA	94.85%	126	94	32	2.8	0.17
Core bond portfolio	100.00	124	124	0	2.8	0.06
FNMA Trust 20 PO	53.46	227	120	−107	14.8	0.90
FNMA Trust 29 PO	56.13	243	102	−141	15.2	0.99
3-year 9% cap	2.91	65	NA	−97	−54.9	6.40
5-year 9% cap	5.20	65	NA	−104	−45.9	4.30
7-year 9% cap	7.56	65	NA	−108	−39.6	3.04
3-year 13% cap	0.26	65	NA	−9	−90.3	30.76
5-year 13% cap	0.83	65	NA	−17	−66.4	16.18
7-year 13% cap	1.64	65	NA	−23	−52.6	9.95

POs are derivative mortgage-backed securities that receive 100% of principal payments from the underlying mortgage collateral and 0% of interest payments.[14] They are highly bullish securities (i.e., very long duration) that benefit when interest rates fall and underlying mortgage prepayment speeds accelerate. In fact, POs behave much like long call options. Currently, they have high option-adjusted spreads and large positive convexity. POs also have a high static spread and hence don't negatively affect book yields. We should note, however, that for all their current attractiveness, these instruments depend upon the realization of a prepayment assumption. As such, POs add the risk of misestimated prepayments to the portfolio. While the duration of the PO, on its own, is too long versus that of the SPDA, a small holding of POs in a portfolio as a substitute for other long-duration assets has a beneficial impact on convexity.

Table 5.6. *Optimized portfolio*
(Prices as of 9 December 1988)

	Present value	OAS	Static spread	PV Weighted Duration	PV Weighted Convexity
Core bond portfolio	$90.35	124	124	2.53	0.05
FNMA Trust PO	3.78	243	102	0.57	0.04
Interest-rate caps	0.72	65	NA	−0.43	0.09
Total assets	$94.85	128	108	2.67	0.19
SPDA	94.85	126	94	2.67	0.15
Net	0	2	14	0	0.04

Interest-rate caps[15] are long-dated agreements that result in payments to the owner to the extent that a short-term interest-rate (e.g., LIBOR) exceeds an agreed-upon level (on the reset dates). As such, caps increase in value and provide cashflow when interest rates rise. This combination of benefits is ideal for the insurer, for it is under exactly the same circumstances that insurers may suffer from policyholder lapse. Caps have large negative durations (varying according to term and strike level) and large positive convexities. Caps currently are considered inexpensive, as short-rate volatility has dropped to around 16% and the yield curve has inverted.

We can combine these two security classes in a portfolio optimized with the core asset portfolio, as shown in table 5.6.

The optimized portfolio contains the original core bond portfolio (95.2% on a market value basis), FNMA Trust 29 PO (4.0%), and a combination of five-year and seven-year 13% interest-rate caps (0.8%). The POs and caps address the convexity issue. While the combination of the these two securities has no impact on the optimized portfolio duration, the convexity increases from 0.06 to 0.20. This gain in convexity is accompanied by a 4 bp increase in portfolio OAS (124 to 128). While the accounting (static) cost of adding the PO/cap combination is 16 bp (124 for the original portfolio versus 108 for the optimized portfolio after amortizing the upfront cost of the cap), the optimized portfolio provides a higher expected spread (OAS) and performs well against the SPDA for large interest-rate shocks. This can be seen from the price/interest-rate curves in figure 5.10, where the surplus drain is removed from the increasing interest-rate scenarios.

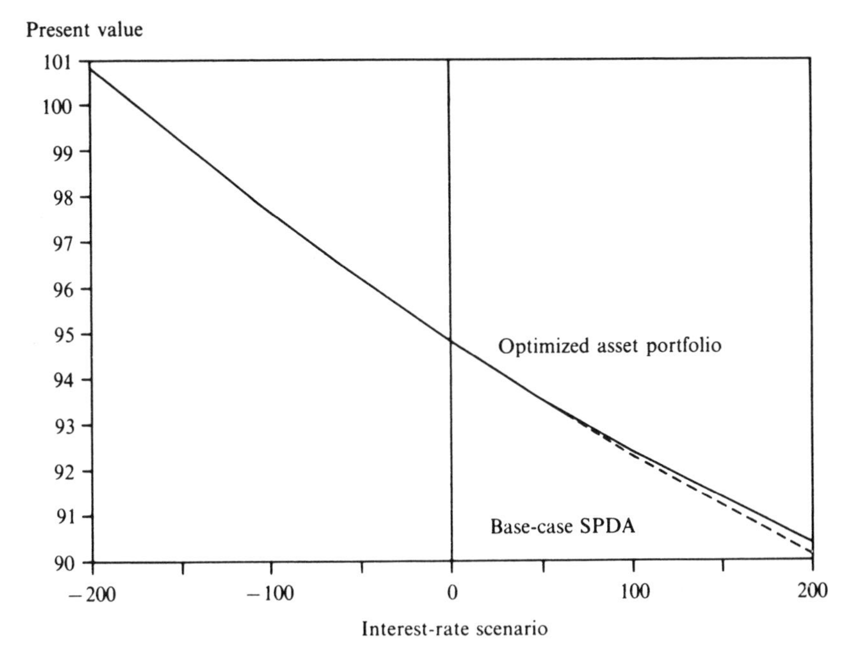

Figure 5.10 Present value of base-case SPDA versus optimized asset portfolio

7 Conclusions

This chapter has presented an option-pricing approach to the valuation of single-premium deferred annuities. This approach has allowed us to calculate an expected cost for the SPDA (the option-adjusted spread) and the value of options granted to the policyholder, and to perform an economic analysis of various asset and liability strategies. We may draw several conclusions:

1 SPDAs can be expensive liabilities for highly rated companies. This is due to a combination of high acquisition costs, crediting rates that are often above (intermediate) Treasury levels (for tax advantaged funds), and the ability of policyholders to select against the insurer when the insurer can least afford it. The option-adjusted spread and option cost highlight the expected cost of the SPDAS, while the option-adjusted duration, option-adjusted convexity, and price/interest-rate curve shown in this chapter highlight the risk.
2 Understanding policyholder lapse behavior is a crucial element in assessing risk and expected cost. Unfortunately, significant data are not

easily accessible. However, since profitability can be greatly affected by lapses, extensive empirical analysis is warranted.

3 The SPDA gives the insurer a powerful tool, the ability to change the crediting rate, which itself can be thought of as an option. This option allows the insurer to capitalize on policyholder inertia by lowering rates and to affect policyholder behavior by judiciously raising rates. An optimal crediting strategy can reduce the cost and riskiness of the SPDA. Policyholder lapse behavior is, again, the key to this analysis.

4 Typically, investment strategy has had an important impact on the insurer's crediting strategy. But because the expected cost and riskiness of the SPDA vary with this crediting strategy, the optimal asset portfolio to be deployed against the SPDA is unclear. We recommend first solving for an optimal crediting strategy independent of the asset portfolio by, for example, minimizing the OAS, subject to marketing constraints. Then the insurer can define the properties of the SPDA and determine an optimal asset portfolio.

5 Capital market instruments, such as interest-rate caps and principal-only strips, can be used to improve the risk-return profile of the asset portfolio relative to the SPDA. These products help deal directly with the lapse risk inherent in the SPDA.

Appendix: a brief discussion of the valuation model

Our valuation model evaluates a wide range of interest-sensitive securities. We can provide a brief overview of the model by separately describing its three major components: stochastic interest rates, cashflow generation, and option pricing. While not an exhaustive discussion, the appendix should give the reader an appreciation of the model. The methods are standard in option pricing, particularly for mortgages, see, e.g., chapter 14.

Stochastic interest rates

We use a Monte Carlo simulation to generate our interest-rate paths.[16] These paths are samples from a population of paths that would determine the current US Treasury yield curve. We use a one-factor model of short-term Treasury rates to define these scenarios – i.e., the only source of randomness in the model is the short-term interest rate. All Treasury prices are defined given (1) the average movement of the short-term interest rate per unit of time and (2) short-term interest-rate volatility. In practice, we assume a particular volatility and solve for the average movement of the short rate per unit of time, so that we reprice the Treasury yield curve.[17] This method of pricing does not explain why Treasuries of various maturities have the prices they have, but rather takes them as given market data.[18]

Cashflow generation

Once we have produced our interest-rate paths, we generate cashflows for each security. For instruments such as mortgage-backed securities and SPDAs, these cashflows are dependent upon both current and past levels of interest rates. Therefore, we generate the cash flows for these securities within each path by moving forward in monthly steps. These interest-sensitive cashflows are generated by an empirical model of prepayment or lapse behavior.[19] Other cashflows – e.g., SPDA expenses – are likewise generated within each path.

Option pricing

The cashflows for each path are then discounted to the present by the one-month Treasury rate within the path. The average of these present values is the price of the instrument. Because the model has discounted the cashflows with Treasury rates, the model price may differ from the market price. If these cashflows are perceived as free of credit risk, then the difference represents an apparent arbitrage. Otherwise, the difference represents a combination of arbitrage and default risk.[20] To equate these two values, the model adds a constant spread to each discount rate across each path. This spread is called the option-adjusted spread (OAS).

The OAS is the expected interest-rate spread to the Treasury curve, inclusive of the impact of any options on the cashflows of the security. We can value these options by removing the volatility from future interest rates. If we do so, then the only possible interest-rate path is the implied future one-month rates contained in the Treasury curve. We can then calculate the spread that when added to this single path equates the model value to the observed value. This is called the static spread. The difference between the OAS and the static spread is the impact of future interest-rate volatility. This difference is the value of the embedded options, or option cost, in basis points.

NOTES

1 See David Babbel and Robert Stricker, "Asset/Liability Management for Insurers," Goldman, Sachs & Co., May 1987.
2 A brief discussion of the valuation model is included in the appendix.
3 Alternatively, we could present these measures as yields. But because we are valuing securities containing embedded options, yields do not represent the likely return or cost, unless interest rates remain unchanged. Because yields are important for accounting purposes, however, they are calculated in the model.
4 Figure 5.3 does not address profitability per se, but deals only with changes in the value of surplus as interest rates change. Figure 5.2 is concerned with profitability analysis, while figure 5.3 presents risk analysis.
5 Assuming the company is perceived to be solvent.
6 While we may gain important insights by using equation 2, the model is capable of handling more elaborate crediting strategies. For instance, the function could

incorporate portfolio yields as a driver rate or as a constraint (i.e., maximum). Also, as we will describe in section 5, we can define a function that lags the market asymmetrically. For example, we can easily define a function that falls faster than it rises.

7 Since our crediting strategy may vary between floating ("a" = 100% in equation 2) and fixed ("a"=0%), the lower the "a", the more the crediting strategy resembles a fixed-rate bond. If we were to model a bond with the coupon set by equation 2, then we could calculate its economic properties (i.e., duration and convexity).

8 If the company always credited a short-term rate of interest, the lapse problem would disappear. However, such a strategy would probably not allow a high enough initial rate to generate significant business.

9 It is difficult to imagine the crediting function of figure 5.5 as the long-run equilibrium crediting profile of an SPDA. Investors seem to get the best of all worlds: higher rates than Treasuries when rates fall and no less than Treasuries when rates rise. In order for this to occur, there must be (1) long-run arbitrage opportunities for firms – i.e., after adjusting for credit losses and option costs, long-run returns above Treasuries must be available – or (2) some other form of long-run industry subsidy must exist.

10 For companies other than the highly rated ones, SPDAs may be a cheap – indeed perhaps the only – way to obtain unsecured funding.

11 For each shift in the initial yield curve, both cashflows and discount rates change. This can be contrasted with the calculation of Macaulay duration, which measures only the impact of discount-rate changes. Macaulay duration is inappropriate as a measure of price sensitivity for securities with embedded options.

12 See Jess B. Yawitz, "Convexity: An Introduction," Goldman, Sachs & Co., September 1986.

13 While we could test an unlimited number of possible strategies, we have chosen the asymmetrical strategy to highlight the option component of the crediting strategy. In this context, we are not primarily concerned with important marketing considerations.

14 See Michael Asay and Tim Sears, "Stripped Mortgage-Backed Securities," Parts I and II, Goldman, Sachs & Co., January 1988.

15 See David Babbel, Peter Bouyoucos, and Robert Stricker, "Capping the Interest Rate Risk in Insurance Products," Goldman, Sachs & Co., February 1988.

16 See Boyle, "Options: a Monte Carlo approach," *Journal of Financial Economics*, 4 (May 1977), 313–38.

17 See Fischer Black, Emanuel Derman, and William Toy, *A One-Factor Model of Interest Rates and Its Application to Treasury Bond Options Financial Analysts Journal*, 33–9, Jan.–Feb. 1990.

18 See J. Cox, J. Ingersoll, and S. Ross, "A theory of the term structure of interest rates," *Econometrica*, 53, (March 1985), for discussions of issues in the term structure of interest rates relevant here.

19 Because these models represent asymmetric behavior (i.e., for certain portions of the function, prepayments or lapses may speed up faster than they slow down) they incorporate an option component. Also, because these models are empirical representations of consumer behavior, they do not necessarily assume purely rational option exercise. See Richard and Roll (1989) and Kang and Zenios (1992) for a discussion of the issues related to consumer choices in mortgages. Similar analysis applies to lapse behavior.

20 For a discussion of default risk, see Robert Litterman and Thomas Iben, "Corporate Bond Valuation and the Term Structure of Credit Spreads," Goldman, Sachs & Co., November 1988.

REFERENCES

Michael Asay and Tim Sears (1988), "Stripped mortgage-backed securities, parts I and II," Goldman, Sachs & Co., January.

David Babbel, Peter Bouyoucos, and Robert Stricker (1988), "Capping the interest rate risk in insurance products," Goldman, Sachs & Co., February.

David Babbel and Robert Sticker (1987), "Asset/liability management for insurers." Goldman, Sachs & Co., May.

Best's Insurance Reports, *Life-Health*, various editions.

Fischer Black, Emanuel Derman, and William Toy (1990), "A one-factor model of interest rates and its application to treasury bond options," *Financial Analysts Journal*, January–February, 33–39.

Daniel P. Boyce and Elizabeth Tovian (1988), "Flexible premium annuities – a persistency study," Research Report, Life Insurance Marketing and Research Associates, Hartford CT.

Phelim P. Boyle (1977), "Options: a Monte Carlo approach," *Journal of Financial Economics*, 4, May: 313–38.

Michael J. Brennan and Eduardo F. Schwartz (1976), "The pricing of equity-linked life insurance policies with an asset value guarantee," *Journal of Financial Economics*, 3, June: 195–213.

Conning & Company (1985), "Risk and rewards in the SPDA market – management considerations for a leveraged product," Insurance Analytical Services Studies, Conning & Company, March.

John C. Cox, Jonathan E. Ingersoll, and Stephen A. Ross (1985), "A theory of the term structure of interest rates," *Econometrica*, 53, March.

John C. Cox and Stephen A. Ross (1980), "An analysis of variable rate loan contracts," *Journal of Finance*, 35, May: 389–404.

John C. Cox and Mark Rubinstein (1985), *Options Markets*, Prentice Hall.

Kenneth B. Dunn and John McConnell (1987), "A comparison of alternative models for pricing GNMA mortgage-backed securities," *Journal of Finance*, 36, May: 471–84.

Ernst & Whinney (1988), *Accounting for Life Insurance and Annuity Products*, Financial Reporting Developments, June.

P. Kang and S.A. Zenios (forthcoming), Complete Prepayment Models for Mortgage Backed Securities, *Management Science*.

Robert Litterman and Thomas Iben (1988), "Corporate bond valuation and the term structure of credit spreads," Goldman, Sachs & Co., November.

Peter D. Norris and Sheldon Epstein (1988), "Finding the immunizing investment for insurance liabilities: the case of the SPDA," Morgan Stanley Fixed Income Research, March.

James E. Pesando (1974), "The interest sensitivity of the flow of funds through life insurance companies: an econometric analysis," *Journal of Finance*, 29(4): 1105–21.

Scott Richard and Richard Roll (1989), "Modeling prepayments on fixed rate mortgage-backed securities," *Journal of Portfolio Management*, Spring, 73–82.

Michael C. Smith (1982), "The life insurance policy as an options package," *The Journal of Risk and Insurance*, 49(4): 583–601.

James A. Tilley (1988), "The application of modern techniques to the investment of insurance and pension funds," Prepared remarks, International Congress of Actuaries, Helsinki, 14 July.

Jess B. Yawitz (1986), "Convexity: an introduction," Goldman, Sachs & Co., September.

Comment on "An economic approach to valuation of single premium deferred annuities"

DAVID F. BABBEL

Asay, Bouyoucos, and Marciano (ABM) have applied the same technology that has been successful in the financial valuation of callable bonds and mortgage-backed securities (MBS) to the valuation of single premium deferred annuities (SPDAs). The application of this technology to insurance product valuations is natural, and long overdue, as it engenders an understanding of the economic importance of policy options that traditional models have heretofore not captured well.

My remarks concerning the ABM study cover three areas: possible extensions to the ABM approach, practical considerations with regard to the interest rates and associated cashflows used in setting up the binomial tree, and potential misapplications of the ABM approach in portfolio structuring.

People familiar with the MBS valuation models used by Wall Street firms will recognize certain buzz words used by ABM and understand at once the particular version of the model that was used. However, it may prove helpful to offer a clarifying comment for the reader less familiar with the extant models. Two classes of models are in common use: one is based on binomial interest-rate trees and the other is based on simulation. The binomial tree is forward looking, and has the advantage of containing an implicit term structure at each node of the tree, while the simulation approach is backward looking, and has the advantage of mapping cashflow patterns that depend on the interst-rate path by which a particular point was reached. ABM take a hybrid approach, simulating interest-rate paths from among the branches of a binomial tree. Thus, their approach garners the advantages of both classes of models, while maintaining computational efficiency. They lose, however, some of the richness of a pure simulation approach in that interest rates are allowed only to move by discrete, rather uniform steps as specified in the binomial tree. It is doubtful that the loss of this added richness would have any economic importance for the application at hand.

The particular binomial tree that was employed as the basis for their simulations is a special case of the generalized one-factor model of Black, Derman, and Toy (BDT). The BDT model's main attraction is that it allows for a term structure of interest-rate volatility, whereas most one-factor models assume constant volatility. The use of non-constant volatility allows BDT to arrive at a better fit of the current term structure of interest rates than most one-factor models. For illustrative purposes, however, in their use of the BDT model, ABM assume constant volatility, so their valuation model becomes indistinguishable from the ubiquitous lognormal binomial interest-rate trees used by others. However, their approach is easily amenable to the inclusion of a volatility term structure.

Most extant MBS pricing models involve two or more factors – a short-term and a long-term interest rate – each of which moves along random paths that are not perfectly correlated with those paths followed by the other. The long-term interest-rate factor is used to trigger mortgage prepayments, while the short-term interest-rate factor is used to discount cashflows back to the present. Other factors which are sometimes used are random interest-rate volatility, and in the case of callable corporate bonds, credit risk. In the ABM study, there is only one random factor. However, this should not be considered a serious detraction from their work. They have modeled other insurance products using more parameters (e.g., volatility term structure) and multiple factors (short- and long-term rates and random volatility) and found their main results to be largely consistent with those achieved with their one-factor model. The feature of SPDAs that allows them to use a more parsimonious model specification is the lapse function, which depends only on the spread between yields offered by competitors (proxied by the five-year Treasury yield, which in their model is completely determined by the short-term Treasury yield and the assumed volatility) and the current insurer crediting rate (which adjusts toward the new five-year Treasury yield at some hypothetical speed).

My second set of comments relates to the riskiness of the promised cash payment streams to the policyholder. Most life insurance products fall within the rubric of insurance insolvency guarantee programs existing throughout most of the United States. Accordingly, promised cashflows to policyholders can be valued as riskless from the policyholders' viewpoint, and also as riskless from the issuers' viewpoint. The insurer option to default on promised payments can be modeled and valued separately as an asset to "put" its liabilities to a group of surviving competitors. In the case of SPDAs and GICs (guaranteed investment contracts), however, several of the large insurers contend that these instruments do not fall within the spirit of the guarantee programs, under which surviving insurers are assessed an amount to make up the shortfall of an insolvent insurer. For example, when Baldwin United became insolvent, a large number of SPDAs were put at risk. Rather than litigate the question regarding the eligibility of SPDAs for guarantee fund bailouts, the insurance commissioner strong-armed a number of insurers to take "voluntary" actions. A group of approximately 100 insurers active in the SPDA market contributed to the voluntary plan, in an effort to maintain the credibility and image of the SPDA industry at large. Thus, the legal question remains unsettled.

Accordingly, it is uncertain whether SPDAs should be modeled as risky liabilities, in which expected cashflows or their certainty equivalents would be substituted for promised cashflows, or rather modeled as other insurance products (in the spirit of ABM), where expected cashflows are identical to promised cashflows, and the put option is modeled separately as an asset of the SPDA issuer. While the approach chosen may make little difference to the insurer, as the put option will reduce the liability value by an amount similar to that which would be achieved by modeling the liability as an obligation subject to credit risk, it can make a sizeable difference to the consumer, who is either being granted a virtually default-free policy or a risky contract. Moreover, while the modeling treatment may result in similar net liability values to the insurer, the legal treatment can impart a different optimal strategy to the insurer's portfolio design, as the option to default can increase in value by following a risky investment strategy without a commensurate increase in a given insurer's liability cost if SPDAs are covered by state guarantee programs.

My final point is to bring to light a practice common among Wall Street firms that may be little appreciated by outsiders. When one is focusing on option-adjusted spreads (OAS) as a component of expected liability costs, the economics of the SPDA market forces insurers to look for high OAS assets in an effort to fund these costly liabilities. Insurers naturally seek out corporate bonds in a mistaken assumption that the high promised yields are somehow associated with higher OAS. In fact, this has not been the case in recent years. Virtually all of the Wall Street models, when using the same inputs that are consistent with Treasury and MBS pricing, have been showing corporate bonds in the AAA to BBB range to have negative or zero OAS, *often even before subtracting out the credit risk component*! Yet this is not well-known to insurers, perhaps because many of the same Wall Street firms use a lower volatility assumption to look for relative value among the corporates before reporting OAS. This lower volatility assumption prices the securities closer to their traded values. It also produces a lower implied cost to the callability option of corporate bonds, resulting in increased OAS.

But this common practice clearly is not consistent with the Treasury rates and their associated volatilities that underlie the unified binomial pricing trees used for both assets and liabilities. It suggests a widespread phenomenon of overpricing of corporate bonds in today's marketplace. I would like to proffer three explanations for this phenomenon. First, the valuation model for corporates may be misspecified in some capacity. For example, the actual callability of a bond may behave, in practice, differently from the optimal callability function embedded in the pricing model. Second, perhaps markets simply do not generate efficient pricing of fixed-income securities, which calls into question the entire valuation methodology and our assumptions about competitive markets. Third, perhaps there are missing elements in the corporate bond valuation models that, if included, would impart added value.

While it is easy to conceive of missing elements, it is difficult to discover elements that would impart additional (positive) value to the pricing of corporate bonds. Let me posit one such element. Book yield. Suppose there is an industry that has a huge appetite for bonds that can, for regulatory purposes, be booked at high yields. The US insurance industry is a prime example of such an industry. It has assets exceeding $2 trillion, and last year purchased more bonds than all the new corporate bonds issued in America. Regulators traditionally focus on book yields rather than economics, and are more likely to look kindly upon insurance policies that appear to be profitable on a book basis. The demand for these higher yielding corporate securities could force up their prices beyond that which would prevail were the focus restricted simply to the economics. In playing this game, insurers have exposed themselves toward the short straddle risk mentioned at the outset of the ABM study. Clearly this game cannot go on indefinitely. Over time, as insurers understand more about the exposure they face, one would expect the corporate bonds to be priced cheaper relative to Treasuries than is currently the case. An analogous situation existed with MBS pricing a few years ago, whereby many MBS securities exhibited positive OAS for extended periods of time. However, as the pricing technology became more widespread, competition forced down the OAS to levels that are much closer to zero in today's market. One would expect the same thing to happen with corporate bonds, as insurers begin to gain pricing insights from adopting approaches similar to those taken by ABM. In this regard, Asay, Bouyoucos, and Marciano have made a valuable contribution indeed.

6 The optimal portfolio system: targeting horizon total returns under varying interest-rate scenarios

EVDOKIA ADAMIDOU, YOSI BEN-DOV,
LISA PENDERGAST, and VINCENT PICA

1 Introduction

The continuously increasing competitiveness and complexity of the fixed-income markets have been a driving force in the use of sophisticated quantitative techniques in financial planning applications.

Not long ago, investors purchased securities or portfolios of securities with the implicit assumption that the securities would be held to maturity and that interest rates would remain stable. In this environment, investment decisions were based primarily on credit quality and yield-to-maturity. To a large extent, interest-rate risk was disregarded. As a result, when rates rose dramatically in the late 1970s and early 1980s, investors were faced with immense interest-rate risk exposure. In response, immunization techniques, such as cashflow matching, duration matching, and portfolio insurance, were developed and were proved to be sufficient.

However, the dramatic growth of the mortgage sector, the introduction of sophisticated new securities and hedging instruments, and increasing interest-rate volatility have forced portfolio managers to develop more complex mathematical techniques that are capable of addressing long-term investment concerns under a variety of interest-rate environments.

Traditional portfolio-management techniques do not provide an efficient solution to the problems associated with actively managing a portfolio in the current market environment. Portfolios that are simply duration-matched cannot achieve the desired tradeoff between risk and reward in volatile environments. Conventional investment parameters, such as yield-to-maturity, duration, and convexity, are insufficient to evaluate a portfolio adequately. The uncertainty of future cashflow/liability streams and the impact that this uncertainty has on portfolio performance can no longer be overlooked. Concepts such as total return and horizon analysis are now the key determinants to successful investment management.

Optimal structured portfolios

The mathematical model described in this chapter is capable of designing optimal structured portfolios that meet various investors' criteria under different interest-rate environments. The model is based on stochastic-optimization techniques that represent the uncertainty of essential parameters of the market. More specifically, a scenario analysis is proposed that generates an optimal portfolio composed of securities with widely differing characteristics that matches the investor's performance profile under a variety of interest-rate scenarios. This analysis differs from traditional portfolio methodologies in two ways. First, it is dynamic and not static; instead of assuming that interest rates follow one path over the life of the portfolio (deterministic analysis), it accounts for a broad range of interest-rate scenarios that may occur over the life of the portfolio (stochastic analysis). Second, instead of using conventional parameters (Macaulay or modified duration, convexity, yield-to-maturity, etc.) to help predict the performance of the portfolio, the analysis is based on the concept of total return over a specified horizon period; this is also known as horizon total return.

Interest-rate movements are embodied by the different scenarios. A few versions of the underlying optimization problem (one for each scenario) are generated to model the interest-rate uncertainty. A weighting scheme is imposed on the scenarios to reflect the portfolio manager's actual view of the direction of future interest rates. The desired portfolio performance for each scenario is defined by the portfolio manager. Next, an optimization technique is applied that bundles the different scenarios and leads to optimal structured portfolios along the desired range of interest-rate movements. The goal is to generate an optimal structured portfolio that does not rely on hindsight, but that accounts for all the defined interest-rate scenarios and their respective probabilities.

A road map to optimal structured portfolios

Figure 6.1 gives a brief overview of the optimal structured portfolio process as it applies to both the creation of new portfolios and the restructuring of existing portfolios. The structure of the paper follows the general flow of the optimal portfolio process.

The purpose of this chapter is to introduce, in non-technical terms, a scenario analysis optimization technique that is capable of generating optimal structured portfolios that match the investor's performance profile under a number of appropriately weighted scenarios. Section 2 discusses the increasing need for addressing risk when making investment decisions

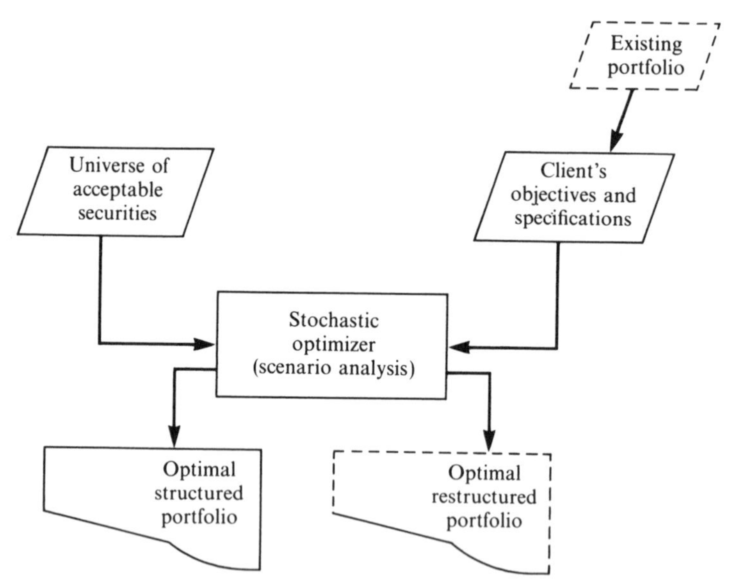

Figure 6.1 The optimal structured portfolio process

and for setting performance objectives *vis-à-vis* a given risk/reward profile. In addition, the section briefly discusses traditional fixed-income securities while focusing on interest contingent, fixed-income instruments. In section 3, frameworks for evaluating the risk/reward tradeoff of various fixed-income instruments are explored. We propose that horizon total return is the most valid framework for the valuation of complex securities. Finally, one-year horizon total returns for various fixed-income instruments are examined and presented graphically.

Section 4 provides a step-by-step analysis of the scenario analysis methodology and the major complications that can arise. Specifically, the stochastic aspect of the problem is explained and a scenario analysis optimization technique is proposed for the generation of optimal structured portfolios over a given time horizon and across a number of interest-rate environments. Section 4 also details the special features of the Prudential-Bache Optimal Portfolio System. Section 5 shows actual applications of the Optimal Portfolio System. We conclude with a summary in section 6.

2 Structured portfolios

Within the context of increasingly competitive and complex markets, portfolio managers place risk control as a primary variable in the acquisition and management of assets. Rather than simply purchasing securities at an attractive spread over Treasuries, managers now purchase structured packages of different securities and hedge instruments that are designed to address asset/liability management concerns such as asset growth, diversification, profitability, and risk management under a variety of interest-rate environments.

Pension funds, banks, insurance companies, savings institutions, and other financial intermediaries have certain common concerns. The financial intermediary must evaluate the interest-rate sensitivity of the existing liability and asset structure. Armed with this information, investment strategies must be developed in such a way that the return-on-assets matches the liability structure under a variety of interest-rate scenarios. For example, if an institution has a neutral outlook on future interest rates, the goal is to maintain a constant spread between the return-on-assets and the cost of the liabilities. However, if the institution has a definite view on the future direction of interest rates, it may wish to "weight" the portfolio so that the returns will be affected positively by the expected change in interest rates.

An added layer of complexity arises from the imposition of the new regulatory guidelines imposed on commercial banks and thrifts. For thrifts, an integral component of these guidelines is the introduction of procedures to measure the institution's interest-rate risk exposure. According to *Thrift Bulletin* 13, the thrift institution's board of directors must establish limits on the level of the institution's interest-rate risk exposure and, most importantly, the board must be aware of the sensitivity of the institution's net-interest income and market value of portfolio equity to interest-rate changes.

In the following section, various portfolio performance objectives are discussed and an efficient means for generating asset growth at risk/reward levels consistent with these objectives is introduced.

Structured portfolio objectives

The most important parameters to consider when structuring a portfolio are the performance objectives of the investor. Portfolios may be bullish, bearish, or neutral; they may be affected either adversely or beneficially by market volatility; and they may be sensitive or insensitive to analytical assumption error.

Depending on the investor's objectives and interest-rate outlook, different types of portfolios can be structured. For example, the conservative investor may want a flat return portfolio that, regardless of interest-rate movements, is expected to provide a constant return. Other portfolio structures may be more appropriate for different investors.

Figure 6.2 shows one-year horizon total-return profiles for flat, bullish, and V-shaped portfolios under seven different interest-rate scenarios. The hypothetical profitability profiles in figure 6.2 illustrate a fundamental risk/reward tradeoff that must be evaluated by all managers. Attempting to eliminate risk may be very costly in terms of profit. For example, the flat return portfolio normally provides substantially smaller base-case returns than bullish or bearish portfolios. The tradeoff must be evaluated within the context of the portfolio manager's expectation as to the likelihood of each potential interest-rate movement.

In the next section, the types of securities that may be included in an optimal structured portfolio are presented. The security types actually selected to be included in an optimal structured portfolio are determined by the portfolio manager's objectives and investment criteria.

Structured portfolio components

There are a number of security types that are candidates for inclusion in a structured portfolio. With the advent of more complex and, at times, uncertain markets, there has been a proliferation of structured securities and hedge instruments that offer unique financial characteristics geared specifically to this environment. Traditional fixed-income securities can be combined with these new structured securities and hedge instruments to create what may be termed structured portfolios.

Structured portfolios can include securities such as Treasury, agency, and corporate bonds; GNMA, FNMA, and FHLMC single-family mortgage pass-throughs; multifamily mortgage securities; strip mortgage-backed securities; CMO bonds and residuals; AA pass-throughs, etc. Hedge contracts may include instruments such as interest-rate swaps, options, futures, and options on futures. While all securities are potential candidates for inclusion in a structured portfolio, there are a variety of types that are more commonly used. Here, we outline the most commonly used securities in the creation of structured portfolios – from traditional fixed-income securities to new, complex securities.

Traditional fixed-income securities

Most fixed-income securities are subject to interest-rate risk; in a rising interest-rate environment, the value of a fixed-income security tends to decline, while the value tends to increase when rates fall. However, if the holder of a standard, fixed-income security holds it until maturity, the initial value of the security is not affected permanently by changes in interest rates. This is due to the fact that, when the bond matures, it still pays the holder its full face value. If the holder intends to sell the bond prior

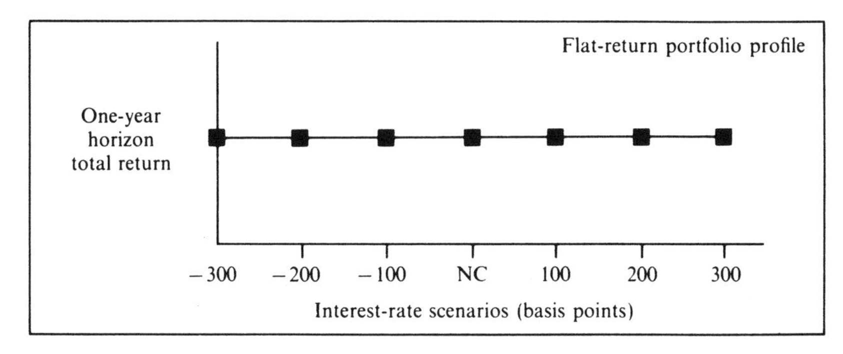

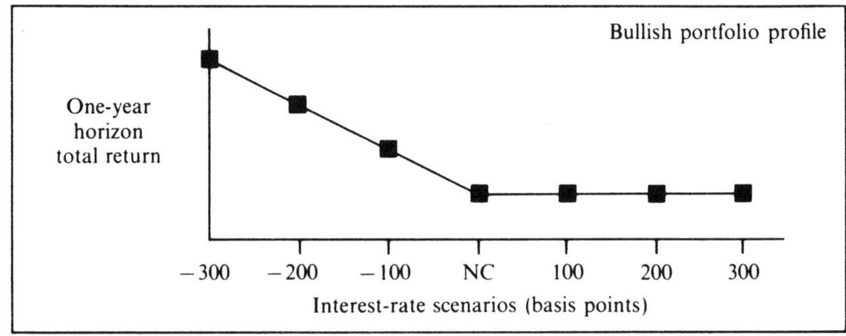

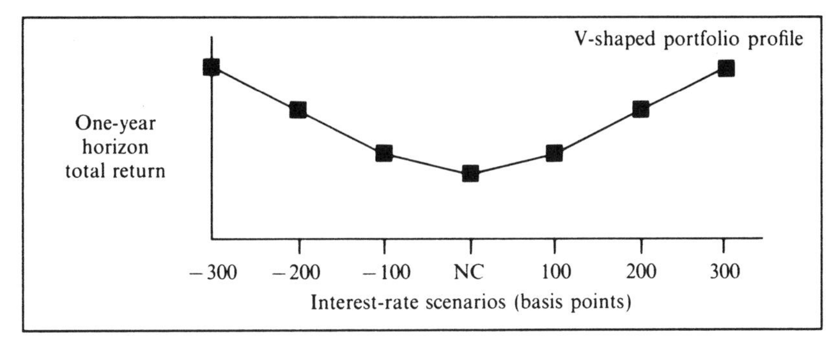

Figure 6.2 Hypothetical portfolio profiles

to maturity and the yields offered by similar bonds are higher, the holder may be forced to sell the bond for less than what was originally paid in order to offer potential buyers a comparable yield to what is currently available in the marketplace.

Treasuries are considered to be free of credit risk as they are guaranteed by the "full faith and credit" of the US government and they are extremely liquid investments. Because of these benefits to investors, their yields are generally lower than those on other more uncertain debt instruments. The major advantage of Treasury STRIPs is that they eliminate reinvestment risk, allowing the investor to lock-in the yield-to-maturity at purchase if the issue is held to maturity. A non-callable corporate bond is equivalent to a long-term promissory note. If the issuer fails to pay either the principal or interest when due, bondholders have a legal claim over common and preferred stockholders as to both income and assets of the corporation.

Interest-rate contingent securities

Interest-rate contingent securities have created the potential for structuring portfolios whose risk/reward profiles are not attainable in the traditional capital markets.

Callable corporate bonds

Callable corporate bonds introduce additional risk – a call provision that gives the issuer the right to fully or partially retire the debt before the stated maturity. In the case of declining interest rates, the possibility of the bond being called becomes more likely. As this occurs, the bond begins to trade to the call date and is treated as a shorter-maturity investment. Yields on callable corporate bonds are higher than on comparable non-callable corporate bonds to offset the potentially negative effects of the call provision.

Mortgage-backed securities (MBSs) and derivatives

Like traditional fixed-income securities, mortgage-backed securities (MBSs) are subject to interest-rate risk. However, there is an added risk associated with investment in MBSs – prepayment risk. Holders of the underlying mortgage collateral have the right, at any time, to prepay part or all of their mortgages, thus introducing a greater degree of uncertainty as to when the cashflows from the underlying mortgages will be received. Prepayment levels depend heavily on interest-rate movements. For example, when interest rates drop sharply, there is a financial incentive for the homeowner to refinance a mortgage at the lower interest rate. However, the new rate must be low enough to more than offset refinancing costs. As a result of the homeowners' option to prepay their mortgages, an MBS's cashflows and other characteristics, such as yield, duration, average life, convexity, option-adjusted spread, and horizon total return, vary with changes in interest rates (Hayre and Mohebbi, 1988, see, also, chapters 5 and 14).

GNMA, FNMA, and FHLMC pass-throughs. Agency single-family pass-throughs provide the basic core investments in structured portfolios. Pass-through certificates are shares issued against pools of specific mortgages. The cashflows from the mortgages are passed through, after subtraction of a servicing fee, to investors, typically with a delay. The payments made to the holders of the security consist of scheduled principal and interest and any unscheduled payments of principal (resulting from prepayments and defaults) that may occur. Agency pass-throughs trade in a highly liquid market. This makes it easy to buy or sell sizeable quantities quickly with minimal impact on the market. Agency MBSs can be financed through repurchase agreements (repos) at attractive rates and bid-ask spreads are minimal. Additionally, a great deal of analytical information (such as that generated by prepayment models (Hayre, Lauterbach, and Mohebbi, 1988 or Kang and Zenios, 1992) is available to facilitate prudent investment.

FHLMC, FNMA multifamily. Multifamily securities also have the credit support of the US government agencies but are available in more limited supply. Multifamily pass-throughs provide the advantages of a four-year to five-year prepayment lock-out and a balloon payment at the end of ten to fifteen years. A disadvantage is greater prepayment uncertainty because of the lack of extensive prepayment experience. However, this uncertainty is moderated by the lock-out and balloon-payment provisions, which single-family pass-throughs do not offer.

Private pass-throughs. Mortgage pass-throughs that do not possess a government guarantee must have some form of credit support in order to obtain a rating. Such credit support may involve the corporate guarantee of the issuer, private mortgage insurance, or the benefit of a senior/subordinate structure. While nearly all private labels carry a "AA" rating or better, there may be significant qualitative differences among them. These securities provide higher yields (approximately 30 basis points or more above FNMA yields), but are less liquid in terms of secondary trading and repo rates.

Interest-only (IO) and principal-only (PO) strip mortgage-backed securities (SMBSs). Interest-only SMBSs and principal-only SMBSs contain either the interest cashflows or the principal cashflows from a pool of mortgage loans. By stripping a mortgage security's interest or principal cashflows, instruments are created that are either very bullish or very bearish (hedge oriented) in nature. Investors who wish to hedge prepayment risk and who have a strong opinion as to future market movements are best suited for investment in IOs and POs. When analyzed individually, IOs and POs are both very volatile securities. However, these securities are best utilized when combined with other types of securities to alter a portfolio's return characteristics or to create a balanced, profitable portfolio.

The primary caution in IO/PO investment is the extreme sensitivity of the security's value to changes in prepayment rates. The amount of interest income earned on an IO SMBS is directly related to the principal balance outstanding at the end of each month. Of course, the principal balance is determined largely by the prepayments on the underlying mortgage collateral. Therefore, in a declining interest-rate environment, increased prepayments reduce the principal balance quicker and thereby cause interest income to decline. Conversely, when interest rates rise and prepayments decline, the principal balance is reduced more slowly allowing for larger interest income. It is said that an IO security exhibits negative duration because, when interest rates fall, the value of an IO falls. When interest rates rise, an IO's value also rises. While there is an active secondary market for SMBSs, the market is not as liquid as the agency pass-through market.

Collateralized mortgage obligations (CMOs)

Since its introduction in 1983, the CMO market has experienced tremendous growth and development. In just a few years, the engineering of CMOs has moved away from its traditional "ABCZ" structure. There is currently a plethora of CMO bond types and it is now very common for issues to contain ten or more classes. The stimulus behind this tremendous growth is the need to design bonds that meet a variety of issuer and investor objectives. CMO engineers can now offer issuers and investors greater choices of maturity, prepayment risk, and performance characteristics. While the cashflows from an underlying pool of mortgage securities are still segmented across time, the new CMO structures employ numerous combinations of classes and stripped cashflows to achieve specific goals. Most importantly, individual classes in a CMO issue must now be analyzed in relation to the entire issue.

In its simplest state, a CMO attempts to reduce a mortgage pass-through security's uncertainty as to the timing and amounts of cashflows by assembling a number of mortgage pools and creating serially maturing bonds from the collateral. CMOs are generally high-quality, AAA-rated securities, yet they offer significantly higher yields than comparable high-quality corporate securities. CMO bonds are in general liquid securities with an active secondary market.

In the traditional CMO ABCZ structure, there are four sequential-pay classes with the last class usually designated as an accrual class (or Z-bond). In such a structure, the first class is given 100% of all principal payments/prepayments until it is completely paid down. The principal payments are then directed to the second class until it has paid down, etc. The last class, the Z-bond, accrues interest at the coupon rate instead of receiving monthly cashflows until all other classes have been paid. As interest on the Z-bond accrues, it is added to the principal balance instead of being paid out to the investor. The Z-bond becomes a current-pay bond once all shorter-maturity bonds in the issue have been retired.

Until mid 1986, most CMOs were designed using this standard structure. However, since that time, the onslaught of CMO bond types has made it increasingly difficult to keep abreast of new developments. While this chapter does not discuss each and every CMO innovation in detail, it examines the major types of CMO bonds and the innovations that have taken place over the past few years (Brehm and Pendergast, 1990).

Current-pay bonds. A current-pay CMO is a bond that begins to receive principal payments from the underlying mortgage collateral once shorter-maturity classes are retired. Prior to the principal-payment period and during principal repayment, interest is paid to the current-pay investor at the given coupon rate based on the outstanding principal balance of the CMO bond. The average life of the current-pay bond is greatly influenced by prepayments on the mortgage collateral. If prepayments increase from the pricing speed, earlier classes most likely will be retired earlier and the current-pay investor will begin to receive principal payments sooner, which will shorten the current-pay bond's average life.

Floating-rate CMOs (FRCMOs). This is a current-pay CMO bond, a Planned Amortization class (PAC), a targeted amortization class (TAC) or a "Companion" bond that is characterized by floating-rate coupon payments, which are calculated using a predetermined spread over an index (typically one-month LIBOR). The coupon on FRCMOs generally adjusts monthly and is subject to a lifetime coupon cap. The performance of an FRCMO is based primarily on two features: the level and behavior of the index on which it is based and the level of prepayments on the underlying mortgage collateral. As the performance of an FRCMO tends to improve in a rising-rate environment, the FRCMO is considered to be bearish. This is because, when mortgage rates increase, a higher interest coupon is earned and, since there is less incentive for homeowners to prepay their mortgages in a high interest-rate environment, the higher interest coupon is paid on a more slowly declining principal balance. Most FRCMOs are priced at or close to par. On a total-rate-of-return basis, in a rising interest-rate environment, an FRCMO provides increasing returns until the lifetime cap is reached, at which point it experiences a slight decline. In a declining-rate environment, the FRCMO is affected by two negative forces, (i) the coupon rate decreases as the index decreases and (ii) prepayments tend to speed up (due to the lower rates) causing the coupon payment to be calculated on a smaller principal balance. The coupon cap is an important FRCMO structural necessity since the index theoretically can increase without limit.

In addition to the standard FRCMO described here, there have been several innovations in the FRCMO market, including Super Floating-Rate CMOs and Inverse Floating-Rate CMOs.

Planned-amortization class (PAC) bonds. The major enhancement offered by PAC bonds is the reduction of prepayment risk. A PAC bond is designed to adhere to a stated schedule of principal payments provided that the underlying mortgage collateral remains within a given PSA range, referred to as a PAC "band." The PAC band tends to span a wide PSA range, providing the investor with stable cashflows under fluctuating interest-rate environments. Since most of the CMO's stability is absorbed by the PAC bond, the non-PAC classes tend to become more volatile as they absorb much of the prepayment risk.

What happens if prepayments fall outside of the PAC band? If prepayments on the underlying collateral exceed the upper PSA limit of the PAC band for a sustained period of time, the PAC's average life will shorten as the faster prepayments will cause the non-PAC classes to pay down sooner thus eliminating the prepayment protection. On the other hand, if prepayments fall below the band's lower PSA limit, the average life of the PAC bond will extend. The degree to which the PAC bond will either extend or shorten depends heavily on the prepayment sensitivity of the underlying collateral, as well as the size of the PAC class in relation to the overall size of the CMO issue.

The PAC innovation has evolved immensely over the last few years with the introduction of PAC II bonds, PAC PO and IO bonds and the spin-off targeted amortization class (TAC) bonds, which provide prepayment protection, albeit less than that offered by the PACs. There are various types of TAC bonds including TAC II bonds, TAC PO/IO bonds, Lock-Out TAC bonds, and Reverse TAC bonds.

Companion bonds. As mentioned above, the presence of a PAC bond increases the amount of prepayment volatility that must be absorbed by the non-PAC classes in a CMO issue. A Companion bond is a coupon-paying bond that absorbs this added prepayment volatility. Because of the added risk, Companions typically are priced at or close to par at higher yields than other types of similar average-life classes. Principal is paid to the companion bond only after it is paid to the PAC bond. When prepayments are high, the added principal cashflows are absorbed by the Companion bond, which tends to shorten its average life. Conversely, when principal shortfalls occur due to a slowdown in prepayments, the Companion receives its scheduled share of principal cashflows only after the PAC bond has been paid its share, thus its average life is extended. Offshoots of the Companion bond include Lock-Out bonds and Super-Principal Only (PO) bonds. The Lock-Out bond is a less volatile version of the Companion bond, while the Super-PO bond is a highly leveraged Companion bond, making only principal payments.

CMO residuals

A CMO residual is the excess cashflow available from the mortgage collateral plus any reinvestment income that exceeds the debt service and operating expenses of the CMO issue. Sources of residual income include the spread between the mortgage collateral and the CMO bond coupons, principal, reinvestment income, overcollateralization, and the bond reserve fund. Residuals are extremely complex to analyze and are relatively illiquid in the secondary market.

Hedge contracts

Various hedge contracts, such as interest-rate swaps, interest-rate caps and floors, options, financial futures, and options on financial futures, may be used to insulate the portfolio from potential losses. In addition, hedge contracts may be viewed as techniques for altering the price volatility and convexity cost of optimal structured portfolios.

Comments

Key concerns for the portfolio manager investing in these complex securities include the high dependency of the securities' cashflows on interest rates, as well as the appropriate evaluation of both individual securities and portfolios. Section 3 on evaluation frameworks reviews some of the more common methods for determining the valuation of fixed-income securities. Within section 3, we propose that horizon total return is the most valid framework for the valuation of complex optimal structured portfolios. Finally, horizon total-return patterns for various fixed-income instruments are examined and presented graphically.

3 Evaluation frameworks

Different frameworks have been used to evaluate the risk/reward tradeoff of various fixed-income instruments. For asset/liability management, immunization techniques tend to be more popular. Immunization's objective is to insure that the present value of the assets is greater than the present value of the liabilities in both rising and declining interest-rate environments. The duration of the liabilities is matched to the duration of the assets and, in order to better match the liability schedule and avoid the single barbell solution provided by duration, convexity requirements are imposed. These traditional risk-reward techniques proved to be sufficient in the late 1970s and early 1980s. However, these techniques are ineffective when applied to the new and complex securities whose cashflows are heavily dependent on interest rates.

The uncertainty in future cashflow/liability streams and the impact that this has on portfolio performance can no longer be overlooked. Specifically, duration does not provide any information about the return of the asset over a specific horizon period or under different interest-rate environments. While duration (Macaulay or modified) appears to be only moderately inaccurate for traditional mortgage securities, it is substantially erroneous for complex securities.

Traditional valuation measures

Investors have many different goals thus making it extremely important to analyze the tradeoffs between risk and reward as they apply to different securities and different investment objectives.

Traditionally, the standard duration yield curve represents the tradeoff between risk and reward. Duration is accepted as a measure of risk and yield-to-maturity is accepted as a measure of reward. In this framework, yield-to-maturity is the performance measure for each security, i.e., the discount rate that equates the present values of all future cashflows to the market price of the security. However, yield-to-maturity has two major drawbacks; it assumes that all cashflows received are reinvested at a single rate and that the security is held to the maturity date. The reality is that securities often are sold prior to maturity and that interest-rate volatility is a critical factor both for securities with interest-rate dependent cashflows and for non-callable securities with fixed cashflows. For example, a fixed-coupon security purchased at par will be worth less if interest rates increase above the coupon rate and will be worth more if interest rates decrease below the coupon rate. Thus, yield-to-maturity is a misleading measure when referring to the total return of a security over a particular horizon. The relationship between the horizon total returns of securities does not correspond to the relationship between their corresponding yields-to-maturity.

Modified duration and convexity also have been used extensively as risk-control measures. Modified duration is a measurement of the sensitivity of the price of a security to changing interest rates, and convexity is a measurement of the sensitivity of modified duration to changing interest rates. The first drawback of both measures is that they do not directly relate to the total return of the asset over a specific period of time. In addition, standard modified duration and convexity formulas lose their economic meaning when applied to securities that experience different cashflow streams under varying interest-rate scenarios. These risk-control measures are effective only for parallel shifts in the yield curve and for securities that do not have interest rate path-dependent cashflows.

Horizon total return

Horizon total return is a security evaluation tool that is superior to yield-to-maturity, modified duration and/or convexity measures. Otherwise known as total return, effective yield, or realized rate of return, horizon total return is an efficient measure for evaluating the performance of fixed-income securities. Given a specific future date (horizon) and an interest-rate environment, horizon total return measures the profit or loss incurred by holding the security until the horizon date. Therefore, total returns of securities maturing at different points in time can be compared. In addition, horizontal total return accounts for the future course of interest rates in the process of determining both the reinvestment of future cashflows until the horizon date and the future value of the security. Horizon total return is determined by three components: capital gains (or losses), cashflow income, and reinvestment income.

Capital gain. The capital gain (loss) component of horizon total return is the increase (decrease) in a security's market value for the specified horizon. The initial price and the salvage value of the security determine the security's capital gain (loss). The initial price is the amount that is paid by the investor to purchase the security. The salvage value is the price of the security on the horizon date. To determine the salvage value of a security, two factors must be considered:

Amortization or premium or accretion of discount. An institution books income in addition to the coupon payments if the security's price is below par or books an expense in addition to the coupon payments if the price is above par. The primary method of accounting for the amortization of a premium or the accretion of a discount on an interest-bearing asset is called the interest method or the level-yield method. The interest method arrives at a periodic interest income (including amortization or accretion) at a constant effective yield to maturity on the net investment, i.e., the full amount of the security plus or minus any amortized premium or accreted discount at the beginning of each period. The difference between the interest income for a given period (as calculated using the interest method) and the stated interest on the outstanding amount of the security is the amount of periodic amortization or accretion and yield adjustment. Standard GAAP and FASB pronouncements describe the appropriate accounting methods used for various financial instruments by different institutions.

Market capital gain. This is the capital gain (loss) associated with interest-rate movements. The market capital gain (loss) is added to the accumulation gain (loss) to determine horizon total return.

Cashflow income. The cashflow income is the total of all coupon payments and principal generated by the asset at the horizon date. For securities with interest-rate contingent cashflows, such as MBSs, this component of horizon total return depends heavily on the interest-rate environment.

Reinvestment income. This is the income earned by reinvesting all cashflows received between the settlement date and the horizon date. Reinvestment income depends on the interest rates available at the time the cashflows are realized. It accounts for a significant portion of a security's horizon total return. The amount of reinvestment income earned depends on the time horizon and the interest rate. The longer the term of the security and/or horizon, the more time there is for the cashflows to be reinvested. The higher the interest rate, the more interest the investor receives and thus the more interest exists for reinvestment.

The examination of the components of horizon total return emphasizes its sensitivity to different interest-rate environments. Horizon total return is not a static tool as is yield-to-maturity, which fails to incorporate market changes. Horizon total return is an efficient, flexible tool that can account for an investor's inclination of market conditions and, based on this preference, can point to appropriate investments. The superiority of horizon analysis rises as the complexity of the investment instrument increases.

Horizon total-return patterns for different security types

As discussed, the value of a security is critically dependent on future changes in interest rates. Different securities react quite differently to changing market environments. Optimal structured portfolios combine individual securities with substantially different characteristics to create portfolios that meet a desired risk/reward profile over a broad range of interest-rate movements. The following section details the one-year horizon total-return profiles of those securities most likely to be included in an optimal structured portfolio.

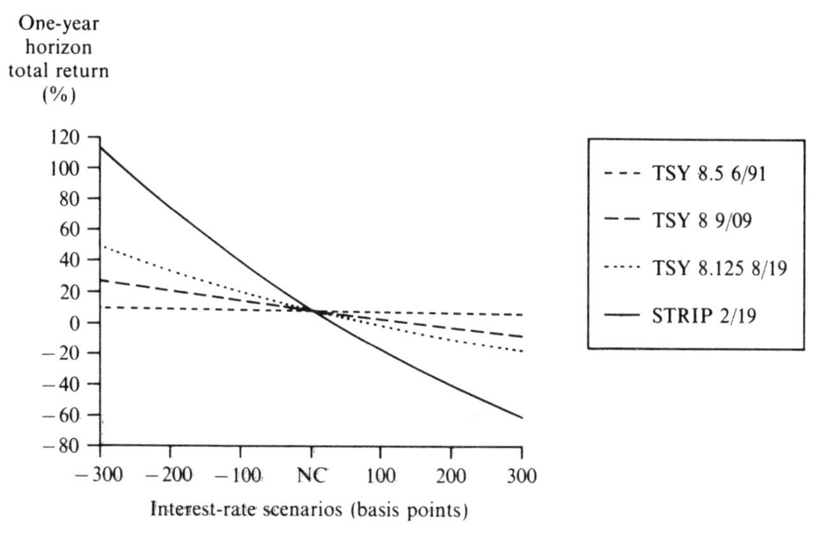

Figure 6.3 One-year horizon total-return profiles on non-callable bonds

One-year horizon total returns for all securities discussed in this section were generated using the Prudential-Bache Structured Portfolio/Synthetic Security (SP/SS) System. All results are stated on a pretax basis. SP/SS total returns are the "all-in" percent return on investment of all interim cashflows at the assumed reinvestment rate. All durations discussed and shown in the tables and figures in this section should be understood as being Macaulay durations.

Non-callable bonds

First, let's consider the horizon total-return patterns of fixed-rate, non-callable bonds. Figure 6.3 illustrates the horizon total-returns of the four non-callable bonds (priced at or close to par with the exception of the government STRIP): an 8.25% two-year government note of June 1991, an 8% ten-year government note of August 1999, an 8.125% thirty-year government bond of August 2019 and a thirty-year government STRIP of February 2019. As can be seen from figure 6.3, the two-year note is least affected by the changing interest-rate scenarios. A basic element of the fixed-income market is that bond prices move in the opposite direction of interest rates. However, the interest-rate risk assumed by an investor only manifests itself if the bond is sold prior to maturity. The sensitivity of a bond's salvage value is linked directly to its Macaulay duration. As the durations of the bonds increase, their price sensitivities also increase.

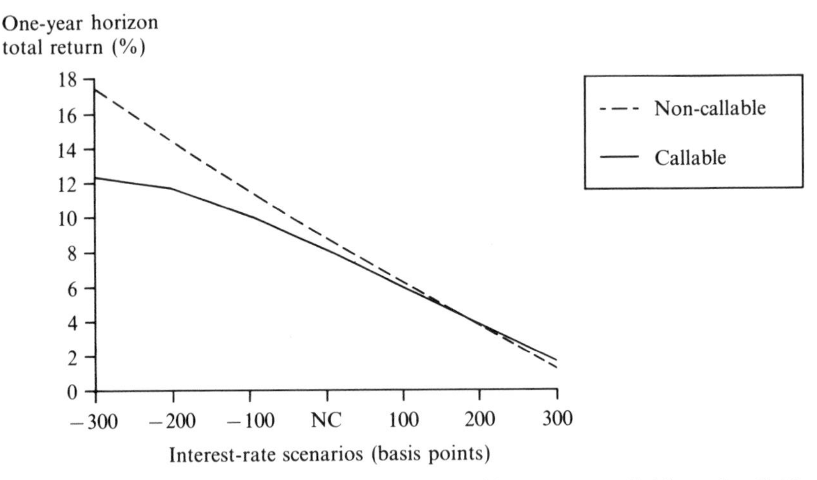

Figure 6.4 One-year horizon total-return profiles on non-callable and callable corporate bonds

Note: ITT Financial: $107.662 (8.30% YMT, +109/10-year); Ohio Power: $104.527 (9.10% YMT, +129/10-year).

The thirty-year government bond displays high price sensitivity – substantially increasing in a declining-rate environment and substantially decreasing in a rising-rate environment. For all four securities, as interest rates fall, their salvage values rise (in some cases, dramatically) and therefore so do their horizon total returns. In a rising interest-rate environment, the reverse occurs. The thirty-year government STRIP displays the greatest price sensitivity.

Callable bonds

Securities with embedded call options have very different horizon total-return patterns. Figure 6.4 shows the returns of two corporate bonds: the ITT Financial 10.125s of 4/5/99 (non-callable) and the Ohio Power 9.875s of 6/1/98 (currently callable at 108.47). In a falling interest-rate environment, the callable bond does not perform as well as the non-callable bond as its duration tends to shorten dramatically.

The callable bond produces superior returns as long as interest rates remain stable. If interest rates fall, the non-callable bond will, in all likelihood, outperform the callable bond. This is generally attributed to the fact that in a declining interest-rate environment the likelihood of a corporation exercising a call on a bond increases. As rates decline below the prevailing rates at the time the security was issued, issuers will want to

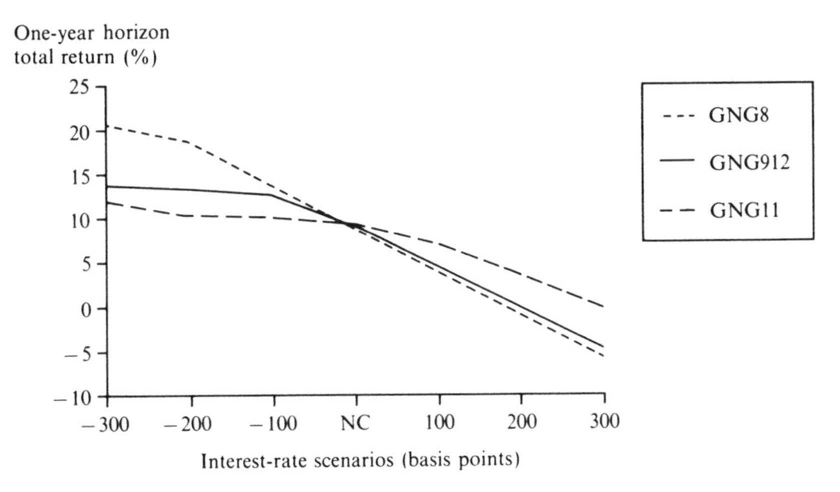

Figure 6.5 One-year horizon total-return profiles on discount, current and premium-coupon GNMA pass-throughs

redeem all of the bonds outstanding and replace them with new securities with lower interest rate. The non-callable bond outperforms the callable bond in a declining interest-rate environment and underperforms the callable in a rising interest-rate environment.

Mortgage-backed securities

Figure 6.5 shows the horizon total-return patterns of discount, current, and premium-coupon GNMA securities under seven interest-rate scenarios. Note that the coupon on the mortgage security is a critical factor in determining the behavior of the MBS under different interest-rate scenarios:

Premium mortgage pass-throughs generally outperform discounts and current-coupon pass-throughs in a rising interest-rate environment. This is due to the benefit of a slowdown in prepayments that allows the investor to receive the high-coupon payments for a longer period of time on a more slowly declining principal balance (due to a slowdown in prepayments). In a declining interest-rate environment, the premium underperforms since principal prepayments are returned at par faster, preventing the investor from receiving the high coupon interest for the period of time originally assumed. Thus, the coupon income may not offset the premium over par paid for the security.

Discount pass-throughs generally perform well in a declining interest-rate environment. As prepayments increase, the security's average life is shortened, which results in the early return of principal at par. The opposite holds true for rises in interest rates.

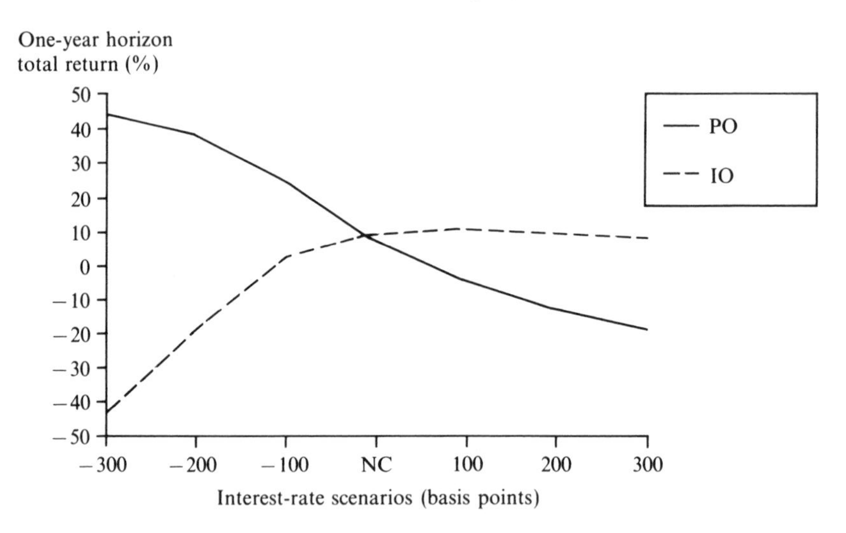

Figure 6.6 One-year horizon total-return profiles on IOs and POs

Interest-only/principal-only (IO/PO) SMBSs

IO and PO SMBSs can have very different horizon total-return patterns. The horizon total return on the interest portion of a high-coupon GNMA SMBS behaves conversely to a non-callable bond. In a falling interest-rate environment, prepayments on a mortgage pass-through security tend to increase causing the security's principal balance to decline. As this occurs, the interest earned on the declining principal amount also declines. However, as interest rates rise and prepayments ease, interest on the principal amount can be earned for a longer period of time. Thus, an IO security is considered a bearish investment since, in a rising-rate environment, its future cashflows and, therefore, horizon total return, increase.

The PO SMBS's response to prepayments is opposite to that of the IO's response. As a PO is a deep-discount security, its value increases when interest rates decline and prepayments increase. When interest rates rise, the value of the PO SMBS decreases as principal is returned slower. Figure 6.6 illustrates the horizon total returns for both an IO and a PO.

Collateralized mortgage obligations

The horizon total-return patterns on different classes in a CMO are highly dependent on the issue's structure. Figure 6.7 shows the horizon total-return profiles for three classes in P-B CMO Trust 2: PB2A, a Companion bond, PB2C, a PAC bond, and PB2D, a Super-PO bond.

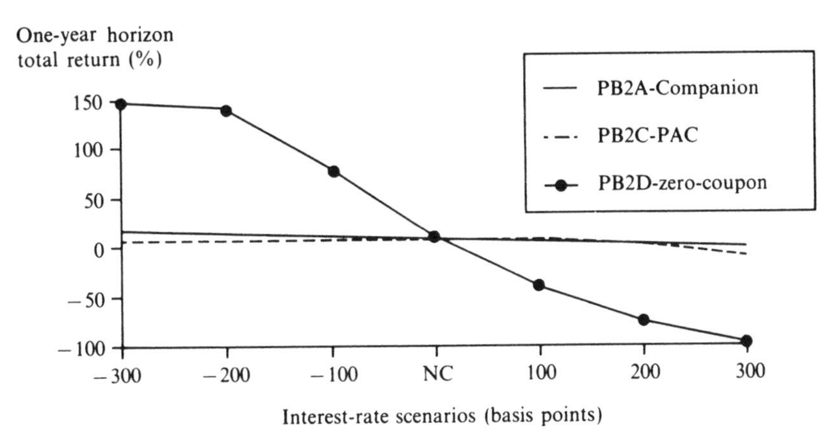

Figure 6.7 One-year horizon total-return profile on a CMO bond

The PAC bond offers the most consistent horizon total-return pattern across all interest-rate scenarios simulated. This is because the PAC keeps to a stable schedule of principal payments as long as prepayments remain within the PAC band. The Companion absorbs any prepayment volatility for the PAC bond. Principal is paid to the Companion bond only after the PAC bond has received its scheduled payment. However, if prepayments increase, the Companion bond absorbs the added principal payments – thus shortening its average life. If prepayments decrease, the average life of the Companion bond is extended.

As figure 6.7 shows, the Companion's performance improves as rates decline (due to an increase in prepayments). If rates rise and prepayments decrease substantially, the Companion's average life extends causing its performance to decline. The Super-PO, like a principal-only SMBS, has powerful bullish profiles. Because Super-POs are priced at deep discounts to par, faster than expected prepayments in lower interest-rate environments cause earlier classes to be retired more quickly. This shortens the average life of the Super-PO and, since it is priced at a deep discount, its return is greatly improved. As can be seen from figure 6.7, the reverse is true when rates rise. If prepayments are slower than expected, earlier classes remain outstanding longer, thereby lengthening the Super-PO's average life and lowering its expected return.

4 Scenario analysis

The interest-rate dependency of a fixed-income investment's future cash-flows cannot be overlooked when developing a realistic model for investment planning. Active portfolio managers are forced to use mathematical programming methods that systemize decision-making processes. Dantzig (1955) and Beale (1955) were the first ones to recognize that the traditional deterministic mathematical programs are not sufficient for most dynamic real-world applications. They proposed stochastic programming as an alternative for the solution of dynamic problems involving data uncertainties. Stochastic problems, however, tend to be very complex and very expensive, in terms of computer time, when an exact solution is desired. Many times an equivalent deterministic problem is formulated which explicitly takes into account future uncertainties of the problem. Research has focused on developing algorithms for the efficient solution of the large-scale equivalent deterministic problems. Decomposition algorithms have been developed that take into account the special structure of each problem. Parallel computing algorithms have been devised to speed up the solution of the problems. The interested reader should refer to Dantzig, Dempster, and Kallio (1981), Wallace (1984, 1986), Wets (1985, 1988), Qi (1985), Ruszczynski (1986, 1988), Rokafellar and Wets (1991), Mulvey and Vladimirou (1989, 1991), Nielsen and Zenios (1992), see, also, chapter 12.

For the purposes of this chapter a multiscenario deterministic approach that sufficiently represents the equivalent, complex stochastic problem will be used. The advantage of the multiscenario approach is that a rate of interest-rate scenarios can be chosen and evaluated based on available statistics and/or the inclination of the investor as to the direction of future rates. In turn, the portfolio manager can avoid being "blind-sided" by any sudden interest-rate shifts and thus insulate the portfolio from the potentially negative effects changes in interest rates may bring. Probabilities can be applied to the scenarios chosen and adjusted so that the effect of different interest-rate environments on the portfolio's expected return can be measured in the most realistic manner. This multiscenario approach is referred to as *scenario analysis*.

Scenario analysis strikes a reasonable compromise between the computationally complex stochastic models and the overly simplistic but unrealistic deterministic models. In scenario analysis, a number of versions (subproblems) of the underlying mathematical problem are created to model the uncertain factors in the analysis. Each scenario is a limited representation of the uncertain elements of the problem and one subproblem is generated for each scenario. Scenario analysis is particularly appropriate for cases in which statistical information is insufficient. Figure 6.8 provides an overview of the optimal structured portfolio process.

The objective of scenario analysis is to create portfolios that follow a user-defined performance profile under different yield scenarios. The

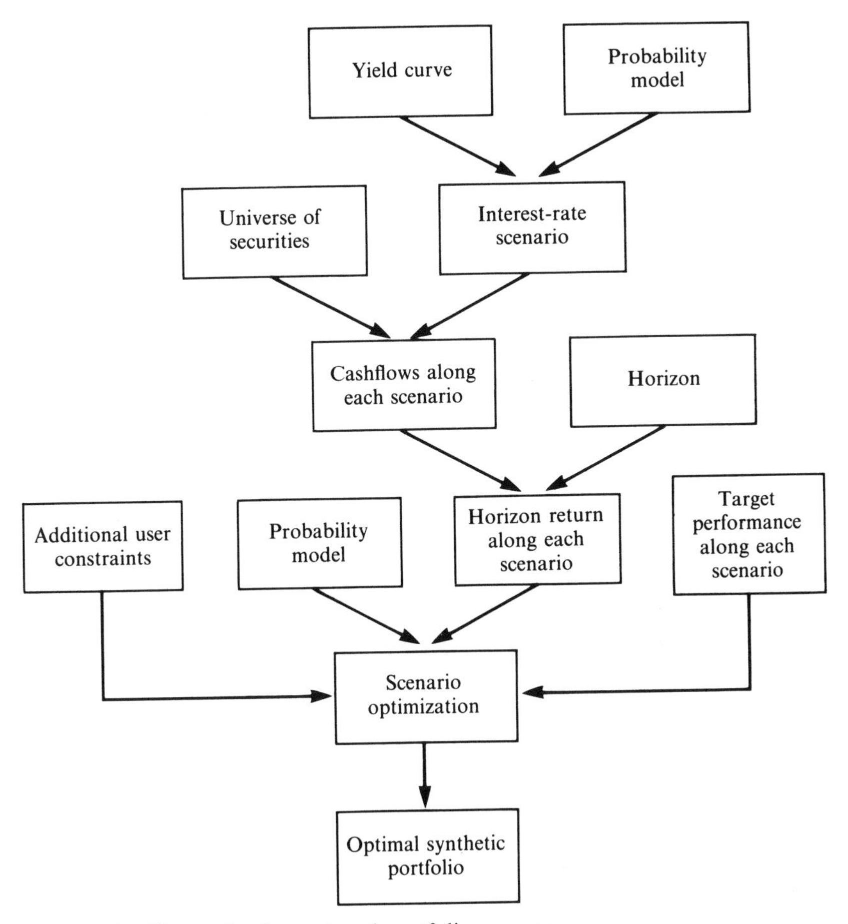

Figure 6.8 The optimal structured portfolio process

uncertainty of the market is reflected in the model by appropriately weighting each scenario. Next, an optimization technique is applied to identify similarities and trends on which an optimal solution to the overall problem can be based. The resulting optimal portfolio meets the user-specified performance target and is protected against a wide variety of probability-weighted interest-rate movements.

In what follows, the various steps of the methodology and the major complications are presented. More specifically, the definition of the appropriate horizon date, the selection of scenarios, and the specification of an efficient objective function as a measure of risk are briefly discussed.

Optimal horizon total-return portfolios

Methodology

The goal of scenario analysis is to design optimal structured portfolios that do not depend on hindsight but follow a user-defined performance pattern under a variety of future interest-rate environments. The following steps are of critical importance in the overall process:

Determination of the horizon date. The first step in the process is for the investor to determine the horizon over which the structured portfolio is expected to perform in the desired way. Depending on the objectives of each specific client, horizons ranging from one day to thirty years can be considered.

Specification of the investor's targets. The next step in the process is for the investor to define the portfolio's required returns under the various interest-rate scenarios selected for the specified horizon. Investors may also select a benchmark security or combination of securities that the structured portfolio must outperform.

Definition of the acceptable securities. Another important step is for the investor to define a universe of acceptable securities. The composition of the optimal structured portfolio depends heavily on the available universe of securities. The quality of the issues chosen to be included in the universe depends on the level of risk the investor is willing to take. Once the type and quality of the issues are chosen, the investor also has to determine, if necessary, what percentage of each type of security should be included in the optimal portfolio.

Determination of future interest rates. One of the most important aspects of asset/liability management is the determination of future interest rates. Since interest rates can take a large number of different paths, it is important to examine the future performance of the securities under different sets of interest-rate assumptions. By analyzing the performance of the securities under a variety of interest-rate scenarios, we minimize the risk that the analysis is merely an opinion about the future.

Theoretically, every possible future interest-rate environment may be considered. In practice, the number of scenarios to be analyzed should be reduced to a subset that would efficiently represent the possible range of future interest-rate environments. The generation of a representative set of scenarios is important in any practical implementation of the multiscenario model since the results are closely dependent on it. Different approaches can be used for the generation of appropriate scenarios. Investors should be particularly careful when defining the scenarios to be considered. The scenarios chosen should be representative of the interest-rate environment and should include extreme cases. In what follows, the different types of interest-rate movements are discussed briefly.

Types of interest-rate changes

In order to study the effect of interest-rate changes on portfolio return, one should look at the three possible components of these changes:

A change in the level of the yield curve (*parallel shift in interest rates*). The most basic type of yield-curve is the parallel interest-rate shift in which the yields of all maturity points on the curve experience exactly the same basis-point move.

A given change in yield affects the security's corresponding price and therefore its horizon total return. The impact of any shift in the yield curve on the security's price is determined by the bond's starting yield, maturity, and coupon rate. A security's maturity usually plays a dominant role when determining the magnitude of the impact. Parallel shifts in the yield curve tend to have a larger price impact on longer-maturity securities. Thus, increasing maturity leads to increased price volatility. Conversely, a parallel shift in the yield curve has a negligible effect at the short end of the yield curve. The effect of a parallel shift decreases as maturity decreases.

A change in the shape of the yield curve (*non-parallel shift*). If one could rely on changes in interest rates to move in parallel, fixed-income portfolio management would be a much simpler task. However, in reality, the yield curve rarely moves in parallel. See figure 6.9 for an illustration of the yield curves over the last four years. When rates move, the yield curve inevitably experiences some change in shape.

There are some basic yield-curve shapes that seem to be the rule-of-thumb in terms of how all yield curves are expected to react. Here, we examine four of these curves.

Normal. Yields rise on a continuous basis with increasing maturity, but with a gentle and continuously decreasing slope.

Flat. A flat yield curve has the same yield over all maturities.

Humped. A humped yield curve at first rises as maturities increase, but then plateaus and begins to decline at the longer maturities. Such a yield curve is usually preceded by a historically high interest-rate environment.

Inverted. Yields fall on a continuous basis with increasing maturity.

A basis change between different security types along the yield curve. Basis risk is associated with different instruments changing yields at differing rates. It is the effect of one instrument (Treasuries, for example)

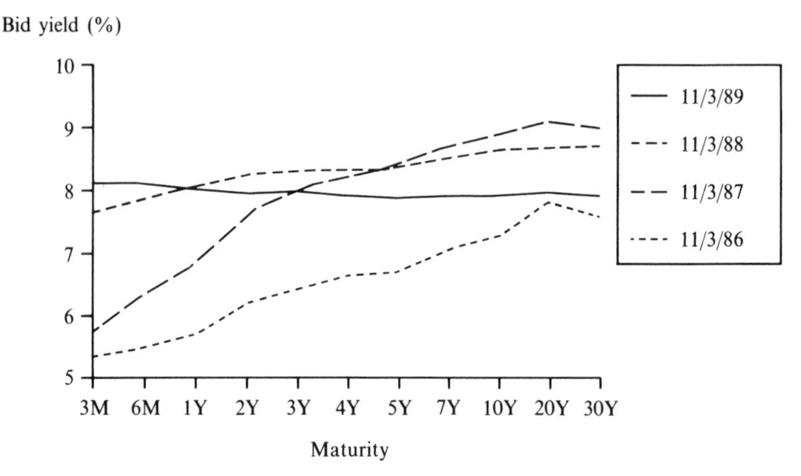

Figure 6.9 Yield-curve shape – a historical perspective, 1986–9

responding to changes in market conditions differently than a closely (but not perfectly) matched instrument (mortgage-backed securities, for example).

The possible changes in the yield curve are far too numerous to define here. However, any portfolio manager must consider the following before making an investment decision: (a) the direction of any possible change in rates, (b) the magnitude of the change across all maturities, and (c) the timing of the change. If the portfolio manager has a strong opinion on any or all of the above, he or she can decide where on the yield curve the portfolio should be positioned.

Selection of the optimal portfolio

After the representative set of scenarios is selected, a weight is assigned to each scenario. The weighting reflects the estimated relative probability of the scenario occurring. For cases in which the appropriate amount of statistical information exists, a stochastic process can be used to assign the probabilities to each scenario. The investor's interest-rate outlook also can be considered when determining the relative weights of the scenarios. For example, if the investor foresees rising interest rates, higher probabilities should be assigned to the scenarios that represent those movements.

Once the representative set of scenarios, the weights, and the horizon date have been defined, one deterministic mathematical problem is formulated for each scenario. Next, an optimization technique is applied that bundles the different scenarios along the desired range of interest-rate movements. The overall goal is to structure a portfolio that accounts for all the defined scenarios (along with their respective probabilities). Ideally a portfolio should be structured in such a way that it generates the maximum return while, at the same time, outperforms the defined target returns for every possible scenario and horizon period. In reality, this is not the case. Not all target returns can be satisfied. Different methodologies can be applied to obtain the desired tradeoff between risk and reward.

Two methodologies are now considered:

Minimizing the total error around target returns,

Minimizing the downside error around target returns.

In the first case, the objective is to minimize the summation of the total square errors surrounding the target return for each scenario. This methodology generates a portfolio that matches the target returns as closely as possible. The disadvantage of this methodology is that it penalizes both the upside deviations and the downside deviations surrounding the target returns. Assume for example that the target return for a specific scenario is 10%. A structured portfolio that generates a 12% return would not be distinguished from a structured portfolio that generates an 8% return for the same scenario. This methodology would penalize the objective function of both portfolios equally.

In reality, upside deviation surrounding the target return is a desired figure and it shouldn't be penalized. Therefore, a better measure would be to minimize the downside deviations around the target returns. Using this methodology, the downside deviations are penalized while the upside deviations remain unrestricted. As a result, the generated portfolio matches or performs better than the target. For purposes of our discussion, minimization of the downside deviation is used as the criteria for optimization. The mathematical formulation for minimizing the downside deviations for both optimal portfolio structuring and restructuring can be found in the appendix.

The Optimal Portfolio System – special features

Based on the methodology described above we have developed the Prudential-Bache Optimal Portfolio System, a viable managerial tool for those concerned with portfolio management or liability financing. Using the Optimal Portfolio System, a number of interest-rate scenarios and horizon periods can be chosen. Once the interest-rate scenarios are chosen, the effect of parallel and non-parallel shifts in the yield curve can be evaluated. In what follows, the special features and capabilities of the Optimal Portfolio System are briefly presented:

Optimal portfolio structuring

It is a particularly flexible tool that can structure optimal portfolios that meet different investor goals, such as:

Outperform a benchmark security or combination of securities;

Perform well in rising interest-rate environments, while maintaining a safety floor in declining interest-rate environments and vice versa;

Track as closely as possible a benchmark security, a combination of securities, or an index;

Depending on the needs of each individual investor, different parameters can be optimized:

Total rate of return	Total dollar gain
Market value	Par amount
Duration	Effective duration
Average life	Convexity
Option-adjusted spread	

It allows for flexibility in the number of scenarios and time periods chosen for consideration. It can account for parallel or non-parallel shifts in the yield curve and specific horizon dates. Thus far, scenarios have been defined in terms of future interest-rate environments. This is not a restriction of the methodology. Scenarios can be defined in many ways. For example, for investors concerned about liability funding and pension plans, scenarios can be defined in terms of mortality rates or projected inflation rates.

It has the ability to capture other investor-defined factors, such as budget, duration, and convexity constraints and it allows the investor to specify the percentage of the portfolio invested in specific sectors of the market.

When cashflow matching is required, it insures that the cashflows of the assets or the cash due from liquidating a portion of the assets are sufficient to meet the corresponding liabilities.

Optimal portfolio restructuring

Changes in the yield curve, client targets, and expectations of market direction and non-economic market forces may require the structured portfolio to be rebalanced in order to achieve its goal. In addition, sound active-management techniques require that a portfolio be reviewed either monthly, quarterly, or semi-annually to insure that the portfolio remains on target. In the next section, we examine the application of the optimal portfolio system and explore the effects of rebalancing.

5 Applications of the Optimal Portfolio System

In this section, an application of the Optimal Portfolio System is presented. The specific considerations and restrictions that have been imposed by a particular client are given and a description of the resulting structured portfolio follows. Finally, the actual holding-period total return of the optimal structured portfolio is reevaluated (two months after the initial structuring) and compared to the benchmark.

Optimal portfolio structuring

The following information was used to structure the optimal portfolio:

Objective: To outperform a GNMA 10.5% that the client considered to be the cheapest bond in the MBS market at the time of execution;

Available universe of securities: governments, corporates, GNMAs, FNMAs, FHLMCs, IOs, POs and CMOs;

Imposed constraints:

- No more than 10% of the portfolio (in market value) can be invested in IOs and POs;
- No more than 25% of the portfolio (in market value) can be invested in one security;
- No security can participate in the portfolio in par amounts of less than 5% of the sum of the par amounts in the portfolio.

In addition to the above constraints, the client had a neutral to slightly bullish point of view with respect to future interest rates. Therefore the objective of the structured portfolio was to outperform the benchmark in a declining interest-rate environment but at the same time sufficiently protect it in a rising interest-rate environment. In the analysis that follows, seven interest-rate scenarios were considered covering a broad range of interest-rate movements (from −300 basis points to +300 basis points in 100 basis-point increments). In all cases the analysis was done based on an immediate parallel shift of the yield curve by the appropriate amount.

From a universe of about ninety securities including governments, AAA corporates, various mortgage pass-throughs, IOs and POs, and a variety of CMO derivatives, the Optimal Portfolio System selected seven securities. More specifically, the Optimal Portfolio System selected high-coupon FNMAs, a premium IO, a Z-bond, a short average-life CMO companion bond, a Super-PO, and a US Treasury STRIP. Table 6.1 presents the composition of the optimal portfolio. Its composition in terms of market value is given in figure 6.10, while its composition in terms of par amount is given in figure 6.11.

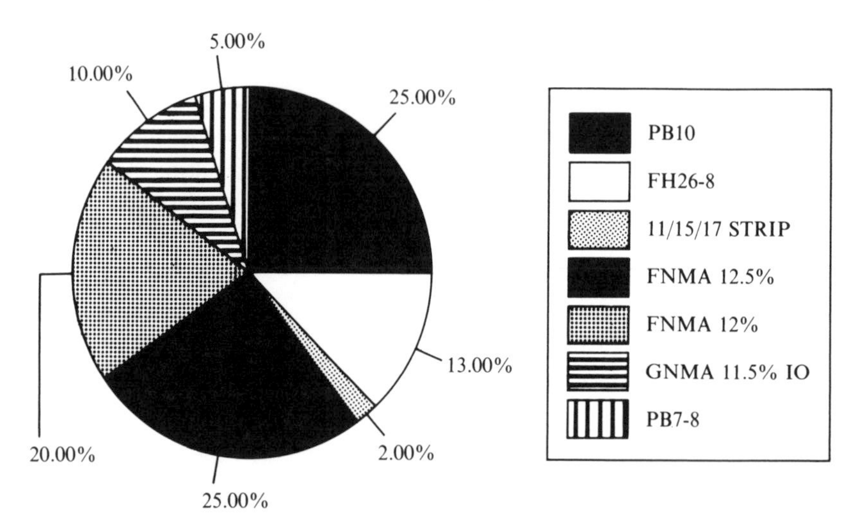

Figure 6.10 Market value of securities selected for the optimal structured portfolio

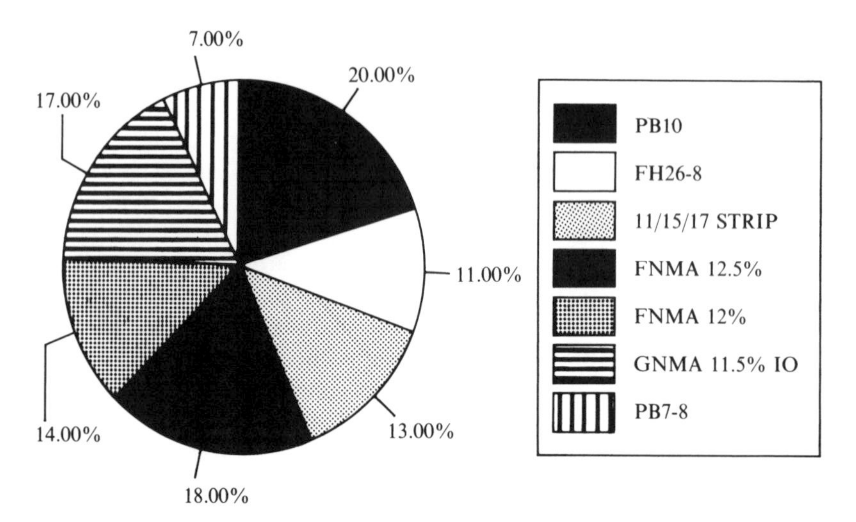

Figure 6.11 Par amount of the securities selected for the optimal structured portfolio

Table 6.1. *A look at the securities selected*

	Par amounts ($)	Security
Sell:	25,000,000	GNMA 10.5%
Buy:	2,150,000	CMO–PB7 (Class 8–Super-PO)
	6,650,000	CMO–PB10 (Class 2–Companion)
	3,600,000	CMO–FHL826 (Class 8–Z-Bond)
	5,688,577	IO GNMA SMBS CMOT26 11.5%
	4,700,000	MBS FNMA 12%
	5,900,000	MBS FNMA 12.5%
	4,355,000	Treasury STRIP (Maturing November 2017)

Figure 6.12 compares the GNMA 10.5% and the structured portfolio in terms of one-year horizontal total returns, while in figure 6.13 the spread (one-year horizon returns) between the GNMA 10.5% and the benchmark is presented. The structured portfolio outperforms the benchmark in all seven interest-rate scenarios. In the worst case (interest-rates increase by 100 basis points), the structured portfolio outperforms the benchmark by 20 basis points while in the best case (300 basis-point rally) it outperforms by 200 basis points.

The analysis performed showed that, in all cases, the client added yield, improved horizon total return, and at the same time shortened duration. Figures 6.14, 6.15, and 6.16 present the one-year horizon total return of the structured portfolio, the components of the structured portfolio, and the GNMA 10.5%, respectively. It is interesting to note that there is not an individual security that would outperform the benchmark in all seven interest-rate scenarios examined.

Retrospective

We reevaluated the structured portfolio versus the GNMA 10.5% benchmark two months after the trade was executed. For the actual two-month holding period, the structured portfolio outperformed the GNMA 10.5% by 20 basis points. Table 6.2 shows the updated prices for the securities included in the structured portfolio as well as for the benchmark.

In addition, our analysis showed that, given the new shape of the yield curve, an increase in interest rates would cause the existing structured portfolio to underperform the benchmark for a one-year horizon total return. A rebalanced portfolio (total proceeds swap) that was projected to

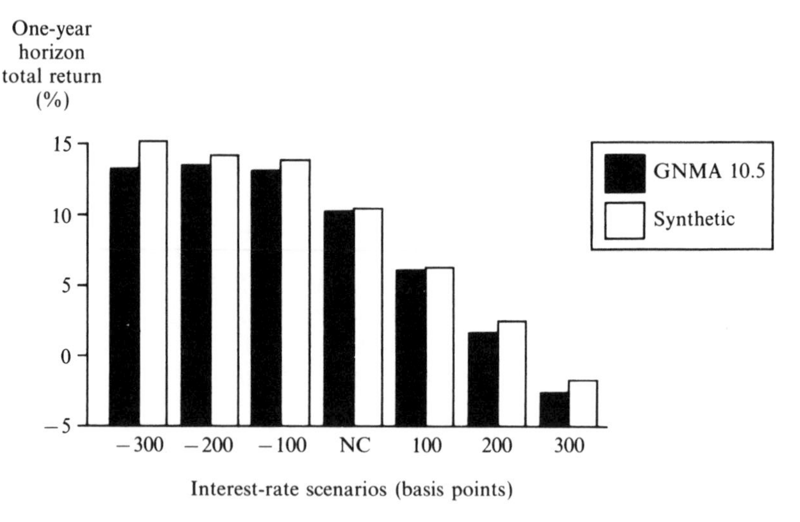

Figure 6.12 One-year horizon total-return comparison of the benchmark (GNMA 10.5%) and the structured portfolio

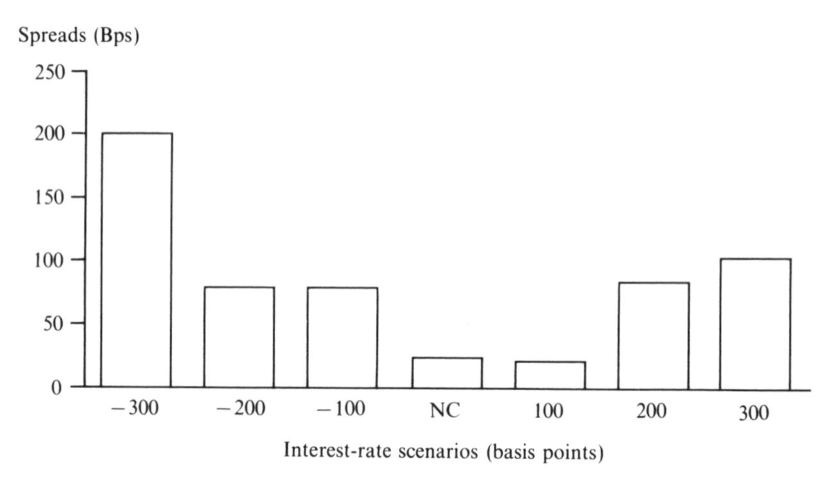

Figure 6.13 Spread comparison of the benchmark (GNMA 10.5%) and the optimal structured portfolio

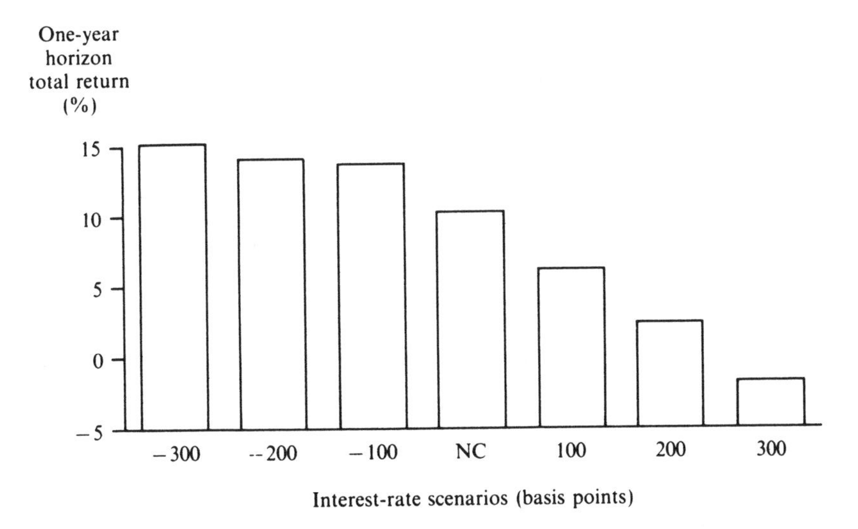

Figure 6.14 One-year horizon total-return profile on the optimal structured portfolio

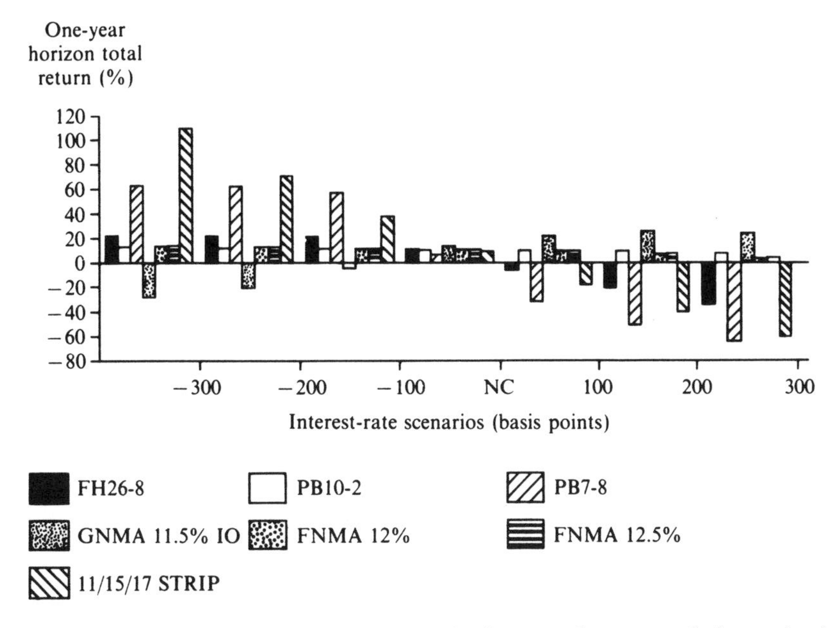

Figure 6.15 Components and one-year horizon total returns of the optimal structured portfolio

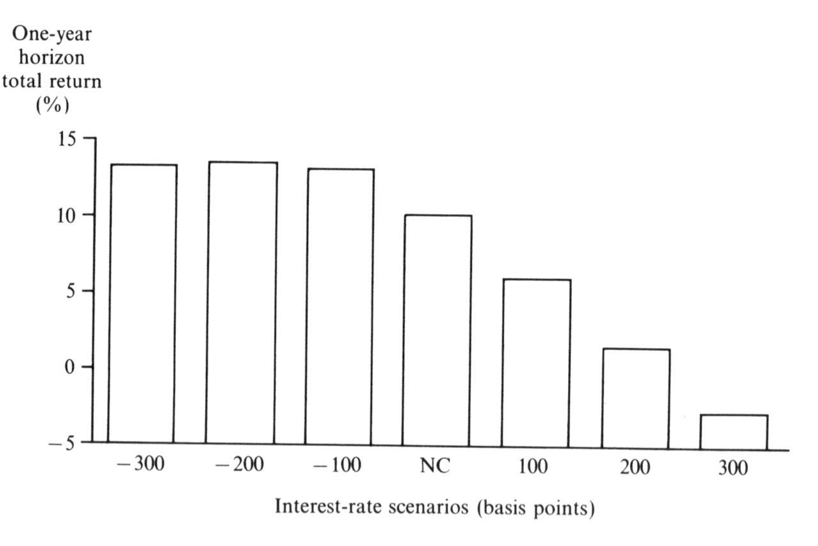

Figure 6.16 One-year horizon total-return profile on the benchmark (GNMA 10.5%)

outperform the benchmark in all seven interest-rate scenarios was suggested to the client. The rebalanced portfolio, however, achieved this performance improvement versus the benchmark by giving up performance versus the existing structured portfolio in the strongly bullish scenarios. After reviewing the rebalanced portfolio with the client, the client decided to remain with the original structured portfolio "as-is" since the client's market viewpoint had moved from "neutral to slightly bullish."

6 Summary

The prudent portfolio manager of the 1990s must look to valuation techniques that enable him to verify the validity of a particular investment strategy in oftentimes volatile and dynamic markets. Rather than looking at a single measurement of risk and return, such as yield-to-maturity and duration or even convexity and volatility, which tend to capture only a portion of the dynamic nature of the securities, a more detailed scenario analysis should be conducted. In scenario analysis, the behavior of the securities, as well as the portfolio as a whole, is examined under different interest-rate environments.

The Optimal Portfolio System is a flexible portfolio management tool that allows its users to actively manage a portfolio of fixed-income

Table 6.2. *Actual holding-period total returns (06/26/89–08/23/89) for the sample optimal portfolio*

	Current portfolio 10.5%	GNMA
Holding-Period Total Return	14.62%	14.42%
Security	**Price on 06/03/89**	**Price on 08/23/89**
GNMA 10.5%	102–05	102–27
CMO PB7 (Class 8–Super PO)	61–00	63–00
CMO PB10 (Class 2–Companion)	93–30	94–16
CMO FHL826 (Class 8–Z-Bond)	91–00	96–00
IO GNMA SMBS (CMOT23 11.5%)	41–26	40–00
MBS FNMA 12%	105–02	105–16
MBS FNMA 12.5%	106–05	106–16
Treasury STRIP (Maturing November 2017)	9.807	10.631

securities. In particular, it is a useful technique for asset/liability management when the liabilities are interest-rate sensitive. By defining a specific horizon date and attempting to maximize the horizon total return for that date, the Optimal Portfolio System enables us to either meet the constraints of a given liability set and get as close as possible to the target performance or to mimic and even outperform a given index.

Appendix: Mathematical formulations

In what follows, two mathematical formulations are presented. The first one (problem P1) can be applied to structure a portfolio that optimally matches specific targets under a variety of interest-rate scenarios. The second one (problem P2) can be applied to optimally restructure an existing portfolio based on prespecified criteria. For simplicity, both mathematical formulations are the ones applied to optimally structure/restructure portfolios that match a user-defined horizon return under a variety of interest-rate scenarios. At the same time, the formulations insure that the portfolio meets specific criteria. However, it is important to note that the Prudential-Bache Optimal Portfolio System is a more flexible decision-support system that can interchangeably handle a variety of different performance measures either as targets (in the objective function) or as additional constraints.

Optimal portfolio structuring

Using the following notation:

S:	set of interest-rate scenarios, $s=1,...,S$;
I:	set of available securities, $i=1,...,I$;
x_i:	decision variables; amount of security $i, i \in I$;
$\bar{x}$:	vector of all $x_i, i \in I$:
$HRR_s(\bar{x})$:	horizon total rate of return of the structured portfolio for scenario s, $s \in S$, as a function of the decision variables $\bar{x}$. This function is in general a non-linear function;
$DUR_s(\bar{x})$:	duration of the structured portfolio (modified, Macaulay, or effective) for scenario s, $s \in S$, defined as a function of the decision variables $\bar{x}$;
$CON_s(\bar{x})$:	convexity of the structured portfolio for scenario s, $s \in S$, defined as a function of the decision variables $\bar{x}$;

$hrrt_s$: constant; target value for the horizon rate of return of the structured portfolio for scenario s, $s \in S$;

w_s: constant; probability weight assigned to scenario s, $s \in S$;

$hrrl_s$: constant; lower bound for the horizon rate of return of the structured portfolio for scenario s, $s \in S$;

a_s, b_s: constants; lower and upper bounds, respectively, of the duration of the structured portfolio for each interest-rate scenario s, $s \in S$;

c_s, d_s: constants; lower and upper bounds, respectively, of the convexity of the structured portfolio for each interest-rate scenario s, $s \in S$;

p_i: constant; ask price of security i, $i \in I$, in the current environment;

e: constant; maximum amount of money, as a percentage of the total amount of money invested in the portfolio, that can be invested in one single security;

ub_i: constant; maximum available amount of security i, $i \in I$.

The problem of structuring an optimal portfolio that matches a desired horizon-rate-of-return-performance objective under a variety of interest-rate scenarios and that, at the same time, meets certain constraints has the following mathematical formulation:

Problem P1

Objective function:

$$\min \sum_s w_s (HRR_s(\bar{x}) - hrrt_s)^-$$

where $(HRR_s(\bar{x}) - hrrt_s)^-$ equals $\begin{cases} 0 & \text{if } HRR_s(\bar{x}) \geq hrrt_s, \\ (HRR_s(\bar{x}) - hrrt_s) & \text{if } HRR_s(\bar{x}) < hrrl_s, \end{cases}$

Subject to:

$$HRR_s(\bar{x}) \geq hrrl_s, \text{ all } s \in S \quad (1)$$

$$a_s \leq DUR_s(\bar{x}) \leq b_s, \text{ all } s \in S \quad (2)$$

$$c_s \leq CON_s(\bar{x}) \leq d_s, \text{ all } s \in S \quad (3)$$

$$p_i x_i \leq \left(\sum_i p_i x_i \right) e, \text{ all } i \in I \quad (4)$$

$$x_i \leq ub_i, \text{ all } i \in I \quad (5)$$

$$x_i \geq 0, \text{ all } i \in I \quad (6)$$

The objective function is an expression of the difference between the horizon rate of return of the structured portfolio and the target horizon rate of return specified by the user and incorporates different probabilities (weights) for each interest-rate scenario. The above objective function minimizes the downside deviation error only. As an alternative, a mathematical formulation that would minimize the total error around the target return could be used.

Inequalities (1) are the lower bound constraints that force the horizon rate of return of the structured portfolio to exceed a specified target.

Inequalities (2) are the upper and lower bound duration constraints.

Inequalities (3) are the upper and lower bound convexity constraints.

Inequalities (4) are the percentage constraints that impose a maximum on the amount of money invested in one security.

Inequalities (5) are the availability constraints that impose an upper bound based on the availability of each security.

Inequalities (6) are the non-negativity constraints.

Optimal portfolio restructuring

Using the following notation:

S: set of interest-rate scenarios, $s=1,\ldots,S$;

I: set of available securities to be purchased, $i=1,\ldots,I$;

J: set of available securities to be sold, $j=1,\ldots,J$;

x_i: decision variables; amount of security $i, i\in I$ to be purchased;

y_j: decision variables; amount of security $j, j\in J$ to be sold;

$\bar{x}$: vector of all $x_i, i\in I$;

$\bar{y}$: vector of all $y_j, j\in J$;

$HRR_s(\bar{x},\bar{y})$: horizon total rate of return of the structured portfolio for scenario $s, s\in S$, as a function of the decision variables $\bar{x},\bar{y}$. This function is, in general, a non-linear function;

$DUR_s(\bar{x},\bar{y})$: duration of the structured portfolio (modified, Macaulay, or effective) for scenario $s, s\in S$, defined as a function of the decision variables $\bar{x},\bar{y}$;

$CON_s(\bar{x},\bar{y})$: convexity of the structured portfolio for scenario $s, s\in S$, defined as a function of the decision variables $\bar{x},\bar{y}$;

$hrrt_s$: constant; target value for the horizon rate of return for scenario $s, s\in S$;

w_s: constant; probability weight assigned to scenario $s, s\in S$;

$hrrl_s$: constant; lower bounds for the horizon rate of return for scenario $s, s\in S$;

a_s, b_s: constants; lower and upper bounds, respectively, of the duration of the structured portfolio for each interest-rate scenario $s, s \in S$;

c_s, d_s: constants; lower and upper bounds, respectively, of the convexity of the structured portfolio for each interest-rate scenario $s, s \in S$;

p_i: constant; ask price of security $i, i \in I$, in the current environment;

p_j: constant; bid price of security $j, j \in J$, in the current environment;

e: constant; maximum amount of money, as a percentage of the total amount of money invested in the portfolio, that can be invested in one single security;

ub_i: constant; maximum available amount of security $i, i \in I$;

eb_j: constant; maximum available amount of security $j, j \in J$, included in the existing portfolio.

The problem of restructuring an optimal portfolio that matches a desired horizon-rate-of-return-performance objective under a variety of interest-rate scenarios and that, at the same time, meets certain constraints has the following mathematical formulation:

Problem P2

Objective function:

$$min \sum_s w_s (HRR_s(\bar{x},\bar{y}) - hrrtr_s)^-$$

where $(HRR_s(\bar{x},\bar{y}) - hrrt_s)^-$ equals $\begin{cases} O \text{ if } HRR_s(\bar{x},\bar{y}) \geq hrrt_s \\ (HRR_s(\bar{x},\bar{y}) - hrrt_s) \text{ if } HRR_s(\bar{x},\bar{y}) < hrrl_s \end{cases}$

Subject to:

$$HRR_s(\bar{x},\bar{y}) \geq hrrl_s, \ all \ s \in S \quad (1)$$

$$a_s \leq DUR_s(\bar{x},\bar{y}) \leq b_s, \ all \ s \in S \quad (2)$$

$$c_s \leq CON_s(\bar{x},\bar{y}) \leq d_s, \ all \ s \in S \quad (3)$$

$$p_i x_i \leq \left(\sum_i p_i x_i \right) e, \ all \ i \in I \quad (4)$$

$$x_i \leq ub_i, \ all \ i \in I \quad (5)$$

$$\sum_i p_i x_i - \sum_j p_j y_j \leq 0, \ all \ i \in I, \ all \ j \in J \qquad (6)$$

$$y_j \leq eb_j, \ all \ j \in J \qquad (7)$$

$$x_i, y_j \geq 0, \ all \ i \in I, \ all \ j \in J \qquad (8)$$

The objective function is an expression of the difference between the horizon rate of return of the optimal restructured portfolio and the target horizon rates of return specified by the user and incorporates different probabilities (weights) for each interest-rate scenario. The above objective function minimizes the downside deviation error only. As an alternative, a mathematical formulation that would minimize the total error around the target return could be used.

Inequalities (1) are the lower bound constraints that force the horizon rate of return of the structured portfolio to exceed a specified target.

Inequalities (2) are the upper and lower bound duration constraints.

Inequalities (3) are the upper and lower bound convexity constraints.

Inequalities (4) are the percentage constraints that impose a maximum on the amount of money invested in one security.

Inequalities (5) are the availability constraints that impose an upper bound based on the availability of each security.

Inequality (6) is the total-proceeds constraint that insures that the money received by selling the selected securities of the existing portfolio will generate proceeds sufficient to buy the additional securities to be included in the optimal restructured portfolio.

Inequalities (7) are availability constraints that insure that the maximum amount to be sold from each security does not exceed the available amount.

Inequalities (8) are the non-negativity constraints.

REFERENCES

E. M. L. Beale (1955), "On minimizing a convex function subject to linear inequalities," *Journal of the Royal Statistical Society*, 17b: 173–84.

P. Brehm and L. Pendergast (1990), "CMO bonds," Financial Strategies Group, Prudential-Bache Capital Funding, New York.

G. B. Dantzig (1955), "Linear programming under uncertainty," *Management Science*, 1: 197–206.

G. B. Dantzig, M. A. H. Dempster, and M. J. Kallio (1981), "Large-scale linear programming," vol. 1. In *IIASA Collaborative Proceedings Series*, Luxenburg, Austria, CP-81-51.

L. S. Hayre, K. Lauterbach, and C. Mohhebi (1988), "The Prudential-Bache prepayment model," Financial Strategies Group, Prudential-Bache Capital Funding, New York.

L. S. Hayre and C. Mohebbi (1988), "Mortgage pass-through securities," Financial Strategies Group, Prudential-Bache Capital Funding, New York.

P. Kang and S. A. Zenios (1992), "Complete Prepayment Models for Mortgage-Backed Securities', *Management Science*, Nov.

J. M. Mulvey and H. Vladimirou (1989), "Evaluation of a parallel hedging algorithm for stochastic network programming," in R. Sharda, B. L. Golden, E. Wasil, O. Balci, and W. Stewart (eds) *Impact of Recent Computer Advances on Operations Research*, North-Holland, pp. 106–19.

(1991), "Solving multistage stochastic networks; an application of scenario aggregation," *Networks*, 1(6): 619–43.

S. Nielsen and S. A. Zenios (1992). "Massively Parallel Algorithms for Nonlinear Stochastic Network Problems", *Operations Research*.

L. Qi, (1985), "Forest iteration methods for stochastic transportation problems," *Mathematical Programming Study*, 25: 142–63.

R. T. Rockafellar and R. J. B. Wets (1991), "Scenarios and policy aggregation in optimization under uncertainty," Mathematics of Operations Research, 35: 309–33.

A. Ruszczynski (1986), "A regularized decomposition method for minimizing a sum of polyhedral functions," *Mathematical Programming*, 35: 309–33.

(1988), "Parallel decomposition of multistage stochastic programming problems," Working Paper 88-094, IIASA.

S. W. Wallace (1984), "A two-stage stochastic facility location problem with time-dependent supply," Working Paper, Department of Science and Technology, Christian Michelsen Institute, Bergen, Norway.

(1986), "Solving stochastic programs with network recourse," *Networks*, 16: 295–317.

R. Wets (1985), "On parallel processor design for solving stochastic programs," Report 85-67, International Institute for Applied Systems Analysis, Luxenburg, Austria.

(1988), "The aggregation principle in scenario analysis and stochastic optimization," Working Paper, Department of Mathematics, University of California, Davis.

7 Optimization tools for the financial manager's desk

MORDECAI AVRIEL

1 Building financial optimization models

The financial work presents interesting challenges to those who build optimization models and wish to implement them by developing the appropriate software. These challenges fall into two interrelated categories: first, the environment in which the model operates is, in most cases, a very risky one and the model's output is expected to be used almost instantaneously. Second, the software based on the model is most often required to be run by the end-user (e.g., a financial manager or a trader), without the intervention and expert advice of optimization specialists. The implication of the risky environment, as far as model development is concerned, is that failing to provide a correct recommendation for optimal action can have very serious financial consequences for the model's user. It is, therefore, vital that the model perform in a foolproof manner in all circumstances. The development of such a model is by no means trivial. For example, without some precautionary measures a very small change in the data input of an otherwise perfectly operating linear program can result in failing to find an optimal (or any feasible) solution and, as far as the end-user is concerned, this means failure. Such failures can be avoided by careful model formulation and by appropriate preprocessing of the problem data that detects inadmissible inputs.

Operating in a real-time environment means that efficient and fast algorithms must be used to find a solution and to make a recommendation for the user. The requirement to get a fast and reliable solution under all circumstances can also mean that it may be preferable to custom build some robust heuristic algorithm to solve the problem and to arrive at a "good" solution, rather than use an algorithm that would stop only at the true optimum (if it finds one), but the time to arrive there may vary considerably from problem to problem. Such situations arise, for example, in problems that would be amenable for solution by non-linear or integer programs.

If the optimization model is to be run by an end-user who is not familiar with optimization theory or with the particular solution algorithm used for the model, special additional steps must be taken by the model developer to ensure the successful use of the model as a decision-aiding tool. These steps concern ease-of-use (user-friendliness) of the model, in addition to all the previously mentioned required characteristics. To make an optimization model implementation user-friendly can be as difficult as, or even more difficult than, building the model itself or developing the solution method. Unfortunately, these user-oriented aspects of modeling and optimization are rarely taught in university courses. Many young professionals starting work in a financial institution are surprised to discover that optimization modeling must also involve a lot of effort to develop the right user interface. In summary, a successful custom-built application of optimization theory intended for a financial manager should be as foolproof and user-friendly as any other good commerical software package (e.g., a spreadsheet). Once these guidelines are observed, there are vast opportunities for optimization experts to develop powerful tools for financial institutions that can considerably improve the decision-making process and increase the firm's profitability. I will describe now two successful applications of such optimization tools and will expand on one of them. For other financial optimization models see, for example, chapters 1 and 2.

2 An asset funding optimization model

The first application has to do with an optimization model for asset funding in a large US bank. Funding traders in the Eurodollar market are continuously faced with the problem of funding newly created loans and renewing existing loans at the lowest available rates without exposing the bank to unacceptable risks of liquidity and interest rates. The funding options available are either immediate borrowings for any length of time period that usually does not exceed six months, or commitments of forward borrowing, commencing on the next or second next business day. Without imposing some limits, the choice of the amounts borrowed (liabilities created) and the periods of borrowing by the trader would mainly depend on current interest rates and on forecasts of future interest rates. One practical limit imposed on the trader is that at the end of every business day the net cashflow for that day must be as close to zero as possible. Since the amounts and tenors of new liabilities created affect future cashflows and money positions, the trader is restricted in the choice of liability alternatives by externally set liquidity limits (on the daily cashflows in the near future) and maturity profile limits (on the aggregate monthly money positions of, say, the next twenty-four months). Consequently, finding the optimal daily funding schedule over a planning period of several months was recognized as a constrained optimization problem that can be best solved by computer. It was decided to develop an appropriate optimization model and the necessary software for its solution.

The custom-built software for the solution of this problem had to play a dual role: First, it was primarily intended to be a decision-supporting tool for the traders in their everyday work of liability creations. Second, it also had to serve the management of the funding group in examining optimal responses to possible future interest-rate and lending position alternatives, as well as in investigating the effects of the imposed limits on the operation and profitability of the group. For both of its roles the model had to be designed along the guidelines discussed at the beginning of this chapter, since neither the traders nor the management of the funding group (the end-users of the model) were expected to be familiar with optimization theory or solution methods. Consequently, an elaborate user interface was developed that enabled the running of the model without the aid of optimization specialists.

The model was formulated according to the dual roles. The traders run the model every morning as the markets open, and also during the day every time a substantial amount of new assets are created for which funding is required. The basic inputs from the users are the amounts, maturities, and rates of new assets created, the amount of funding required, current borrowing rates for "standard" borrowing periods (overnight, one day forward, two day forward, 1 week, 2 weeks, etc.), and, optionally, forecasts of one or two future borrowing-rate scenarios. These user-supplied inputs are combined with cashflow and money position data from one of the bank's central data bases and with a specially developed calendar dating routine (written in Pascal).[1]

More specifically, the optimization model is built with the following steps: first, the current and future cashflow and asset–liability positions from the data base are updated by a user-supplied input on new assets and liabilities created. These updates lead to new current and future liability requirements. Second, borrowing rates are computed for every day of the planning horizon (default: six months) for all "standard" borrowing periods from the user-supplied current and forecasted future rate data, using a special case of cubic spline interpolation technique.[2] Third, the rates are combined with the calendar routine to compute the cost of borrowing for every day of the horizon and for all borrowing periods. Finally, using the current yield curve, discount factors are computed for every day of the planning horizon. All this data processing is an essential part of the model building procedure, since the data computed at this stage serve as the coefficients of a large-scale linear program by which the optimization is carried out.

The linear program represents a multiperiod optimization problem that finds the lowest cost current funding as required by the user, along with the lowest cost funding of all presently outstanding assets in the planning horizon. The optimal funding schedule takes into account present and forecasted future rates, and tradeoffs between creating a single liability for a given length of time and rolling over a series of liabilities for the same total period. In the latter case, on any future intermediate day when previously created liabilities mature, new liabilities are created at the then prevailing (higher or lower) rates. In addition, all externally imposed limits on the funding schedule (liquidity and maturity profile limits) must be observed by the solution. The user (trader or money manager) can further restrict the model by fixing or bounding the amounts of current and forward liabilities for which an immediate decision is sought. The variables of the linear program are the amounts of new liabilities created, their value dates (current or forward) and maturities. The objective function to be minimized is the present (discounted) value of the total cost of funding all currently outstanding assets in the planning horizon.

The linear program representing this problem consists of approximately 1,200 variables and 900 constraints. It is solved by the GAMS/MINOS 386 system on a 80386-based microcomputer. For an introduction to the GAMS system see Brooke, Kendrick, and Meeraus (1988). Since GAMS has no built-in calendar routines at present (a serious deficiency for financial optimization), a special Pascal code was written to prepare the necessary data for GAMS and to write them in a format directly readable by GAMS. This appending of a calendar routine with GAMS proved to be an efficient and time-saving effort. In fact, it is doubtful if an attempt to use only GAMS for the model formulation would have resulted in a correct and usable code. For a similar approach of integrating preprocessing routines with GAMS in a financial application see chapter 6 by Adamidou *et al.*

The output of the model consists of two parts, representing the two operating modes of the model: for the everyday work of the funding traders (the "narrow" mode) a screen showing the recommended optimal actions (current and forward borrowings for immediate execution) is shown upon completion of the linear program. For examining policy alternatives (the "wide" mode), additional output is produced. By using the duality theory of linear programming, marginal rates for all the borrowing periods are computed and presented on the screen. Marginal rates for periods that are not "basic," i.e., no borrowing is recommended for them by the model, are the rates at which it would become optimal to borrow a small positive amount. This information is important to the user in order to examine the sensitivity of the optimal solution to the rates and, in a more operative sense, to see at what rates would it be beneficial to close a deal. Similarly, an output screen showing the cashflows and money positions resulting from the optimal solution, relative to the imposed limits, is also produced. These output screens were built by a Pascal code "post-processing" the output data from the GAMS system such that they are presented in a convenient and easily understandable form for the user.

The time required from the user to prepare a run of this model depends on the needed amounts of updating the data base and specifying bounds. Setting up a typical daily run takes about three to five minutes. Executing the model-building and solution phases by GAMS and to produce the output takes approximately seven minutes.

3 Bank reserve management optimization

3.1 Problem description

The second application concerns optimizing cash reserve management of commercial banks. The following is a description of a model built for Israeli banks, but, with slight modifications that mainly concern regulatory requirements, it can be applicable in other countries as well.

Israeli commercial banks are required by the Bank of Israel (BofI) to maintain certain portions of their daily domestic demand deposits in a liquid form by depositing them in BofI, or alternatively, by investing in certain qualifying short-term government bonds. Fulfillment of these requirements is monitored by BofI on a monthly basis, with further checks performed weekly. In spite of the rather complex set of requirements, money managers or CFO's of commercial banks have ample opportunities to manage the bank's domestic money positions so that the aggregate monthly reserve requirements are fulfilled and at the same time extra profits are earned in the cash and short-term government bond markets.

At the beginning of every business day the money manager of the commercial bank is faced with the following problem: given the levels of demand deposits and customer credits from the previous business day and given some forecasts of these levels to the end of the current calendar month, the options available with regard to the bank's present money position (excluding new asset creations), and in particular its reserve requirements, are as follows: deposit in BofI, keep cash for daily transactions, lend or borrow overnight (interbank transactions), borrow overnight from BofI (discount windows),[3] trade qualifying short-term government bonds (called Makam in Hebrew) in the bond market, buy new issues of qualifying short-term government bonds from BofI. The basic reserve requirement is to deposit a portion (say 10%) of the daily demand deposit in BofI. Interest is received on this deposit according to a variable scale. Of the required 10% of demand deposits, a first step (say the first 5%) earns no interest at all, a second step (the remaining 5%) earns 9% interest, and a third step of any deposits over the required 10% earns 6% interest (the actual rates of interest are subject to change and vary from one type of demand deposit to another). Although the required levels of deposit are computed on a daily basis, the fulfillment of the requirements is first checked on an aggregate weekly basis. On this aggregate basis, banks must fulfill at least the first segment of their reserve requirements, or face (weekly) penalties. The second, and final check is done by BofI on the aggregate monthly level. At this level banks must reach all their required reserve levels, or else they face (monthly) penalties. The penalties are actually mandatory borrowings from BofI to reach the required deposit levels for the period of violation at a very high rate. However, in lieu of a part of the deposits, commercial banks are allowed to keep certain types of short-term government bonds, and are encouraged to buy new issues of such bonds at a discount.

The money manager's problem can be formulated as a multiperiod optimization problem with a horizon consisting of the remaining days in the current calendar month, since reserve requirements are computed on a monthly basis and at the end of each month all reserve account transactions are settled.[4] The problem is to maximize the bank's asset position at the end of the month, subject to satisfying the regulatory requirements.

3.2 Data inputs and specifications

As mentioned above, the solution of the money manager's problem is obtained by an optimization model formulated from input data, forecasts, regulatory and financial considerations, and user-imposed limitations. Some of these elements remain unchanged, but many others are subject to daily variations. In order to assist the money manager to run the model without relying on outside help from optimization experts, a menu-driven user interface was built that performs all the necessary tasks for building and solving the model. The main input menu screen is shown in figure 7.1. Preparing a typical run involves the following steps:

(i) Updating the bank's demand deposit levels for the previous day and, if necessary, the forecast of these levels to the end of the current month.

(ii) Updating the reserve account levels for the previous day. Simply

BANK XXXX - RESERVE ACCOUNT MANAGEMENT PROGRAM 14:32:23
Copyright 1989
Wednesday, August 8th, 1990 by M. Avriel, R. Avriel and I. Zang

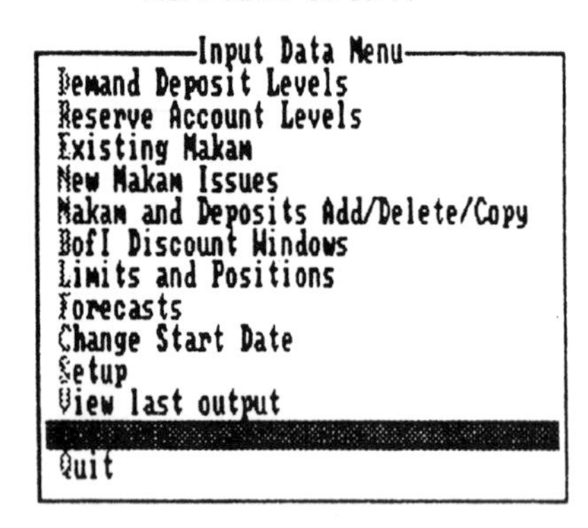

Figure 7.1 Main menu for reserve account management

stating this means listing the uses of the bank's funds (the difference between demand deposits and customer credits). Among the uses are deposits in BofI to fulfill reserve requirements, vault cash, short-term government bond positions, interbank transactions, overnight borrowings from BofI discount windows, etc.

(iii) Updating current yields of the qualifying short-term bonds (Makam) traded on the stock market and, if desired, specifying trading limitations.

(iv) Inputting data on new Makam issues announced or expected in the current month.

(v) Since the model takes into account all regulatory and user-imposed limitations, as well as current and forecasted interest rates and bond yields, updating or specifying of these parameters may be needed. The easiest way to accomplish this step is to call up the remaining items on the main input menu and to change the values of the parameters on the screen.

(vi) Finally, choosing OPTIMIZE on the main menu, a preprocessing and solution program is commenced that terminates in an output screen.

3.3 *Model formulation*

For the solution of the money manager's problem, a linear programming model was formulated. Here we present the mathematical formulation of the model:

Let us define first the following index sets:
R = remaining business days in current month,
W = weeks in current month,
RW = remaining weeks in current month,
MK = government bond (Makam) issues,
WI = BofI discount windows (increasing rates).
Now the model equations.

Daily cashflow equation for day R

$$CashEnd(R-1)+DepChg(R)+\Sigma_{WI}Cr(WI,R)+IntBB(R)+\Sigma_{MK}SellSec(MK,R)*SPr(MK,R)+\Sigma_{MK}MatSec(MK,R)=Dep(R)+Cash(R)+\Sigma_{MK}BuySec(MK,R)*BPr(MK,R)+IntBL(R)$$

where:

$CashEnd(R-1)$ = reserve account balance at the end of day $R-1$,
$DepChg(R)$ = change in demand deposit and credit levels on day R,
$Cr(WI,R)$ = credit to be taken from BofI discount window at rates WI on day R,
$SellSec(MK,R)$ = par value of bond MK to sell on day R at price $SPr(MK,R)$,
$BuySec(MK,R)$ = par value of bond MK to buy on day R at price $BPr(MK,R)$,
$MatSec(MK,R)$ = par value of bond MK maturing on day R,
$IntBB(R)$ = amount to borrow overnight in the interbank market on day R,
$IntBL(R)$ = amount to lend overnight in the interbank market on day R,
$Dep(R)$ = amount to deposit in BofI on day R,
$Cash(R)$ = amount of cash to keep in the bank on day R.

Security position of bond issue MK at the end of day R

$$SecPos(MK,R)=SecPos(MK,R-1)+BuySec(MK,R)-SellSec(MK,R)-MatSec(MK,R)$$

where:
$SecPos(MK,R)$ = par value of bond MK position at the end of day R.

Total amount deposited in BofI in week W

$$ActDep(W)+\Sigma_{R\in RW}Dep(R)*Pl(R)=Dep_1(W)+Dep_2(W)+Dep_3(W)$$

where:

$ActDep(W)$ = amount deposited in BofI in week W before starting day for the model,

$Dep_1(W)$, $Dep_2(W)$, $Dep_3(W)$ = deposit levels in BofI in interest steps 1, 2, 3, respectively,

$Pl(R)$ = number of calendar days from business day R to business day $R+1$.

Fulfillment of reserve requirements in week W

$$WeekFul(W)=\Sigma_{R\in RW}[Dep(R)+Cash(R)]*Pl(R)+ActDep(W)+ActCash(W)$$

where:

$WeekFul(W)$ = fulfillment of reserve requirement in week W,

$ActCash(W)$ = amount of vault cash in week W before starting day for the model.

Note that vault cash kept in the bank is equivalent to deposit in BofI (first step as explained above).

Compute deposits in the first two steps in week W

$$Dep_1(W)+ActCash(W)\Sigma_{R\in RW}Cash(R)*Pl(R)Pen(W)\geq Weekly_1ReqDep_2(W)\leq Weekly_2Req$$

where:

$Pen(W)$ = penalty for violating reserve requirements in week W,

$Weekly_1Req(W)$, $Weekly_2Req(W)$ = reserve requirements in steps 1 and 2, respectively (given).

A portion of the bank's short-term government bond securities position is recognized as a reserve deposit. This recognized portion is computed on a weekly basis as follows: it is the smaller of the bank's total qualifying bond position and a certain percentage (currently 20%) of the bank's weekly reserve requirement. We obtain the following constraints:

Recognizing bondholdings as reserve deposit in week W

$$Recog(W)\leq 0.2*WeeklyReq(W)$$

$$Recog(W)\leq ActPos(W)+\Sigma_{R\in RW}[\Sigma_{MK}SecPos(MK,R0]$$

where:

$Recog(W)$ = portion of bondholdings recognized as reserve deposit in week W,

$WeeklyReq(W)$ = reserve requirement in week W (given),
$ActPos(W)$ = qualifying bondholdings in week W before the starting day of the model.

Fulfillment of reserve requirements in current month

$$\Sigma_W[WeekFul(W) + Recog(W) + Pen(W)] + PenM \geq MonthlyReq$$

where:
$PenM$ = penalty for violating reserve requirements in current month,
$MonthlyReq$ = required reserve level for current month (given).

Compute cash balance at the end of day *R*

$$CashEnd(R) = Cash(R) + Dep(R) + IntBL(R)*ILRate(R) - IntBB(R)*IBRate(R) - \Sigma_{WI}Cr(WI,R)$$

where:
$ILRate(R)$, $IBRate(R)$ = interest rates on interbank lending and borrowing on day R, respectively.

Objective function

$$Return = CashEnd(LR) + \Sigma_{MK}LastPr(MK)*SecPos(MK,LR) + TotInt - \Sigma_W Pen(W) - PenM$$

where:
$Return$ = variable to be maximized,
LR = index of the last day of the month,
$LastPr(MK)$ = market price of bond MK on the last day of the month,
$TotInt$ = interest received from BofI on reserve deposits during current month.

The above relationships represent the main blocks of constraints in the linear program. There are, of course, many additional details that have to be included in the model. Among these, we may mention here constraints limiting the turnover rate of the bank's bond portfolio, and various bounds on the variables.

The size of the model depends on the starting day within the current month and the number of qualifying government bonds in which the model can trade. On the first day of the month there are about 1,000 constraints and 2,000 variables. On the last day of the month, these numbers are reduced to about 150 each. The use of this model has very similar features to those of the previous application: it is run at least once a day by persons unfamiliar with optimization methodology, without intervention by optimization specialists.

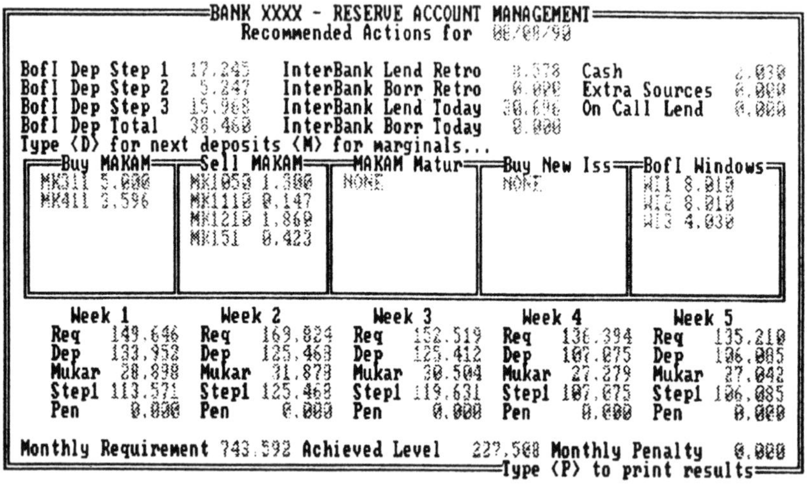

```
═══════════BANK XXXX - RESERVE ACCOUNT MANAGEMENT═══════════
               Recommended Actions for  08/08/90

BofI Dep Step 1  17.245  InterBank Lend Retro   3.378  Cash            2.030
BofI Dep Step 2   5.247  InterBank Borr Retro   0.000  Extra Sources   0.000
BofI Dep Step 3  15.968  InterBank Lend Today  30.696  On Call Lend    0.000
BofI Dep Total   38.460  InterBank Borr Today   0.000
Type <D> for next deposits <M> for marginals...
 Buy MAKAM      Sell MAKAM      MAKAM Matur     Buy New Iss     BofI Windows
 MK311 5.000    MK1050 1.300    NONE            NONE            W11 0.010
 MK411 3.596    MK1110 0.147                                    W12 0.010
                MK1210 1.860                                    W13 4.030
                MK151  0.423

   Week 1          Week 2          Week 3          Week 4          Week 5
Req   149.646  Req   169.824  Req   152.519  Req   136.394  Req   135.210
Dep   133.952  Dep   125.463  Dep   125.412  Dep   107.075  Dep   106.085
Mukar  28.898  Mukar  31.873  Mukar  30.504  Mukar  27.279  Mukar  27.042
Step1 113.571  Step1 125.463  Step1 119.631  Step1 107.075  Step1 106.085
Pen     0.000  Pen     0.000  Pen     0.000  Pen     0.000  Pen     0.000

Monthly Requirement 743.592 Achieved Level   227.508 Monthly Penalty   0.000
                                          ══Type <P> to print results══
```

Figure 7.2 Main output screen for reserve account management

3.4 *Solution and output*

Because of the rather complicated calendar dating functions, and forecasted yield curve computations needed for selecting the optimal solution, a Pascal "preprocessing" routine was written that provides the necessary input for the GAMS/MINOS program. The Pascal code in turn receives its inputs from the interactive screen program described in section 3.2. The results of the optimization are in turn "postprocessed" by another Pascal code in order to provide the user with an easily understandable set of recommended (optimal) actions. Figure 7.2 shows the main output screen.

In the upper portion of the screen the user can read the recommended BofI deposit, vault cash and interbank lending/borrowing amounts for the current day. In the middle portion of the screen the recommended trades of the qualifying bonds are shown, together with the optimal levels of borrowing from the BofI discount windows. The lower portion of the screen shows the recommended fulfillment of the weekly and monthly reserve requirements. Auxiliary output screens are also available to the user. One such screen shows the optimal deposit levels in BofI for the next five days (information that reflects the recommended deposit policy beyond the current day). Another screen shows marginal rates for the short-term bonds.

The whole modeling and optimization system is successfully implemented on 80386-based microcomputers. Preparation time for a typical run takes about five minutes, while the model-building, compilation, and optimization stages last between five and six minutes at the beginning of the month and two and three minutes at the end of the month.

ACKNOWLEDGMENTS

The author would like to thank Prof. Israel Zang for his contributions to the reserve management model, and Ron Avriel for his assistance in developing the Pascal codes for both applications mentioned in this paper. Support received from the Fund for the Advancement of Research at the Technion is gratefully acknowledged.

NOTES

1 The routine is needed to compute future cashflows resulting from borrowings for specified periods. The definitions of "periods" in finance are not always trivial. For example, a "one month" liability created on the last business day of February (say, the 28th) is due to be repaid on the last business day of March (say, the 31st).

2 The technique fits a smooth polynomial curve of borrowing rates through the date points supplied by the user. It was adapted from a Pascal routine in Press *et al.* (1986).

3 Discount windows are sources available to banks to borrow from BofI. Banks are given quotas within which they can borrow limited amounts on an increasing rate scale. For example, a bank can borrow up to a certain amount at the lowest prevailing rate, an additional amount at a higher rate, etc.

4 The planning horizon for deposits in BofI is the current month, defined as starting on the last Thursday that falls on or before the first of the current calendar month and ending on the last Wednesday of the current calendar month. Thus the planning horizon is always exactly four or five weeks. Since bond trading activities should take into account forecasts beyond the current month, for these activities the planning horizon is extended by another two weeks.

REFERENCES

E. Adamidou, Y. Ben-Dov, and L. Pendergast (1992), "Optimal horizon portfolio return under varying interest-rate scenarios," in this volume.

A. Brooke, D. Kendrick, and A. Meeraus (1988), *GAMS: A User's Guide*, The Scientific Press.

H. Dahl, A. Meeraus, and S. A. Zenios, "Some financial optimization models: I Risk management, and II Financial engineering," in this volume.

W. H. Press, B. P. Flannery, S. A. Teukolsky, and W. T. Vetterling (1986), *Numerical Recipes: The Art of Scientific Computing*, Cambridge University Press.

8 A flexible approach to interest-rate risk management

HENRIK DAHL

1 Introduction

Bond portfolio immunization and dedication are two widely used methods for controlling interest-rate risk, see, e.g., Platt (1986), Bierwag (1987), Fabozzi and Pollack (1987) and chapter 1 in this volume. The two approaches are seemingly very different. Under the immunization scheme, interest-rate risk is controlled by matching durations on assets and liabilities. When the interest rate moves, both sides of the balance sheet are affected in the same manner, leaving net present value virtually unaltered. Under the dedication scheme, the aim is to match cashflows to a degree which is economical under particular reinvestment and borrowing assumptions. As long as reinvestment rates do not fall below the scenario rates and borrowing rates do not rise above the ones assumed, the strategy ensures that enough cashflow is generated to gradually withdraw liabilities as they occur.

Which strategy should a particular investor choose? The traditional answer is that an immunized portfolio is riskier than a dedicated portfolio, hence also promises higher returns. If the investor is willing to accept some risk, he should establish an immunized portfolio, otherwise he should dedicate. However, this answer leads to additional questions. For example, to decide on the strategy, the investor may wish to know which risks are assumed in immunization, how large they are, and whether they agree with the investor's market views.

Similarly, if the investor believes interest rates will drop, neither of the two strategies may be useful to him. Instead, the investor may wish for a portfolio with a net positive duration, which promises higher returns if rates should in fact drop but experiences losses if rates should increase. On the other hand, the investor may not want a duration mismatched portfolio, if this simultaneously increases exposure to other, undesired, risks.

Thus, to enable an investor to choose a portfolio, it is necessary to be a bit more explicit about the multidimensional nature of fixed-income risks.

Below, we shall describe an approach to fixed-income risk management, which is based on the arbitrage pricing theory (APT, cf. Ingersoll (1987)). Under this theory, expected excess returns compensate for variations in multiple systematic risk factors. If we are able to isolate these risk factors, we will also be able to quantify how they affect returns of a particular portfolio. Consequently, it will be possible to manipulate the sensitivity of the portfolio returns to variations in the risk factors, to achieve a particular desired tradeoff between risk and return.

The first step is therefore to isolate the risk factors and quantify their impact on bond returns. We present a linear factor model of returns on pure discount bonds, i.e., bonds that make only a single but certain payment at a specified future date. Using this model, we derive return sensitivities to factor variations for default-free, fixed-payment bonds. We do so using a no-arbitrage condition: a regular bond may be thought of as a portfolio of pure discount bonds. To prevent arbitrage, the bond should therefore promise the same returns as the portfolio of discount bonds. We also present empirical estimates of the model.

Having quantified risks, we can proceed to controlling them. Under the factor model, we show how an immunized portfolio can be constructed by setting all net factor sensitivities equal to zero. In general, such a multifactor immunized portfolio can be established in many ways, so it is natural to select the portfolio composition which maximizes some criterion. We therefore specify an optimization model for constructing factor immunized portfolios, and discuss how this model relates to the immunization and dedication techniques mentioned above. It will be seen that the factor approach is more flexible than either of the two standard procedures, and that, in the limit, it covers both of them as special cases. Finally, we discuss how the factor-based approach could be used for selecting portfolios with a specified risk exposure to some factors while other factors are hedged. Along the way, we present some numerical examples to illustrate which kind of portfolios will result from applying this flexible approach.

2 Measuring interest-rate risk

In this section, we will describe how interest-rate risk may be measured. We first present a no-arbitrage pricing formula for bonds using the term structure of interest rates, and second see how variations in the term structure will affect bond prices. Third, we present a factor model of returns on pure discount bonds, and use this to derive simple risk measures for bonds. Fourth, we present an empirical estimate of the factor model for the Danish bond market, and discuss the implications of this estimate for risk management.

2.1 Arbitrage-free bond pricing

A pure discount bond is a security which pays one dollar with certainty at a specified future date. Suppose we know the prices of all such pure discount bonds. Then we could derive a fair price of a given bond by noting that we could reproduce the bond cashflow by a portfolio of pure discount bonds with nominal amounts of pure discount bonds maturing at time t equal to the size of the bond cashflow at that date. Since the bond and the portfolio of pure discount bonds have the exact same cashflow, they should have the same price to avoid arbitrage.

More formally, let:
d_t be the price of a pure discount bond maturing at time t,
C_t be the bond cashflow at time t, and
P be the bond price.
Then the value of the bond should be:

$$P=\sum_t C_t d_t \tag{1}$$

The prices of the discount bonds are given by the yields on those bonds, i.e., the yield y_t of a payment of 1 dollar to be received at time t. We will assume this yield to be continuously compounded, so:

$$d_t=e^{-y_t t} \tag{2}$$

We will refer to the collection of prices of discount bonds (as a function of term to maturity) as the discount function, and the corresponding collection of yields as the term structure of interest rates. Clearly, we can express the fair bond price using the term structure of interest rates:

$$P=\sum_t C_t e^{-y_t t} \tag{3}$$

In practice we cannot observe the term structure. However, we can use prices of traded bonds, including zero coupon issues if tax effects are not dominant, to derive or estimate the term structure of interest rates, assuming that these bonds are priced fairly. More importantly, given that any bond price depends on the term structure of interest rates, it is easy to see that any change to the shape or level of the term structure will affect bond prices. Using a first-order approximation, we find the impact on the bond price of an instantaneous change to the term structure to be:

$$dP=\sum_t -tC_t e^{-y_t t} dy_t \tag{4}$$

This expression serves as the basic means to measure interest-rate risk. Since yield changes occur randomly, bond present values will change randomly. Yield increases reduce present value, whereas yield reductions increase present value. The impact depends on the size and timing of the bond cashflow.

2.2 *Duration*

Equation (4) is too general to be of much practical value, since it gives us bond price sensitivities to a continuum of yields. To be useful, we need to make simplifying assumptions about the term structure changes. The most common assumption is that all yields change by the same amount. In that case, we find the traditional bond duration, D, to be a measure of interest-rate risk:

Set $dy_t = dy$. Then we have that the percentage change in present value to a term structure shift is:

$$D = -\frac{\partial P}{P}\frac{1}{\partial y} = \frac{\Sigma_t t C_t e^{-y_t t}}{P} \tag{5}$$

viz. simply the present value weighted term to the cashflow, with present value weights computed at the term structure rates. It may be noted that this model implies that a short maturity bond has a lower duration than a long maturity bond with the same coupon rate, hence will be found to be less risky than the long bond. The implication for risk management is straightforward. Two bonds with the same duration have the same risk. If you want less risk, i.e., you expect rates to go up, you buy a short maturity bond, and if you want more risk, you buy a long-term bond.

Although the assumption of parallel shifts to the term structure of interest rates is a convenient one, it is not realistic. First of all, it can be shown that parallel shifts will imply the existence of arbitrage possibilities. The reason lies in the convex nature of the discount function. We can compute the convexity of a pure discount bond as the relative second-order derivative of the present value with respect to the term structure.

Let q_t denote the convexity of a pure discount bond maturing at time t. Then:

$$q_t = \frac{\partial^2 d_t}{\partial y_t^2}\frac{1}{d_t} = t^2 \tag{6}$$

It is seen that convexity is positive for pure discount bonds (and hence also for any bond with fixed payments). This means that the duration measure above will understate the true impact on bond returns of yield changes. Also, since convexity grows by the square of the term to maturity, or, equivalently, by the square of the duration of the pure discount bond, the estimation error will be higher, the longer the maturity.

Two portfolios with equal duration may have different convexity, depending on maturity. For example, we could construct a portfolio of a one-year and a thirty-year pure discount bond with a duration of ten years. This portfolio will have higher convexity than a ten-year pure discount bond. If all yields now increase by the same amount, the value of the ten-year pure discount bond will be reduced by more than the value of the equal duration portfolio, because the higher convexity of the portfolio offsets some of the loss of present value. Similarly, if all yields drop by the same amount, the value of the portfolio increases by more than the value of the ten-year pure discount bond. Therefore, no matter what happens to yields, as long as they move in parallel, the portfolio will always give a higher return than the pure discount bond. Consequently, we have an arbitrage possibility: buy the portfolio and sell an equal amount of the ten-year pure discount security. The net outlay today is zero, but, as soon as rates change, we can reverse the trade at a profit. This implies that parallel shifts generate arbitrage possibilities and theoretically they can therefore not be expected to describe the real world movements of the term structure.

2.3 A linear factor model

Apart from this purely theoretical reason to object to the assumption of parallel shifts, it is also unnecessarily simplified. In addition, casual observation of the term structure shows that it does not shift only in parallel. Without any significant additional complexity, we can generalize the assumptions of term structure variations to the case where more than just a single type of shift occurs. In particular, we can consider the class of linear factor models of returns. Let:

f denote an index of factors, $f=1,\ldots,n$, where n is the number of factors.

T denote calendar time,

r_{tT} denote the rate of excess return at time T of a pure discount bond maturing at time t,

dy_{tT} denote the change at time T of the yield of a pure discount bond with constant term to maturity t,

a_{ft} denote the impact of a change of factor f on the returns of a pure discount bond maturing at t, and

dF_{fT} denote the change to the level of factor f at time T.

Then, we could write the linear factor model as:

$$r_{tT}=\sum_{f} a_{ft}dF_{fT}+\varepsilon_{tT} \tag{7}$$

Here, ε is an error term, which we assume to be independently normally distributed with zero mean and constant variance.

The model states that, apart from a non-systematic error, the returns on pure discount bonds move as a result of a linear combination of several independent risk-generating factors. If factor 1 is increased by one unit, the model states that the one-year rate of return increases by a_{11}, while the two-year rate of return increases by a_{12}, etc. We will refer to the coefficients a as the *factor loadings*, and to dF as the factor level changes. Notice that the loadings are assumed to be constant over calendar time, whereas the factor-level changes are time dependent.

We can easily translate the above model of returns into one describing movements of the term structure by:

$$dy_{tT} = \sum_{f} b_{ft} dF_{fT} + \omega_{tT} \tag{8}$$

with:

$$b_{ft} = \frac{a_{ft}}{t} \tag{9}$$

We have not specified what the factors F are. Several approaches have been proposed in the literature. Thus, for instance, Brennan and Schwartz (1979), develop a model of term structure variations based on variations of the short- and the long-term rate. Other explanatory variables could be introduced as well to enhance that model. However, it seems that most theoretical models of the term structure variations are empirically deficient. Instead, we could attempt to determine the factors empirically. This is the route taken in Garbade (1986) and Litterman and Scheinkman (1988), and which we shall follow here.

The idea is that, in order to measure and control interest-rate risk, you do not need to be able to interpret the factors as real economic entities like the money supply or an exchange rate, etc. All you need to know is how a change to the factor affects the term structure and therefore the values and returns of traded bonds. This is in fact the same approach used above to derive the duration as a risk measure. There, we said that a parallel shift to the term structure would affect bond returns in a predictable manner, but we did not say what caused the parallel shift to occur.

To estimate the model in (7), we first generate the matrix of historical returns of pure discount bonds – either by observations of traded zero coupon issues or by using estimated term structures. Next, we find that set of loadings of the first factor which explains most of the variance of historical returns on pure discount bonds, subject to the normalization condition that the variance of factor 1 changes is 1. We can find these loadings as the first eigenvector of the matrix of historical returns. Given these loadings, we proceed to find the loadings of the second factor which explain most of the residual variance of historical returns, and given that the changes of factors 1 and 2 should be uncorrelated. This will be the second eigenvector of the matrix of historical returns. In this manner, we can continue with factor 3, etc., until we reach a stopping criterion, for example that we have explained 99% of historical returns.

This estimation technique will give us three results. First, we find how many factors we need to use to explain historical returns to the desired degree of accuracy (n above). Secondly, we find how each of these factors have moved over time (dF above). And thirdly, we find out how the movements of each factor affect returns of pure discount bonds and therefore the shape of the term structure (the factor loadings a or b above).

Based on US data, it appears that three factors are sufficient to explain excess returns of Treasury bonds (see Garbade (1986, 1989) and Litterman and Scheinkman (1988)). These are a *level factor*, which implies a roughly parallel shift to the term structure, a *steepness factor* which changes the slope of the term structure, and a factor which changes the *curvature* of the term structure.

The factor loadings can be different for other countries. Thus, in Denmark, we find three to four factors, explaining 99.9% of the term structure variation (see Dahl (1989)). These are a steepness factor which explains 86% of historical variation; a curvature factor, accounting for about 11% of historical term structure movements; a factor which affects the term structure up to ten years but does not affect longer yields. This third factor explains about 3% of total variation. The fourth factor is very short term in nature and implies a twist of the term structure only up to maturities of four years. It is not very important since the first three factors explain 99.6% of historical changes to the Danish term structure. The term structure loadings of the three Danish factors are shown in figure 8.1.

We have tested the stability of these factors over time and with respect to the specification of the model used to estimate the term structure of interest rates, which is used to generate the implied historical returns on pure discount bonds, and found them to be very robust.

2.4 *Factor-based risk measures*

Having found the factor loadings b, above, we can compute factor modified durations, i.e., the impact on bond returns from a marginal change in each factor. The idea simply is that a factor change affects the shape and level of the term structure, which implies a bond price change. The computation

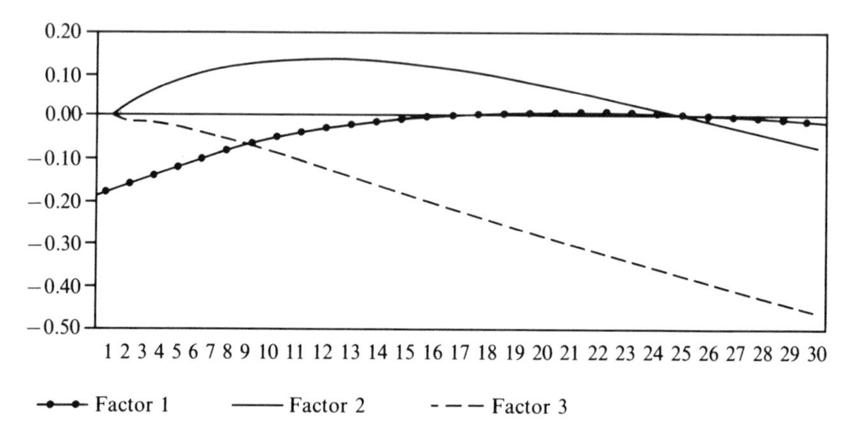

Figure 8.1 Factor loadings

uses (4) and the factor model in yield space (8) to find that:

$$D_f = -\frac{\partial P}{\partial F_f}\frac{1}{P} = \frac{\Sigma_t t b_{ft} C_t e^{-y_t t}}{P} \tag{10}$$

since the factors are constructed to be independent by our estimation procedure.

This expression deserves three comments. First, to understand the expression, note that the term $tC_t e^{-y_t t}/P$ is the portion of returns of the bond which result from a unit change to the yield for that maturity. The factor model tells us that a unit, i.e., standard deviation, change to the factor does not change yields by 1%, but rather by b_{ft}%. Therefore to find the total impact on bond returns from a factor change, we need to scale the above return portions accordingly.

Second, note that for a purely parallel shift to the term structure, $b = 1$ for all t. Therefore, in that case, the factor duration in (10) is identical to the standard measure in (5). The standard modified duration is therefore a special case of the factor model based measures, although, as was indicated above, since parallel shifts induce arbitrage opportunities, it is not likely that we find a parallel shift among the empirical factors.

Third, while the modified duration measure normally implies that longer bonds are riskier than shorter bonds, that is not necessarily the case under the factor model. For example, consider a factor 3 shift above. This affects short-term yields without any impact on long-term yields. When this happens, short-term instruments will be more affected than long-term instruments. Short-term bonds are therefore riskier than long-term bonds in that case. Similarly, when curvature changes, we may see an increase in short-term and long-term yields, but a drop in medium-term yields. We may therefore find that, while short-term and long-term bonds lose value, medium-term bonds give positive excess returns.

These observations have important implications for interest-rate risk management, since it demands a more nuanced response to expected interest-rate changes. Under the standard assumptions of parallel shifts, when you expect rates to increase, you reduce the duration, i.e., maturity, of your portfolio to offset the potential losses. Under the factor model, however, you need to be more specific about which rates are going to increase. A reduction of standard duration might expose the portfolio to increases in short-term rates, i.e., factor 3 shifts.[1] Whereas the standard duration model does not allow for an easy measurement of risks due to changes in the shape of the term structure, the factor model can be used to compute risk measures that automatically uncover such exposures.

In addition to computing factor durations, we can also compute factor convexities, i.e., the impact on bond returns from volatility of the factors around an unchanged mean. We find that:

$$Q_f = \frac{\partial^2 P}{\partial dF_f^2}\frac{1}{P} = \frac{\Sigma_t t^2 b_{ft}^2 C_t e^{-y_t t}}{P} \tag{11}$$

Again, when parallel shifts are the only type of term structure movements, the factor convexity reduces to the standard measure of bond convexity.

3 Factor-risk management

We are now ready to turn to the question of how you can control portfolio interest-rate risk under the factor model above. We shall first discuss multifactor hedging and next see how this approach relates to the more traditional techniques for bond portfolio dedication and immunization. Finally, we shall briefly indicate how a portfolio manager could use the factor approach to selectively hedge particular risks while maintaining a desired exposure to other risks.

3.1 Factor hedging

Note that, for a given change to the term structure, the impact on any bond price is systematic. An increase in the one-year yield will reduce the value of any payment made in one year, irrespective of which bond makes this payment, although the size of the present value change depends on the size of the actual bond cashflow in one year. The fact that bond returns are systematically affected by factor movements implies that it is possible to hedge the impact of factor changes.

To see how the impact of factor changes on bond returns can be eliminated, let us first note that hedging and immunization always aim at protecting value relative to some benchmark. This benchmark could be an index of securities, in which case we talk about indexation, or it could be a portfolio of liabilities, in which case we talk about asset-liability management, or it could be target horizon return, which is the traditional focus of bond portfolio immunization or asset management. Although these areas may appear to be very different, we can treat them in the same manner.

When attempting to lock in a given return over a particular investment horizon, we may think of the horizon as the time when we wish to withdraw a certain amount. Conceptually, this simply amounts to satisfying a hypothetical liability, consisting of a single payment on the horizon date. Therefore, we may think of traditional immunization as a particular form of asset-liability management, where the liability side simply consists of a given nominal amount of a zero coupon bond maturing on the horizon date.

Similarly, when indexing a portfolio, we may think of the target index as a hypothetical liability portfolio. The aim of indexation is then to make sure that the net portfolio, i.e., assets minus the hypothetical liabilities, should never present losses.

For these reasons, it is convenient to cast the question of how to hedge portfolio returns under the factor model in the framework of asset liability management. In what follows, we therefore assume that we have a well-defined set of liabilities, whether actual or hypothetical ones. We then attempt to find an asset portfolio composed of bonds which will always have at least the same value as the present value of liabilities, no matter what happens to the level or shape of the term structure.

Let:

i denote the set of securities in the asset universe,
X_i denote the nominal holdings of each security,
P_i denote the security full prices (flat price plus accrued interest),
D_{if} denote the factor loading of factor f in security i,
Q_{if} denote the factor convexity of factor f in security i,
P_L denote the present value of liabilities,
D_{Lf} denote the liabilities' factor durations, and
Q_{Lf} denote the liabilities' factor convexities.

We then have that the value of assets should be equal to the value of liabilities:

$$\sum_i X_i P_i = P_L \tag{12}$$

In addition, we want the assets and liabilities to perform the same when factors change. To arrive at a practical expression, let us totally differentiate (12) to find that:

$$\left(\sum_i X_i P_{iT} - P_{LT}\right) dT + \sum_f \left(\sum_i X_i P_i D_{if} - P_L D_{Lf}\right) dF_f +$$

$$\sum_f \left(\sum_i X_i P_i Q_{if} - P_L Q_{Lf}\right) dF_f^2 = 0 \tag{13}$$

Here, $P_{.T}$ is the partial derivate of the price with respect to calendar time.

The first term of (13) measures the impact on net portfolio value as time goes by without anything else happening in the market. The derivative of value with respect to calendar time is the time value of the portfolio. This term is (locally) deterministic, and we may choose not to consider it for hedging purposes. Instead, we may desire to eliminate the influences of stochastic movements of the factors precisely in order to secure higher time values, i.e., yield, on the assets than on the liabilities.

The second term of the expression measures the impact of a directed change of each factor, while the last term measures the impact of factor volatility on net portfolio value.

In order to eliminate the influence of any combination of factor changes, we can see from the expression that we need to select an asset portfolio which has the same duration and convexity for each factor as the liability portfolio. In that case, the coefficients to each dF and dF^2 are zero, and factor movements will not affect net portfolio value. Therefore, the conditions for factor hedging become:

$$\sum_i X_i P_i D_{if} = P_L D_{Lf} \;\forall f \tag{14}$$

$$\sum_i X_i P_i Q_{if} = P_L Q_{Lf} \;\forall f \tag{15}$$

Equation (14) states that the dollar duration of assets and liabilities should be set equal for each factor, and (15) states that, for each factor, dollar convexities should be set equal on assets and liabilities, in order to establish the hedge. It is easy to see that we may choose to allow for greater factor convexities on assets than on liabilities. The reason is that, as can be seen from (11) above, factor convexities are uniformly positive for bonds. Therefore, when setting the net portfolio convexities non-negative whilst keeping net factor durations at zero, we have the conditions for a global minimum on portfolio value at current yields. Once factors move, the net portfolio value will become positive.

In general, we can select many portfolios that satisfy these requirements. We may therefore want to optimize the portfolio composition. A natural formulation of the optimal factor immunization problem would be to maximize asset portfolio time value, subject to the conditions that present value and factor durations on assets be equal to those of liabilities while asset factor convexities should be at least as large as those of liabilities:

$$\underset{X}{\text{Maximize}} \sum_i P_{iT} X_i$$

subject to

$$\sum_i P_i X_i = P_L$$

$$\sum_i D_{if} P_i X_i = D_{Lf} P_L, \forall f$$

$$\sum_i Q_{if} P_i X_i \geq Q_{Lf} P_L, \forall f$$

$$X_i \geq 0$$

This optimization model is a linear programming problem which is easily solved for even very large selection universes. It can also be enhanced with various constraints on the portfolio composition, and be solved as a mixed-integer programming problem to give tradeable solutions. It is seen that normally, the resulting portfolio will consist of a number of bonds equal to the number of factors plus one (for the present value equality condition).

As an example of the type of portfolios that result from applying the factor immunization model, consider the following case where we attempt to immunize a given liability stream. The universe for portfolio selection is composed of Danish government bonds and investment grade bonds from the Danish Agricultural Board (DLR) and Ship Building Credits (SKE), sixty bonds in all. All bonds have a high outstanding volume, and are either bullets or serial bonds, which amortize a fixed percentage of the initial outstanding volume at each payment date. All are non-callable.

Figure 8.2 shows the resulting cashflow from the asset portfolio together with the target liability cashflows, when solving the above factor immunization problem without any side restriction on portfolio composition. It is easily verified that the cashflows are very closely matched. This may be quite surprising given that the optimization model does not include any

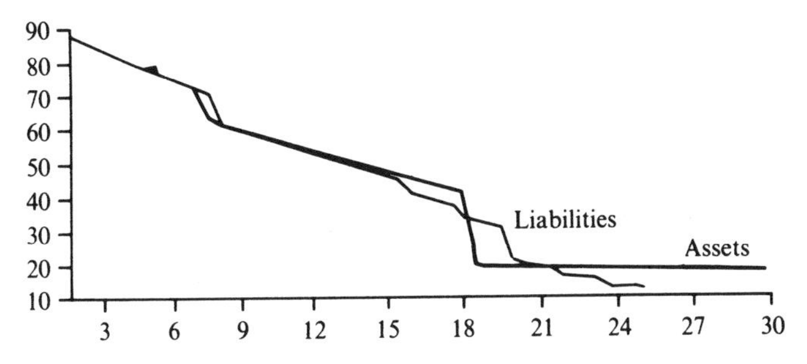

Figure 8.2 Cashflow pattern (full factor hedging)

information about cashflows. However, the result is quite general, and mostly you will find factor immunized asset portfolios to have roughly the same cashflow as that of liabilities.

3.2 Relationship to bond portfolio dedication

To underline how closely matched the factor immunized portfolio cashflow is figure 8.3 shows the cashflows from a bond portfolio dedicated to the same liability stream. We have used the formulation of dedication of Zipkin (1989). In this model, we have assumed that excess cashflows can always be reinvested to cover the next liability at current forward rates less 500 bps, while any cash deficits can be covered by borrowing at forward rates plus 500 bps. These assumptions are quite conservative, and the resulting portfolio would therefore normally be thought of as virtually risk free.

Comparing figures 8.2 and 8.3, it appears that the factor immunization model gives results quite close to those of the dedication model. Although there is not a particularly good reason why this should always be the case when hedging three factors, there is a good reason why we should expect a convergence between the two models when sufficiently many factors are added.

Consider the following argument. The equilibrium pricing equation for a portfolio of fixed-payment securities is given by (3) above. Now consider a shock to the term structure:

$$y'(t) = y(t) + h(t)$$

where $h(t)$ can be any function depending on the term to maturity. Total instantaneous dollar returns can be found by differentiating:

$$dP(h(t)) = -\sum_t tC_t e^{-y(t)t} h(t) \tag{16}$$

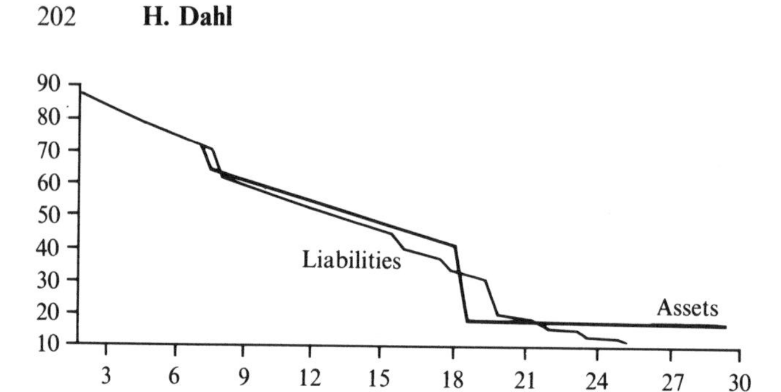

Figure 8.3 Cashflow pattern (conservative dedication)

Since the shock function can take on any shape, it can be verified that full hedging, i.e., a net price sensitivity of zero for any realization of the shock, requires a perfect cashflow matching of assets to liabilities. Note that (16) is really just another way of expressing our factor sensitivities. Thus, in the limit where infinitely many factors are hedged, each corresponding to a realization of the shock function h, we will end up with cashflow matching. In practice, of course, only a finite number of factors are used, namely those which are found to be empirically important (and for fixed-payment securities only those which are independent). The argument holds, however, that the more factors are hedged, the more the resulting portfolio will converge to cashflow matching. In this sense, dedication is contained in the factor model.

It may be worth mentioning that the factor model allows for an economical way to choose which h-functions to hedge. Certain realizations of the h-functions never occur in practice (more precisely, they occur with a negligible probability), and are therefore not worth hedging. Since the factors are found by an estimation technique which maximizes the explanatory power of each factor, the analysis also tells you the empirical probability of movements of each factor. Then you may decide whether to hedge 95%, 99%, or 99.99% of empirical term structure variations by choosing the number of factors to hedge.

3.3 Relationship to standard immunization

At the other extreme, it may be useful to compare factor immunization with standard duration based immunization. As we have discussed above, standard duration is obtained from the factor model when only a single factor which shifts the term structure in parallel is considered. It should therefore not be surprising that the standard immunization techniques are a special case of factor immunization.

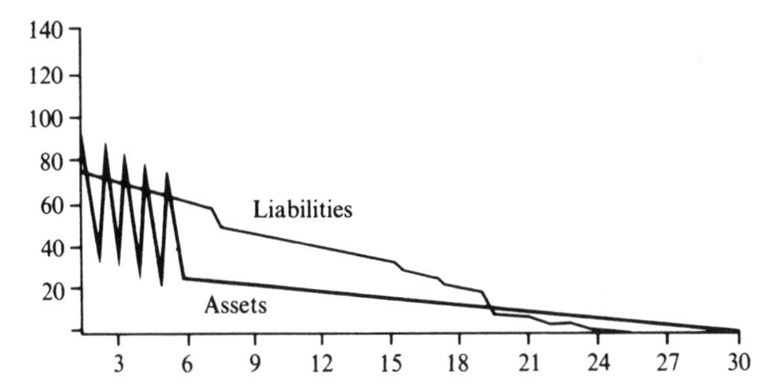

Figure 8.4 Cashflow pattern (standard immunization)

It is well known that without any side restrictions on the portfolio composition, optimally immunized portfolios are barbells, that is portfolios with payments spread out relative to the target cashflow. Economically, the reason is that if parallel shifts are the only source of risk, we can arbitrage by buying a portfolio with as high a convexity as possible, as discussed earlier. The higher convexity is obtained by spreading out the cashflow around the target duration as much as possible.

In our numerical example, we find exactly this kind of optimal portfolio. The immunized portfolio cashflow is shown in figure 8.4.

In practice, of course, nobody would accept this portfolio cashflow as a hedge. The problems are twofold. First, because of the cashflow mismatch, the immunized portfolio is highly exposed to non-parallel shifts to the term structure. Secondly, over time, the portfolio will need frequent rebalancings, because durations of assets and liabilities will drift apart over time. Therefore, apart from the shape risk, the portfolio is subject to liquidity risk, i.e., the risk that some securities cannot readily be bought or sold without incurring high transaction costs. It is easily verified that the factor immunized portfolio will not need as frequent rebalancings, since the cashflow of that portfolio is much closer to liabilities.

The problems of immunized portfolios are normally solved by adding side constraints to the optimization problem. Especially, the desired cashflow pattern is obtained by forcing certain proportions of the bonds of the portfolio into particular maturity buckets. Such diversification constraints certainly do reduce the shape risk and liquidity risk inherent in immunized portfolios, but it is not clear that they do so in an economic manner. Instead, such constraints may be viewed as *ad hoc* repairs on the basic assumption of single-factor parallel shifts. The multifactor model on the other hand explicitly utilizes the multidimensional nature of term structure movements to eliminate these risks.

3.4 Selective hedging

Consider the definitions of factor durations and convexities above. They describe the rate of instantaneous return on a bond or on a portfolio from movements in the factors. We have seen that to eliminate influences from factor movements on portfolio returns, all net factor durations and convexities should be set to zero. If, on the other hand, factor durations are mismatched, the portfolio returns will be subject to the risk that factors move.

In particular, assume that the net duration of factor 1 is negative. In that case, an increase in factor 1 will increase portfolio returns, while a reduction in factor 1 will reduce returns. This means that, if you believe the term structure to become steeper (a factor 1 increase), you should attempt to reduce net portfolio factor 1 duration. Conversely, if you believe the term structure to flatten, you should increase factor 1 duration in order to enhance returns. If you do not have any view but wish to hedge, you should set net factor duration equal to zero.

This rule, of course, is simply the multifactor version of the traditional rule that an expected interest-rate reduction should be met by an increase in portfolio duration and an expected rate increase should be met by a reduction of duration. However, in the multifactor case, you can express your views for each type of term structure movement in isolation.

Easy as this sounds, in practice, it is a bit more complicated than that. First, normally securities are not single-factor instruments. Rather, they are sensitive to most or all factors. Thus, switching one security for another will not only affect exposure to a single factor but to most of them: exchanging a security with a low steepness sensitivity for one with a high steepness sensitivity may simultaneously change exposure to curvature even though this is not desired.

The problem of portfolio risk management therefore becomes a simultaneous one where desired factors are over- or underweighted, and the rest are hedged. We can formulate an optimal-factor risk-management model to guide portfolio selection in these cases.

Suppose that of the n risk factors, h should be hedged, while the portfolio manager has specific views for the rest, g, of them. Also assume that the portfolio manager is able to specify how much these factors are going to move, i.e., he knows the size of dF_g. Finally, assume that the present value, factor durations, and factor convexities of liabilities are known.

To simplify matters, the portfolio manager could now construct a new term structure from the existing one by adding the factor shocks that he expects, whilst assuming zero changes to the one he wishes to hedge:

$$y_t^* = y_t + \sum_g b_{gt} dF_g \tag{17}$$

He could use this new hypothetical term structure to find an expected price of each security in the asset portfolio:

$$P_i^* = \sum_t C_{it} e^{-y_t^* t} \tag{18}$$

Using this information, we could formulate the optimization problem as one which attempts to capture the highest gain from the expected factor movements, while hedging the influence of the remaining factors:

$$\underset{X}{\text{Maximize}} \sum_i (P_i^* - P_i) X_i$$

subject to

$$\sum_i P_i X_i = P_L$$

$$\sum_i D_{ih} P_i X_i = D_{Lh} P_L$$

$$\sum_i Q_{ih} P_i X_i \geq Q_{Lh} P_L$$

$$X_i \geq 0$$

It may be mentioned that, although the model will position the portfolio optimally for the foreseen scenario, it will also cause the portfolio to fail miserably, should the opposite scenario occur. If for instance the investor expects a steepening of the term structure and positions his portfolio optimally for this scenario by reducing the steepness duration, and the term structure actually inverts instead, the investor loses. This, of course is due to the active risk-taking.

Figure 8.5 shows an example of a portfolio cashflow generated by this model, where the investor expects an increase in the curvature of the term structure, i.e., where short- and long-term rates are reduced while medium-term rates are increased. The investor wishes to maximize his gain from this scenario but does not want to expose his portfolio returns to changes in factors 1 (steepening) and 3 (short-term rate movements).

It is seen that the asset cashflow is now clearly mismatched relative to the liability stream. The resulting portfolio resembles a barbell, where excess payments are generated in the near and in the distant future, while a cash deficit is carried in the interim period. Were it not for the simultaneous hedging against other types of term structure movements, this would

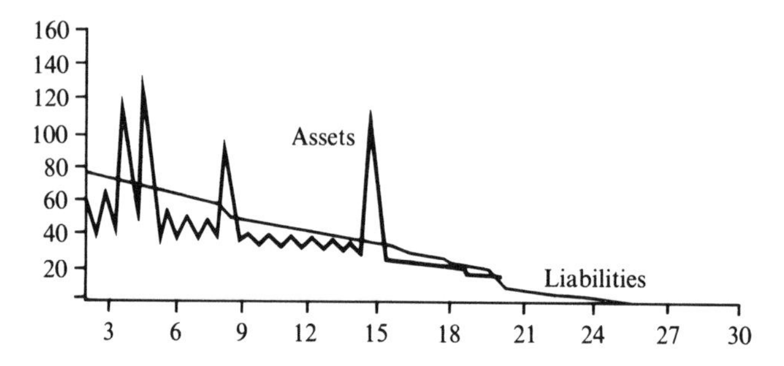

Figure 8.5 Cashflow pattern (speculation in curvature)

indeed have been the cashflow pattern. The portfolio would consist of cash and the longest possible instrument. However, we see that the asset cashflow also has a spike around nine years maturity, and that the overall cashflow is not more mismatched than was the case for the immunized portfolio in figure 8.4 above. The nine-year surplus cashflow serves to prevent losses due to increased or reduced steepness of the term structure. In addition, up to maturities of five years, the cashflow is first net negative and next net positive. This structure of payments insures against movements of short-term rates.

In summary, the factor approach allows for flexibility in creating asset portfolios. Some risks can be eliminated while others are actively sought. The principles for doing so are similar to and almost as simple as the standard trading rules based on duration mismatches. But whereas duration mismatching can be seen to change exposure to other types of term structure movements, the factor approach allows you to control exposure to each type of movement separately. This may be quite valuable when establishing index portfolios, since it allows for controlled return enhancement without unwanted exposures.

It may be worth mentioning that dedication techniques do not allow for a similar flexibility. In the dedication model, the degree of risk exposure is controlled through the choice of reinvestment and borrowing rates. Setting the reinvestment rate to zero and the borrowing rate to infinity will force a close cashflow match and very low risk. Setting the reinvestment rate and borrowing rate equal to the current forward rates, the dedication model simply results in a present value match, with no control over risk. It is not clear how to set rate assumptions to obtain a cashflow mismatch which exposes the portfolio to exactly the desired risks.

4 Conclusions

In this chapter, we have outlined a flexible risk-management model based on factor analysis of empirical term structure changes. We have argued that the model covers immunization and dedication as special cases but also allows for selective risk-taking with simultaneous hedging of undesired risks. Also, we illustrated the model by numerical examples. It was argued that the factor model converges to cashflow matching of bond portfolios. Even when hedging just three factors, it appeared that hedging could be performed quite accurately. Since the resulting cashflow was quite closely matched to the liability stream, it could be argued that the portfolio would not need frequent rebalancing due to duration drift. The excess returns assumed to be locked in by the hedge should therefore not be expected to be eroded by transaction costs.

We have focused exclusively on management of interest-rate risk in bond portfolios. However, it appears that the approach could be used more broadly. First, Knez, Litterman, and Scheinkman (1989) have applied factor analysis to the money market and have isolated different sector factors, i.e., factor affecting returns on commercial paper or certificates of deposit relative to Treasury Bills, etc. Using these sector factors, it should be possible to enhance the above framework to also control sector risk.

Secondly, it may be expected that the approach could be applied to portfolios with fixed-income derivatives as well, including swaps, options, futures, and mortgage-backed securities. All these products are sensitive to changes in the shape of the term structure, and most likely even to a higher degree than regular bonds. The main difficulty is to compute the option-adjusted factor durations and convexities of these products. Today, option-adjusted durations and convexities are generally computed using the option-adjusted spread methodology. However, it is not clear that the stochastic models of the term structure that are used in these calculations are consistent with the modes of variation found by factor analysis, and it appears that some research is needed to answer the question of how to select a set of suitable stochastic processes of interest rates to generate term structure movements that agree with the empirical ones.

NOTES

I would like to thank an anonymous referee for excellent comments on a previous version of this chapter. Also, the chapter has benefitted from discussions with my colleagues Simon Schultz, Per Soegaard-Andersen, and Torben Visholm, at SimCorp. Any errors or omissions are of course exclusively mine.

1 It appears that the term structure inversions in many European countries during 1989 and 1990 were to a large extent due to such variations in short-term yields. As a response to expected rate increases Danish banks uniformly reduced asset portfolio maturities during 1989. This implied an involuntary exposure to short-term rate movements, and as a result of such factor 3 shifts, many Danish banks lost significant amounts over this period.

REFERENCES

G. O. Bierwag (1987), *Duration Analysis*, Ballinger Publishing Company, New York.

F. Black, E. Derman, and W. Toy (1990), "A one-factor model of interest rates and its application to treasury bond options," *Financial Analysts Journal*, January–February, 33–39.

A. J. Brazil (1988), "Citicorp's mortgage valuation model: option-adjusted spreads and option-based duration," *Journal of Real Estate Finance and Economics*, 1.

M. J. Brennan and E. S. Schwartz (1979), "A continuous time approach to the pricing of bonds," *Journal of Banking and Finance*, 3.

J. C. Cox, J. E. Ingersoll, Jr., and S. A. Ross (1981), "A re-examination of traditional hypotheses about the term structure of interest rates," *Journal of Finance*, 36.

(1985), "A theory of the term structure of interest rates," *Econometrica*, 53.

H. Dahl (1989), "Variationer i rentestrukturen og styring af renterisiko," *Finans/Invest*, 1 (in Danish).

H. Dahl, A. Meeraus, and S. Zenios, "Some financial optimization models: I Risk management," this volume.

S. Diller (1989), "The yield surface – a three dimensional approach relating yield to duration and convexity, Bear Stearns, fixed income strategies," February, New York.

F. J. Fabozzi and L. M. Pollack (eds) (1987), *The Handbook of Fixed Income Securities*, Dow-Jones, Irwin, New York.

K. Garbade (1986), "Modes of fluctuations in bond yields – an analysis of principal components," Bankers Trust Company, Money Market Center, June, New York.

(1989), "Polynomial representations of the yield curve and its modes of fluctuation," Bankers Trust Company, Money Market Center, No. 53, July New York.

J. E. Ingersoll, Jr. (1987), *Theory of Financial Decision Making*, Rowman & Littlefield, New York.

P. J. Knez, R. Litterman, and J. Scheinkman (1989), "Explorations into factors explaining money market returns," Technical Report, Goldman, Sachs & Co., 1 March, New York.

R. Litterman and J. Scheinkman (1988), "Common factors affecting bond returns," Goldman, Sachs & Co., Financial Strategies Group, September, New York.

R. B. Platt (1986), *Controlling Interest Rate Risk, New Techniques & Applications for Money Management*, John Wiley and Sons, New York.

P. Ritchken, P. (1987), *Options: Theory, Strategy and Applications*, Scott, Foresman and Company, New York.

S. M. Schaefer (1984), "Immunisation and duration: a review of theory, performance and applications," *Midland Corporate Finance Journal*, Fall.

P. Zipkin, (1989), "The structure of structured bond portfolio models." Technical Report, Columbia University, Graduate School of Business, January, New York.

9 Currency hedging strategies for US investment in Japan and Japanese investment in the US

WILLIAM T. ZIEMBA

1 Introduction

Investment to and from Japan can be very profitable. For example, a dollar invested in the Nikkei Stock Average in 1949 was worth over $500 at the end of 1989. In recent years there has been much Japanese investment in the US, particularly in bonds, stocks, and real estate. The drop in the yen dollar rate from the 260 range in the fall of 1985 to the 120 range at the end of 1987 and its sharp rise back to 160 in mid 1989 and back under 125 in early 1992 shows the extreme risk involved in these investments. This chapter investigates currency hedging strategies for Japanese investors making investments in US assets and Americans investing in Japan. The traditional approach is to fully eliminate the currency risk using forward or futures contracts to offset the long exposure to the foreign currency. For the American investing in Japanese stocks the hedge often provides a bonus: an essentially risk-free gain of 1–4% per year due to the difference in interest rates between the two countries. Although improvements adding risk are conceivably possible this approach is a very satisfactory resolution of this problem. The situation has been much more difficult and complicated for Japanese investment in the US. The forward/futures hedge eliminated the currency risk but at a cost of 1–4% per year until 1990. Strategies that do not lose this interest-rate differential and in fact collect positive premiums are available by selling and buying yen put and call options. Simulations using an investment in three-year Treasury bonds investigate this using recent currency movements and Black–Scholes estimated prices for plausible future scenarios to evaluate the risk–reward tradeoffs and the worst possible outcome using the various strategies. The added risk seems well worth taking for many investors to achieve substantial expected gains.

Futures and options can be used for hedging risk reduction purposes as well as for speculation. For the yen, there are active options markets as well as futures markets around the world. Long-term yen currency warrants are traded on the American Stock Exchange. Yen futures markets in Japan began in June 1989. For the stock market, only index futures were available for use in Japan until the new NSA index options contracts began trading in Osaka in June 1989. Futures options contracts on the TOPIX on the TSE and the Nagoya 25 began trading in October 1989. However, long-term puts and calls on the NSA and Topix have been trading for some time in markets outside Japan. Except for long-term equity warrants there are no futures or options contracts in existence or planned regarding individual stocks or for small stocks. Hence, hedging strategies involving individual stock portfolios using the index derivative instruments will be subject to tracking error, which must be considered in the design of such strategies. Bailey and Ziemba (1991) and Ziemba and Schwartz (1992) discuss these index options and warrant contracts.

When the yen strengthened from 260 in the fall of 1985 to its low near 120 in 1987 and 1988, the *seibo* (life insurance companies) lost some 50 billion dollars from their US Treasury bond investments by not hedging. During 1989 and 1990, the yen has fallen back to the 150–60 range and then to the 125–35 range. In early 1992 the rate was about 125. Hence they have made up some of these losses. The strategy discussed here is a simple one that utilizes short-term puts and calls. Other instruments that could be used for currency hedging are put warrants as well as currency swaps, foreign exchange annuity swaps, foreign currency borrowing, interest-rate caps and floors, investment or deposit in foreign currencies (such as dollar deposits with reverse floating rates against option contracts), hybrid capital market instruments (such as index bonds), etc.

The *seibo* have generally neglected the currency risk for investing long-run in the higher interest rates in the US and for international diversification of their portfolios. The main reason is that their capacity to take large risks has come from the soaring prices of their accumulated real estate, stocks, and bonds in the domestic markets since 1985. The decline in stock and land prices in 1990 and 1991 has led to much less Japanese investment abroad (see Stone and Ziemba (1993)). Altman and Minowa (1989) discuss risks and returns from Japanese investment in high yield corporate (junk) bonds using futures hedging strategies. Obviously with so much at stake, strategies to protect this investment against unfavorable currency movements is of considerable interest.

2 Hedging a US stock portfolio against a possible increase in the value of the yen for a Japanese investor

To begin our discussion, let's consider the case of a Japanese individual or institutional investor with a US denominated portfolio that is either owned or contemplated. To make the case simplest, suppose the investor has 10

Table 9.1. *20 October, 1988 futures for Japanese yen*

Japanese yen (mm) 12.5 million yen; $ per yen (0.00)

						Lifetime			
	Open	High	Low	Settle	Change	High	Low	Open interest	Yen/ US$
Dec. 78	0.7931	0.7948	0.7908	0.7911	+0.0013	0.8530	0.7115	46,621	126.41
Mar. 89	0.8007	0.8019	0.7981	0.7982	+0.0013	0.8590	0.7439	2,094	125.28
June 89	0.8093	0.8096	0.8065	0.8065	+0.0013	0.8400	0.7500	785	123.99
Sept. 89	0.8150	0.8180	0.8145	0.8140	+0.0014	0.8180	0.7690	174	122.85

Est. vol. 28,556; vol. Wed. 27,380; open into 49,674, −1,104.

million dollars in yen or about 1,271.5 million yen at the spot rate of 127.15 yen per dollar in late October 1988. It does not matter if the money is in dollars from earnings in the US, or in yen because this sum can be immediately converted at very low cost from one currency to the other.

Strategies for Making and Keeping Excess Profits in the Stock Market (Ziemba, forthcoming) provides strategies and suggestions for investment in the US, so we will not discuss how to wisely invest there. Assume that you are about to purchase the 10 million dollars in US denominated stocks, bonds, and treasuries. How can you protect against the chance that the yen will keep rising and possibly go to 100 against the dollar?[1]

A hedge involves the purchase of yen futures to match the sale of yen on day 0 – the conversion day – into dollars and the sale of these futures on day τ – the day you sell all or part of your stock in dollars and convert it back into yen. On Thursday, 20 October 1988 the futures displayed in table 9.1 were available. The spot price was 0.7865 US dollars per 100 yen, or 127.15 yen per dollar. The futures price of the yen is such that it is expected that the dollar will fall more than one yen in each three-month period. Over a twelve month period, the dollar will fall in the futures market hedge by about 4.75 yen or about 3.75%. This difference is the amount that it must fall in the futures market to account for the difference in interest rates in Japan and the United States. For example, the US discount rate in late October 1988 was 6.5%, and the Japanese was 2.5% for a 4% difference.[2] If the futures were not priced this way then risk-free arbitrage would be possible by buying dollars with borrowed yen and investing them in US dollars at the higher interest rates, meanwhile guaranteeing no loss by selling these dollars at high rates in the futures market.

The point is that an investor will lose about 3.75% per year by using this type of hedge. One buys:

$$\frac{\$10 \text{ million US} \times 127.15\text{¥}/\$}{12.5 \text{ million ¥/contract}} = 101.7$$

or about 102 futures contracts and sells them when the stock is sold and the proceeds are converted back into yen. Since the 101.7 is not exactly 102 and one may not get exactly the same selling price for the futures contract as the spot price equivalent, except on an exact delivery date (because the futures price involves expectations, the so-called basis risk, as well as interest-rate differentials) there may be some additional small costs or benefits from this transaction.

The percentage gain on the portfolio in yen will be approximately:

$$\left(100 + \begin{matrix}\text{Rate of increase} \\ \text{in the portfolio in} \\ \text{US dollars in percent}\end{matrix}\right)(0.9625) - 100$$

So if the portfolio gains 25% in dollars it will gain only about 20% in yen. Hence a full 5% of the gain is lost just to manage the hedge. Can one avoid this loss? Yes, to some extent, but not without bearing some currency risk.

3 A partial hedge: the basic idea

One way is to partially hedge by selling put options on the yen. The short position in puts is equivalent to a long position in yen that pays a premium. This strategy eliminates the 3.75% annual loss because the yen is projected to strengthen in the futures markets and indeed provides a similar premium. So instead of losing 3.75% per year you gain about 7.60% per year (using the data below). The actual premium depends upon the usual factors that affect option prices: current volatility, time to expiry, the difference between the current yen price and the nearest strike price, the interest rates, and current market expectations regarding the yen/dollar exchange rate.

But this gain does not come for free, because you are only partially protected against falls in the dollar to the tune of the 7.60% or 11.35% per year in comparison to the futures approach. The spot price was 78.65 US cents per 100 yen on 20 October 1988. The options were priced as listed in table 9.2. With options one has many choices of how to do the partial hedge but let us sell December puts.[3] They are the most liquid and have about two months to expiry. Since the yen is expected to strengthen and the spot price is already 78.65 we can sell the 79 puts for about 1.001.

Table 9.2. *20 October 1988 options on Japanese yen*

Strike price	Calls-Settle Nov-c	Dec-c	Jan-c	Puts-Settle Nov-p	Dec-p	Jan-p
77	2.15	2.43	—	0.06	0.34	0.42
78	1.27	1.70	2.46	0.17	0.61	0.67
79	0.60	1.11	—	0.48	1.00	—
80	0.22	0.69	1.29	1.09	1.57	—
81	0.07	0.41	0.90	—	2.27	—
82	0.02	0.23	0.60	—	3.09	—

Let's see how this works out. Assume that in the ensuing two months the portfolio, including dividends goes up 3% – a roughly 20% yearly rate of return which is similar to the historical returns from the TOPIX or NSA. The investor will leave some of this stock as collateral for the margin required and invest the premium proceeds in T-bills and interest-bearing accounts with the brokerage firm to cover losses on the short calls.

The investor will collect the premiums, which, with rolling over every few months, amount to the roughly 7.6% per year and does not lose the 3.75% interest premium decay in the futures markets. So if the time horizon is long, say three years, then the investor has more than 25% to play with in possible drops in the dollar against the yen in comparison with the futures hedge. So assuming one rolls up and dynamically updates this hedge you are better off with the partial hedge over the three-year horizon as long as the dollar stays above 95 yen.

It's not so simple, though, because whipsawing volatility is hazardous to this strategy. For example, when the dollar falls our investor loses all the gains in the yen above about $\frac{2}{3}$% per month premiums and the about $\frac{1}{3}$% per month from the interest-rate differential. Also the investor does not participate in the gains in the dollar because the protection is against falls in the dollar. So, if there are sharp falls, such as the 5 yen fall in table 9.3 it will take a while to recoup the losses. However, the 5 yen fall or 4% in only two months still results in a gain of 1.42% on the partially hedged yen, which is less than 1% worse than the straight hedge.

This strategy is appropriate for a market with a yen that is slightly on the increase against the dollar. After the fall to the lower end of the Group of Seven target area for the yen of a reputed 120 to 140 this may not be too risky a strategy. At some stage, the already grossly overvalued yen in price-parity terms will, once the US twin deficits are held in check, start falling and then one will want to participate in its rise more fully than you

Table 9.3. *Results of the short put, partial hedging strategy on the US portfolio for a Japanese investor, gains and losses in US $ on December 1988 expiry (on second Friday)*

Day 0	Hedge comparison	Dollar remains constant @ 127.15¥/$ or $0.7865/100¥	Dollar rises 1.75¥ to 129¥/$ or $0.7750/100¥	dollar rises 5¥ to132.15¥/$ or 79.90¢/100¥	Dollar falls 2¥ to 125.15¥/$ or 79.90¢/100¥	Dollar falls 5¥ to 122.15¥/$ or 81.87¢/100¥
Convert 127,150,000¥ into $10 mil US @ $0.7865/100¥ or 127.15 ¥/$	$300,000 (gains on stock with dividends less 0.625% loss on hedge or $62,500	$300,000 (gains on stock with dividends)	$300,000 (gains on stock with dividends)	$300,000 (gains on stock with dividends)	$300,000 (gains on stock with dividends)	$300,000 (gains on stock with dividends)
Sell 102 Dec. 79 puts @ 1.00¢/100¥ Collect $125,000 less commissions of about $2,040 (assuming $20 contract)		$122,800 (in premiums $1,228 (interest on premiums at 6%/year)	$122,800 (in premiums $1,228 (interest on premiums at 6% year)	$122,800 (in premiums $1,228 (interest on premiums at 6%/year)	$122,800 (in premiums $1,228 (interest on premiums at 6%/year)	$122,800 (in premiums $1,228 (interest on premiums at 6%/year
		−0, since short puts expired worthless	a loss of $161,400, short puts 0.0125¢ in the money, with commission	a loss of $426,595, short puts 0.0333¢ in the money, with commission	−0, since short puts expired worthless	−0, since short puts expired worthless
Total gain is in $	$1,237,500	$424,028	$262,628	($2,567)	$424,028	$424,028
% gain in $	2.38%	4.24%	2.63%	(0.026%)	4.24%	4.24%
% gain in ¥	2.38%	4.24%	4.12%	3.91%	2.60%	1.4%

can with the covered call strategy. You can then simply go long in the stocks or protect the other way by selling yen futures options puts.

As with all option positions they must be carefully monitored because sudden movements in the spot price are magnified in the options markets. Still, with proper management, the short put partial hedge strategy with dynamic adjustments can provide a useful way to hedge against possible future dollar drops against the yen.

4 Testing out the partial hedge idea: Japanese investment in US Treasury bonds[4]

The short put strategy discussed above is useful for an investor especially if the dollar does not fall too much. But it does not participate in possible rises in the dollar. We discuss some related strategies that do participate in dollar advances and limit the loss for yen gains. These strategies are then compared by simulation to ascertain how good they really are.

First let us look at the various strategies. For the context and calculations, we move to Wednesday, 23 November 1988, when the yen was trading at 82.41 cents per 100 yen or 121.34 yen per dollar, near its low for the year. At this time US interest rates had risen to protect the dollar so the spread in the futures markets was about $4\frac{1}{2}\%$ per year. Figure 9.1 compares the various strategies. In these graphs the profits or losses from the option trading are kept separate from the asset returns. You may think of the underlying asset as being in US currency with zero mean return and no variance. This simplifies these diagrams. The most risky strategy, called N and shown in figure 9.1a is simply to not hedge at all. This, of course, is what is done most frequently by investors. By not hedging one does not lose the roughly $4\frac{1}{2}\%$ annual futures discount but there is no protection if the yen rises. All gains in the dollar against the yen are captured. But all losses from gains in the yen are lost. If the yen rises less than $4\frac{1}{2}\%$ per year this strategy beats the hedge.

Figure 9.1b shows the futures hedge, called H. This strategy is essentially riskless. One receives the same return, a loss of about $-4\frac{1}{2}\%/12$ per month or about -0.75% over two months, namely -0.91 yen or 0.618 cents per 100 yen.

Strategy O, shown in figure 9.1c is the short put partial hedge described in the previous section. This strategy is the next most risky procedure. The investor sells yen puts near the money. The investor collects the premium of say 7% per year and does not pay for the $4\frac{1}{2}\%$ discount. So if the yen rises less than $11\frac{1}{2}\%$ per year this strategy provides higher mean returns than the futures hedge. This strategy does not participate in dollar gains. It is only protected against yen gains by the roughly $11\frac{1}{2}\%$ per year.

The five new strategies limit the downside risk of yen rises and participate in dollar rises in various ways. Figure 9.1d describes the strategy: A1 Sell put near the money, buy put 2 cents out of the money. The related strategy was also tested: A2 Sell put near the money, buy put 3 cents out of the money.

These strategies are less risky than strategy O. They participate in the dollar rise once it moves 2 cents or 3 cents, respectively, but the protection against the yen's rise is limited to the premium. This strategy is the opposite of strategy C described below and has similar risk, assuming one worries about yen falls as well as rises. It is more risky against yen rises than C. The net premiums collected are less, say 4–5% per year than with strategy O because the deep out put must be purchased.[5]

Strategies B1 and B2 are the least risky. B1, which is shown in figure 9.1e, has the investor buy a call 2 cents out of the money in addition to strategy A1. So the full position is to sell a put near the money, buy a put 2 cents out, and a call 2 cents out. The related strategy B2, which corresponds to A2 but with the extra 2 cents out call with the at the money short put and the 3 cents out long put was tested as well.

The investor is now protected on both sides and participates in dollar rises once it moves through the strike price and the losses on the yen's rise are limited to 2 cents less the net premiums. The net premiums collected are from the short put minus the out of the money puts and calls. This will not be very much, maybe $\frac{1}{2}$% to 3% per year depending on volatility, execution, etc.

The final strategy, C is shown in figure 9.1f. In this strategy the at-the-money put is sold and a deep out call is purchased. This is a cheaper strategy cost wise than B1 since the deep out put is not purchased. What you gain is to limit the maximum loss in any two-month period if the yen rises sharply to a 2 yen change in the yen/dollar rate. So you have protection against the yen rising. But you do not participate in dollar gains. The risk of this strategy is less than with A1. For this strategy you collect less premium because a call is bought as well as the put sold. The options sold provide more money than those bought so the total return is about 4–5% per year.

These strategies are compared by investigating how they would have performed in the previous two or more years and simulating their likely future performance. The context used for the future simulations was investment in US Treasury bonds with a fixed rate and a three-year maturity. The specifics being:

> Nov. 1991 $8\frac{1}{2}$% Bid 98-02 Ask 98-06 yield 9.23%, and
> coupons received on stream were assumed invested at 8% per year.

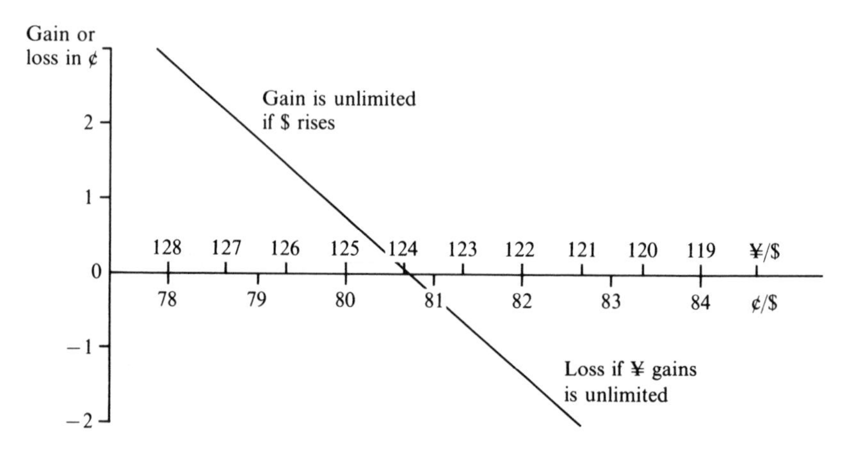

(a) Case N: no hedge

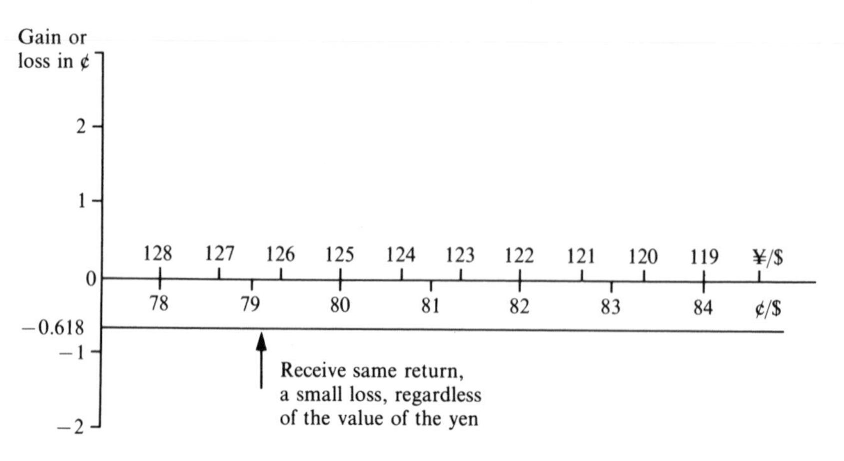

(b) Case H: futures hedge

Figure 9.1 The various hedging strategies compared: N, H, O, A, B, and C

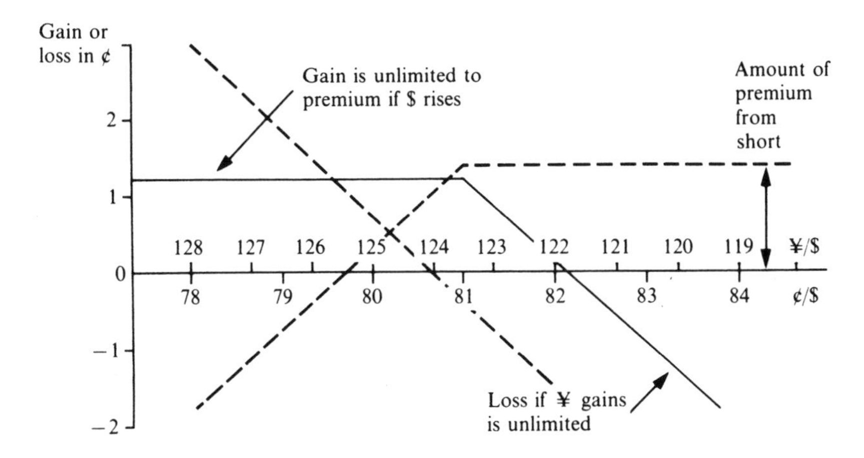

(c) Case O: the partial hedge with put sold near money, at 81¢/100¥

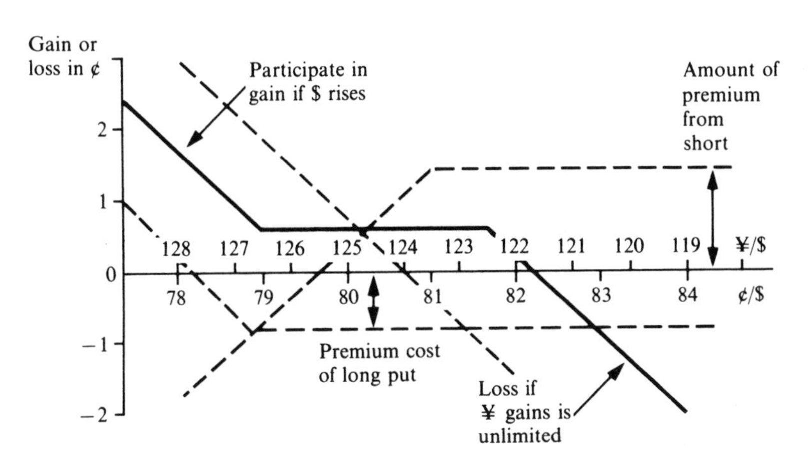

(d) Case A1: the partial hedge with put sold near money at 81¢/100¥ and put bought out of money, at 79¢/100¥

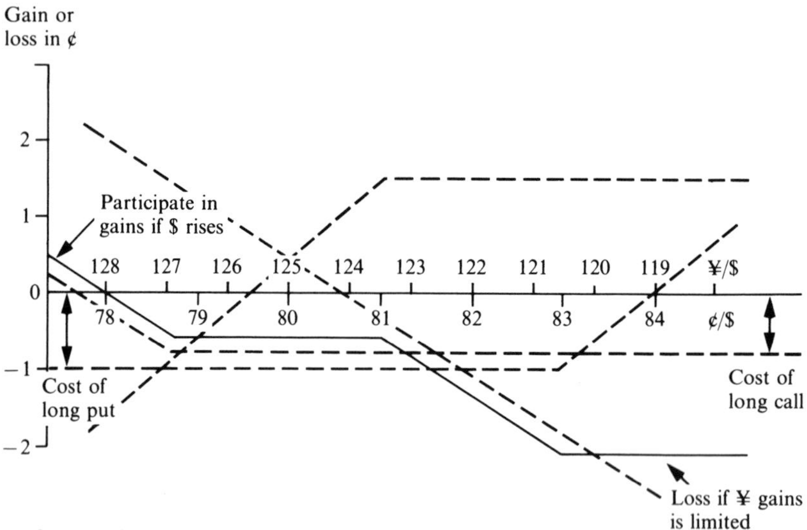

(e) Case B1: the partial hedge with put sold near money, put bought 2¢ out of money and call bought 2¢ out of the money

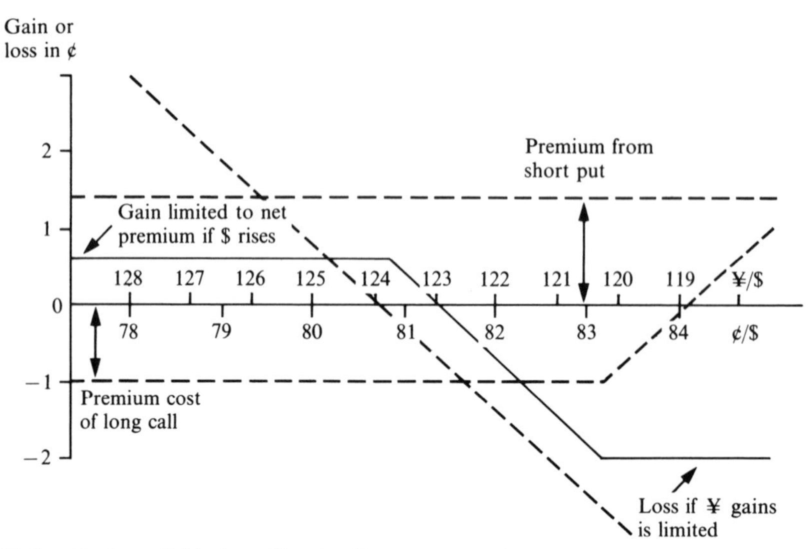

(f) Case C: the partial hedge with put sold near money, at 81¢/100¥ and call bought out of money, at 83¢/100¥

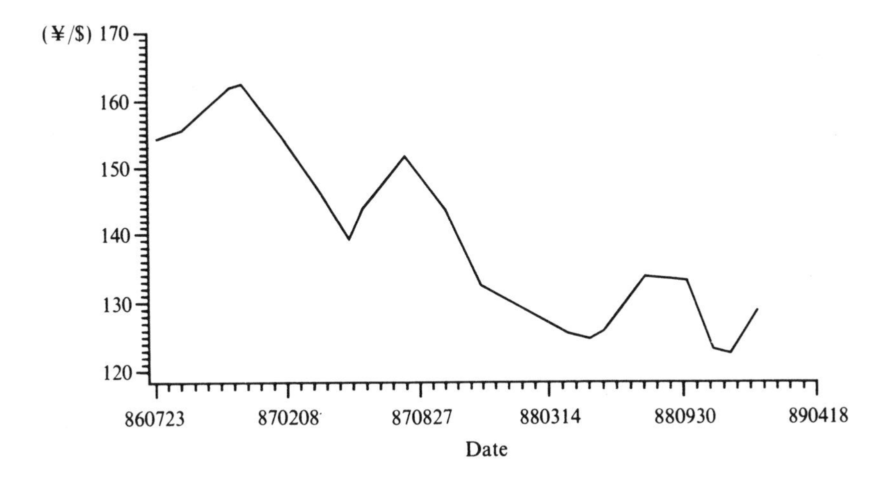

Figure 9.2 ¥/$ exchange rate, 31 July 1986–19 January 1989

An amount A yen = 100 yen is converted into US dollar at time 0 and invested in US assets at time 0. At time τ the proceeds are converted back into B yen. The data were from 23 July 1986 to 19 January 1989. The test used a T-bond with a yield of 6.7% and it was assumed that the coupons were reinvested at 6%. This was a period of sharp increase for the yen except during 1988. As shown in figure 9.2, the yen started at about 154 and ended up about 128. The first year and a half the yen rose over 25% (154→120). Then it stayed in a volatile trading range of 120–35. With such a large gain in the yen one would expect the futures hedge to work the best. It did, returning 110.83 without risk. Strategies B1, B2, and C were only about 1% behind, but they had more risk. The other strategies were 5–10% worse. Figure 9.3 shows the results.

The conclusion is: if the yen rises 15–20% in a year or so, the futures hedge will be the best strategy. You have to have the yen rise no more than say 10% per year for the other strategies to possibly be better. So over three years the rise in the yen must be less than, say, 30–35%. Given the dollar's sharp fall since 1985 a further 30% drop would take it under 90 yen/dollar, an event that may or may not occur. Obviously, the yen cannot keep rising forever. Indeed the dollar rose to the 150 yen level in 1989 and to 160 yen in April 1990 before it fell to the 125 yen to 135 yen range in late 1990 and then to the 125 yen range by early 1992. Hence, there may well be use for the partial hedge strategies that are now investigated with the simulation of possible futures for the yen/dollar exchange rate.

A Black–Scholes simulation shows how the strategies might do in

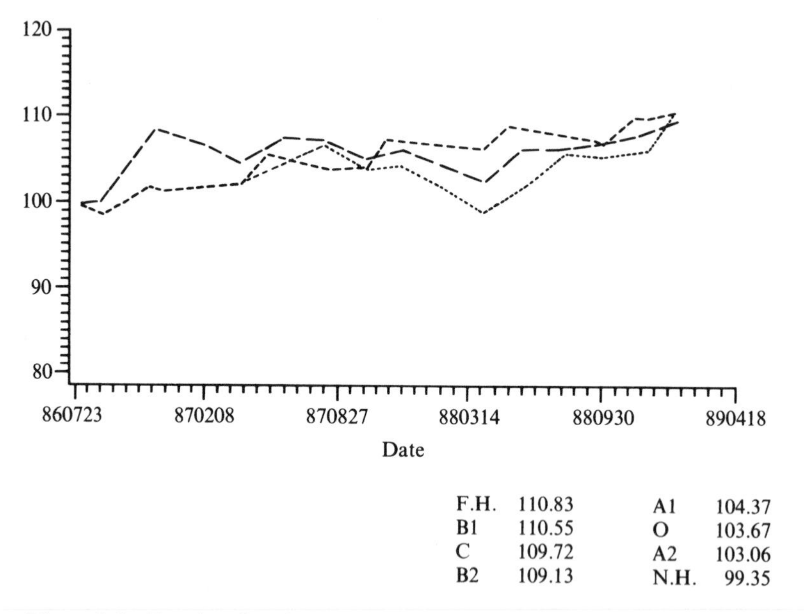

Figure 9.3 Results of various strategies, 23 July 1986–19 January 1989

various possible futures. The idea was to assume that the put and call prices were well estimated by their Black–Scholes options prices. Volatility was assumed to be 11.8% which was that of 1988.[6] For interest rates we used 9% for the US (the federal funds rate) and 3.9% for Japan (the call rate). The yen/dollar rate evolves day by day as:[7]

$$(\text{¥}/\$)t+1=(\text{¥}/\$)t\exp\left(-\frac{d}{252}+\sigma\delta\right)$$

d = the assumed yearly drift of the scenario over the 252 trading days
σ = the assumed volatility = 11.8% but double for the weeks with shocks
$\delta=\begin{cases}+1\\-1\end{cases}$ for increases and decreases, for the binomial model with drift.

The two-month options are bought and sold at their Black–Scholes prices and held to maturity when they either expire worthless or are covered at the then in the money price assuming no bid-ask spread.

The following scenarios were tested:

1 The yen rises at 4% per year in each of the three years. This is roughly what the futures market expected ($d=0.04$).

1′ The same as 1 but every three months there is a shock and for one week the volatility doubles.

2 The yen has a zero ($d=0$) drift in the first year. In the second and third

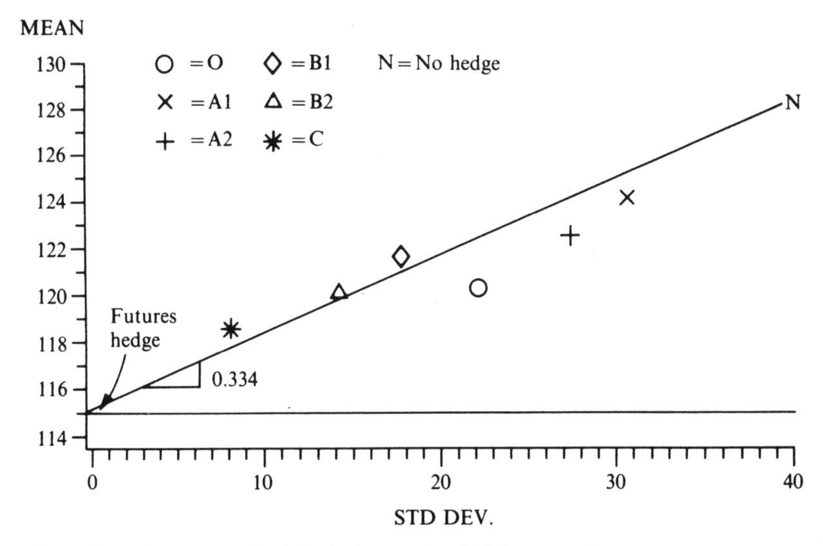

Figure 9.4 Mean standard deviation tradeoff of the scenario results for the various categories

years the yen falls with a drift of 5% per year ($d=0.05$). This is a mildly stronger dollar scenario.

3 The yen drifts up 4% in year 1 ($d=-0.04$) but then the dollar drifts up 15% in the second year ($d=0.15$) and 20% in the third year ($d=0.20$). This is the US trade and budget improvement scenario moving close to price parity in three years.[8]

4 Then yen drifts up 10% in the first year then 5% more in the second year and 10% more in the third year. This is the hard landing for the dollar with the $/¥ rate falling to less than 100.

These five scenarios seem to span the range of possibilities reasonably well. There could be more violent moves in the market, but it seems doubtful that the yen would fall much below 95–100 (scenario 4) or rise much above about 165 (scenario 3). For some of the simulation outcomes, the yen/dollar rate is outside these bands so we do have estimates of the strategic performances of these cases also.

Observations:

1 The mean yen/dollar rate was 120.4 with a minimum of 47.6 and a maximum of 254.1.

2 The future hedge returns 114.96 without risk.

3 The other strategies return more on average but they have more risk. Additional mean return is approximately linear in standard deviation risk with a slope of 0.334 units of mean return per unit of standard deviation. Figure 9.4 shows this tradeoff. B1 is slightly above the curve

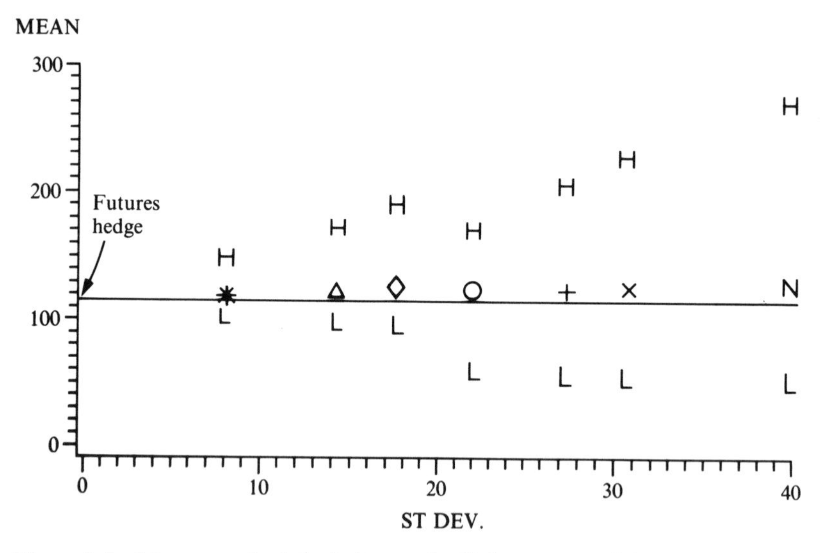

Figure 9.5 Mean-standard deviation tradeoff plus ranges of the scenario results for the various strategies

but more or less you get what you pay for in a Markowitz mean-standard deviation sense. Another measure of risk is the lowest payoff. Figure 9.5 shows the minimum payoffs. The specific numbers appear in table 9.4.

The strategies N, O, A1, and A2 have very low minimum values. These are with very low dollar values. Assuming that the dollar cannot fall below 90 yen then the minimum values are (starting with 100 in year 0):

N	96
O	89
A1	99
A2	99

So you could lose money with these strategies.

B1 and B1 and C have minimum payoffs of 95, 97, and 102, respectively, with all scenarios. With the yen no lower than 90, the minimum payoffs for these strategies are:

B1	102	with ¥/$ at 93	¥/$=74 before loss of 5
B2	104	with ¥/$ at 94	¥/$=60 before loss of 2
C	105	with ¥/$ at 102	Never loses, its minimum is 102 when the exchange rate is 83¥/$.

So C compares favorably with the hedge. Its mean is 3.5% higher. Its

Table 9.4. *Summary of the 125 simulation runs, 25 per scenario times 5 scenarios for each of the eight strategies, both in total and individual*

Variable strategy	Sample size	Mean return	Std. dev. of return	Min. return	Max. return	Std. error of mean return
Exchange rate	125	120.42	37.46	47.60	254.08	3.35
N	125	128.29	39.91	50.71	270.67	3.57
H	125	114.96	0.00	114.96	114.96	0.00
O	125	120.24	22.13	60.01	170.47	1.92
A1	125	124.16	30.68	54.15	226.77	2.74
A2	125	122.48	27.35	55.21	205.31	2.45
B1	125	121.61	17.71	95.02	189.80	1.58
B2	125	120.12	14.31	97.66	171.69	1.28
C	125	118.53	8.06	102.24	148.48	0.72

Scenario	N	O	A1	A2	B1	B2	C
1	113.14	115.13	113.25	113.37	113.31	113.48	115.6
1′	114.05	111.69	113.29	112.50	117.44	116.68	116.7
2	140.96	129.18	135.11	132.46	126.46	124.05	121.1
3	173.92	138.57	157.25	151.22	143.36	137.88	126.3
4	99.37	106.61	101.88	102.88	107.44	108.53	113.0
Total	128.29	120.24	124.16	122.48	121.61	120.12	118.5

Scenario	STD N	STD O	STD A1	STD A2	STD B1	STD B2	STD C
1	25.69	18.79	22.07	20.48	10.10	8.31	6.12
1′	28.58	19.62	24.34	22.58	12.01	10.15	6.35
2	32.01	19.27	24.85	22.16	13.72	10.82	7.33
3	39.50	18.76	26.94	22.64	17.22	13.05	7.70
4	22.57	18.41	20.28	19.41	8.54	7.42	5.53
Total	39.91	22.13	30.68	27.35	17.71	14.31	8.06

lowest return with the yen at 92 is 105 versus the 115 with the futures hedge. Its standard deviation is low. Strategies B1 and B2 look promising as well. These strategies seem to promise a 2% minimum return per year and get more mean return 5.7% with B1 and 4.4% with B2 versus the futures hedge.

Table 9.5 gives the minimum returns for the various strategies given a variety of low final values for the dollar/yen exchange.

Table 9.5. *Minimum returns for the various strategies for various minimum final values for the dollar in yen*

Value of $ in ¥	NH	O	A1	A2	B1	B2	C	Futures hedge
90+	96	89	99	99	102	104	105	115
95+	100	107	105	105	106	105	110	115
100+	106	108	109	110	108	112	112	115
105+	112	108	112	111	108	112	112	115
110+	117	101	115	113	108	110	111	115
115+	121	115	120	118	118	112	112	115
120+	127	126	126	124	119	117	118	115
125+	132	128	130	129	121	120	119	115
130+	138	132	139	137	126	126	119	115
135+	142	131	142	139	127	125	118	115
140+	147	130	138	138	130	130	121	115
145+	154	131	144	144	144	139	135	115
150+	159	148	145	144	125	123	127	115
160+	170	132	153	142	148	138	131	115
170+	176	150	168	159	140	132	123	115
Mean returns	128	120	124	122	121	120	119	115

This study was oriented toward the concern that the yen would strengthen to provide poor returns on the investment. Should the dollar neither fall nor strengthen, then all of the partial or no hedge strategies will beat the futures hedge. Hence at final exchange rates of 120+ yen for the dollar one sees this effect. Not surprisingly, the no hedge strategy is then the best but the partial hedge strategies that were good protection against a fall in the dollar are pretty good should the yen fall.[9]

This simulation used twenty-five histories for each strategy. Each result is path dependent so that the outcome depends upon how the yen/dollar rate reached its final value. For professional application one might do several things in addition to more model development and simulations to decide on a preferred strategy, such as:

1 Using a market-maker, prices can be obtained that are better than those

estimated from the Black–Scholes equations. One can sell at the ask and buy at the bid. That is worth a few percent over the three years.

2 These calculations use a two-month rollover period with no dynamics. There is room for some improvement by altering slightly the strategy as the outcomes unfold. One also might be able to buy back shorts if prices are low and combine a little market timing on the margin.

3 The interest on the shorts is worth more than calculated here as are the coupons. The latter benefits the futures hedge as well.

More mathematical analyses emphasizing continuous time strategies using futures are discussed by Adler and Detemple (1988), Duffie and Jackson (1986), Duffie and Richardson (1989), Eaker and Grant (1987), Solnik (1989), Svensson (1988) and references therein. The goals of these authors differed from those here. We are searching for a way to attempt to minimize the effects of the deep futures discount on the yen, while they were more concerned with minimizing the variance of returns over time. See also Jorion (1989), Stone and Hensel (1989), and Perold and Schulman (1988).

The analysis here and in most other literature assumes that stock prices and currency movements are uncorrelated. This is definitely not the case with the yen and the Tokyo market. Indeed the Tokyo market generally reacts favorably to higher yen values and negatively to the reverse. This was especially true in the 1990 decline when the yen and the indices fell sharply, and then turned around together in late April and May and then fell again from July to October 1990. It is not simple though. During 1989 there was a sharp rise in the NSA coupled with a falling yen. It was higher interest rates that triggered the stock market fall in 1990. The point of this for a complete analysis is that one needs to consider the relationship between currency and stock price movements and their relationship with interest rates and other macro variables such as inflation. Such is the depth of difficulty of this problem.

5 Hedging a Japanese portfolio against a possible drop in the value of the yen for a US investor

This situation is the reverse of the previous case, except now it is assumed that the interest-rate differential works in the investor's favor. Hence, the futures prices of the dollar are higher than the spot prices. In addition, because the dollar is expected to fall, the prices of the puts you would have to sell to produce the partial hedge are not as high as the calls. Hence, the futures hedge seems like the preferable way to protect against possible gains in the dollar. Let's look at it with some numbers.

Assume that 10 million US dollars has been invested in Japanese

Table 9.6. *Results of the short call, partial hedging strategy on a Japanese portfolio for a US investor, gains and losses in yen on December 1988 expiry (second Friday in December)*

Day 0	Hedge comparison	Dollar remains constant at 127.15¥/$ or $0.7865/100¥	Dollar rises 1.75¥ to 129¥/$ or $0.7,750/100¥	Dollar rises 5¥ to 132.15¥/$ or 79.90¢/100¥	Dollar falls 2¥ to 125.15¥/$ or 79.90¢/100¥	Dollar falls 5¥ to 122.15¥/$ or 81.87¢/100¥
Convert $10 million US into 127,150,000¥ @ $0.7865/100¥ or 127.15¥/$	3,814,500¥ (gains on stock with dividends) gains on hedge of 0.625% or 794,688¥	38,145,000¥	38,145,000¥	38,145,000¥	38,145,000¥	38,145,000¥
Sell 102 Dec. 79 calls @ 1.11¢/100¥. Collect $140,250 less commissions of $2,040		$1328,210 or 17,573,402¥	$138,210 or 17,573,402¥	$138,210 or 17,573,402¥	$138,210 or 17,573,402¥	$138,210 or 17,573,402¥
		−0, since short calls expired worthless	−0, since short calls expired worthless	−0, since short calls expired worthless	a loss of $116,790 or 14,849,849¥ short calls 0.009¢ in the money with commission	a loss of $367,965 or 44,946,625¥ short calls 0.0287¢ in the money with commission
Total gain in $	46,091,880¥	55,894,123¥	55,894,123¥	55,894,123¥	41,041,274¥	10,944,198¥
% gain in $	3.63%	4.40%	4.40%	4.40%	3.23%	0.86%
% gain in ¥	3.63%	4.40%	2.90%	0.45%	4.88%	4.99%

securities at the 20 October 1988 spot exchange rate of 78.65 cents per 100 yen or 127.15 yen/dollar. The dollar can fall about 3.75% or about 4.75 yen to 122.40 yen/dollar before the advantage of the hedge ceases to give you an edge over the status quo considering a one-year horizon.

The futures hedge allows you to change your yen back into dollars at the currently favorable rate of 122.40 yen/dollar. So your 10 million US dollars is really worth 10,375,000 dollars as long as the stock market performance in yen matches the performance in the US stock market. This 375,000 dollars edge provides a cushion for the yen tracking error risk as well as the market risk and transactions costs.

A short call partial hedge is illustrated in table 9.6 and compared with the futures hedge. Assuming the stock portfolio plus dividends increases by 3% in the two-month period from 20 October 1988, until the options expire on the third Friday in December, the hedge gains 3.63% – the 3% gain plus the two-month bonus of 0.63% from the interest-rate differential. If the exchange rate stays constant then the partial hedge returns 4.40% in yen as well as in dollars. You do not gain much extra with dollar falls because you have sold the calls short. The gains of 4.88% and 4.99% with dollar falls of 2 and 5 yen, respectively, differ slightly from the 4.40% of the constant exchange rate case because of transactions costs, the fact that the 102 short calls is not the exact cover for the partial hedge and the basis difference between the spot and strike prices of the options.

The investor is pretty well fully protected on this side but does not participate in any further gains. When the dollar rises, the investor is only partially protected to the tune of the options premium for, unlike the yen into dollars play, the investor does not gain the interest differential in comparison with the futures hedge. In addition, because the yen is projected to increase in value in the futures market, the calls are not as valuable to sell as the puts were. Still, our investor makes 2.90% when the dollar rises 1.75 yen and 0.45% when it rises 5 yen in this two-month period. On balance the straight hedge seems preferable for this investor. The advantage of the essentially riskless 3.75% per year makes this strategy a hard one not to employ. The US investor in Japan does not need to use the complicated strategies that the Japanese investor in the US might wish to utilize.[10]

The Canadian investor or those from countries such as England, France, Italy, and Switzerland, that have higher interest rates than in the US have even more advantages investing in Japan. Figure 9.6 shows this with four different approaches. In figure 9.6a the portfolio value in yen is displayed with its typical volatility assuming an upward trend in prices. The portfolio value in Canadian dollars is much more volatile as shown in figure 9.6b. Huge jumps in the value of the portfolio have occurred often in these markets. For example, on Friday, 11 May 1990, the Canadian dollar fell 5 yen from 135 to 130. Since currency and stock price movements are correlated one will have large moves in both directions from the trend line in yen. Figure 9.6c hedges the currency and the Canadian investor is able as of December 1990 to eliminate the Canadian/yen risk and gain over 5% essentially risk free. Given this edge the investor may well be wise to spend some of it on portfolio insurance protection through Nikkei put warrants or some other scheme. The situation is shown in figure 9.6d where there is a floor on the minimum portfolio return that is above the nominal investment with a positive drift of 1–2% per year. Given the poor performance of Canadian portfolios and managers over the past twenty years they would likely benefit from strategies like figure 9.6d. The difference between hedged and unhedged returns is explored more fully after we discuss adjusting the amount of the hedge.

6 Adjusting the amount of the hedge

The US investor in Japan does need to adjust the number of contracts in the hedge as the portfolio of Japanese stocks varies in value. Since each yen currency contract is for 12.5 million yen the number of contracts required to hedge the position when the currency is exchanged is

$$\text{int}\left[\frac{S_0 A}{\yen 12.5\text{M per contract}}\right] = c_0$$

where S_0 is the current spot exchange rate say 140 yen/dollar, A is the investment, say \$10 million, and int means that one takes the largest integer that is less than the quantity in the brackets. With these data $c_0 = 112$ contracts exactly but usually one will have to round down so as not to over hedge. To update the hedge, one simply keeps the number of contracts sold on day t equal to

$$\text{int}\left[\frac{S_t A(1+r_t)}{\yen 12.5\text{M per contract}}\right] = c_0$$

where S_t is the yen/dollar spot exchange rate on day t and r_t is the gross rate of return on the portfolio from the start up to day t.

$$\text{int}\left[\frac{150(10\text{ million})(1.16)}{12.5\text{ million yen}}\right] = \text{int}(139.2) = 139$$

Hence one needs twenty-seven more contracts. This day-by-day adjustment procedure works well in general. Problems can occur from basis risk when the futures gets out of whack with the spot rate or when there is

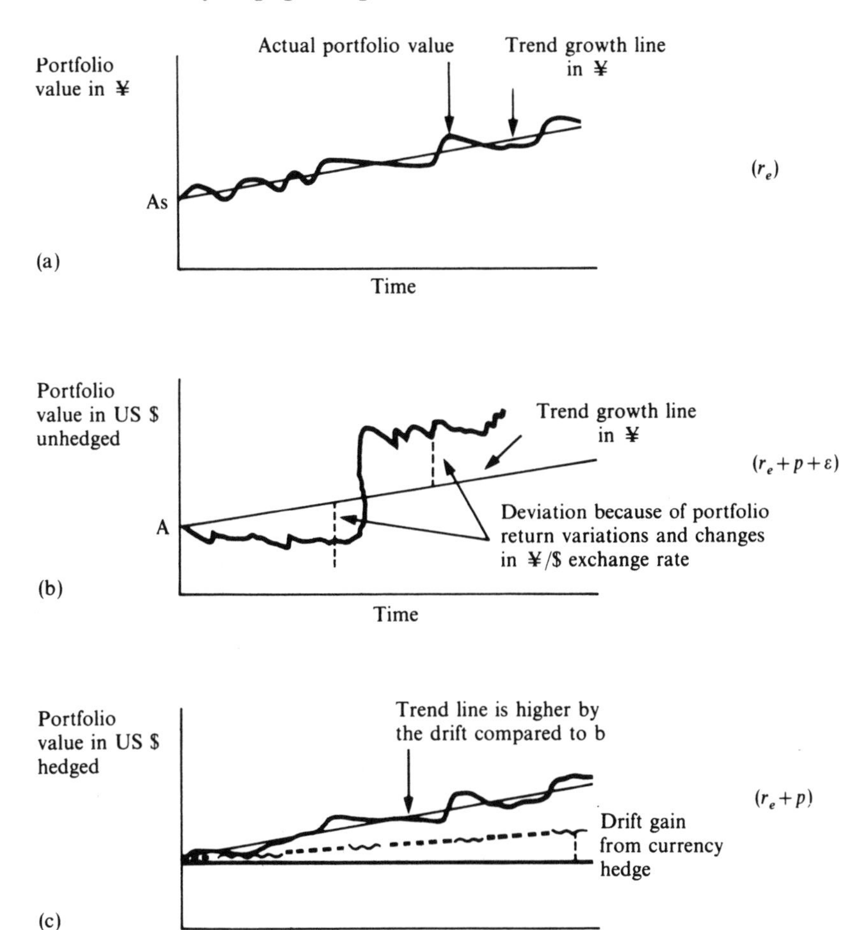

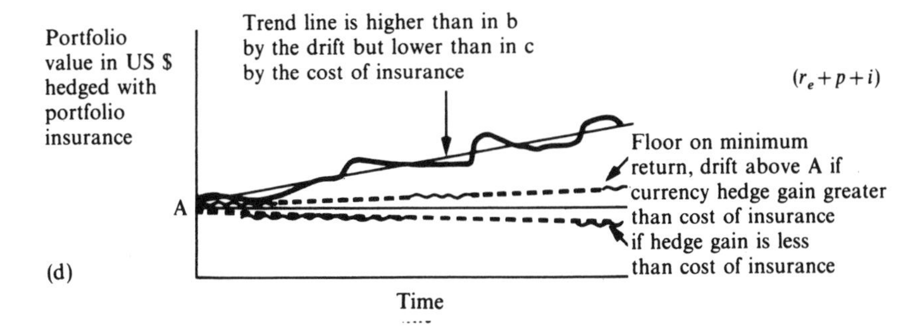

Figure 9.6 Typical portfolio values in yen and dollars over time

a sharp move in stock or currency values. The former rarely amounts to more than one half yen so the tracking error is less than 0.5% which is small in comparison with the interest differential futures discount gain. The latter, a sharp move in the stock portfolio value or the yen/dollar rate can cause hedging losses. Like portfolio insurance strategies used in the October 1987 stock-market crash, one must dynamically adjust in a continuous fashion. Since one is not dealing with stock futures premiums that can get 5% or more out of whack with the cash indices but with currency futures the risk from the potential basis change does not seem that great.

7 Hedged and unhedged returns[11]

The mean return from hedged and unhedged returns is the same for an investment from one country to another as long as the returns from investments and from currency changes are uncorrelated. The difference in actual returns is then equal to this mean return plus the return in the local currency plus the currency surprise. This is shown as follows. Let:

r_u = the unhedged return
r_h = the hedged return
r_c = the currency return
r_e = the return in the local currency
S_t = the spot currency exchange rate to change local into foreign returns
F_t = the forward/futures rate local to foreign currencies
V_t = the asset values in the local currency

$$1+r_u=\frac{S_tV_t}{S_{t-1}V_{t-1}}=\left(\frac{S_t}{S_{t-1}}\right)\left(\frac{V_t}{V_{t-1}}\right)=(1+r_c)(1+r_e)$$

$$\begin{aligned}1+r_h&=\frac{F_tV_{t-1}+(V_t-V_{t-1})S_t}{S_{t-1}V_{t-1}}\\&=\frac{(F_t-S_{t-1})V_{t-1}-(S_t-S_{t-1})V_{t-1}+S_tV_t}{S_{t-1}V_{t-1}}\\&=1+r_e+\underbrace{\left(\frac{F_t-S_{t-1}}{S_{t-1}}\right)}_{p}-\underbrace{\left(\frac{S_t-S_{t-1}}{S_{t-1}}\right)}_{r_c}.\end{aligned}$$

The actual return of the hedged portfolio equals that of the unhedged portfolio plus the futures hedge premium p minus the currency return. If we let $\varepsilon=r_c-p$ then the return of the hedged portfolio equals that of the unhedged portfolio plus the currency surprise ε. Taking expectations yields:

$$E(r_h) = E(r_u) + E(\varepsilon) = E(r_u)$$

since currency surprise has mean zero.

In terms of local returns and currency returns one has:

$$\begin{aligned} 1 + r_h &= 1 + r_u - \varepsilon \\ &= (1 + r_c)(1 + r_e) - \varepsilon \\ &= 1 + r_e + r_c + r_c r_e - \varepsilon \end{aligned}$$

So $r_h = r_e + p + r_c r_e$ and $E(r_h) = E(r_e) + p + \text{cov}(r_c, r_e)$. If $\text{cov}(r_c, r_e) = 0$, then $E(r_h) = E(r_e) + p = E(r_u)$.

In practice, the currency returns and the local returns have low correlation[12] so the expected hedged return equals the expected local return plus the futures hedge premium.

Figure 9.6 shows the typical behavior of the various portfolios for a Canadian investor. For a US investor the situation is similar except that the drift in figure 9.6d is about 1% and the floor in figure 9.6d has a slight negative drift of 1–2% per year.[13] The return in local currency r_e shown in (a) has the same volatility as the hedged portfolio (b), namely $\sigma(r_e)$. The hedged portfolio has return $r_e + p$ with volatility $\sigma(r_e)$. The unhedged portfolio has return $r_e + p + \varepsilon$ which has the same mean return as the hedged portfolio but higher volatility assuming $\sigma^2(\varepsilon)$ is positive but $\text{cov}(r_e, \varepsilon) = \text{cov}(r_e, r_c) = 0$.

NOTES

Without implicating them I would like to thank Warren Bailey, Yuko Beppu, Andy Turner and my colleagues at the Yamaichi Research Institute particularly A. Komatsu and H. Shintani for their help and useful discussions. Mr H. Maruyama performed the simulation calculations in the empirical part of the paper. Part of this research was conducted at the Yamaichi Research Institute with the advice of William T. Ziemba. I thank the Yamaichi Research Institute for permission to publish those results. This research was also partially supported by the Social Sciences and Humanities Research Council of Canada, and the Centre for International Business Studies, University of British Columbia.

1 Most of this paper was written in the fall of 1988 and the winter of 1989 when the yen was in the 125–30 range and many predicted that it would increase further due to the US twin trade and budget deficits. The fear then was about a 100 yen

per dollar rate. In May 1990 we had a 150–60 yen and the possibility of further falls. Hence, in recent times hedging may not have been the best strategy *ex post*. At these and higher levels it would have been very wise again. Indeed the yen was about 125 when we went to press in January 1992.

2 In January 1992, the US discount rate was 3.5% and that in Japan 4.50% for a −1% difference.

3 We could buy calls, such as the 79s for 1.11 cents but that would get us back into the interest-rate loss situation as in the futures markets at roughly similar costs. At-the-money calls would eliminate the currency risks as the futures hedge did and out-of-the-money puts would partially protect against large losses with lower costs.

4 H. Maruyama of the Yamaichi Research Institute assisted in this test by performing the simulation calculations that follow.

5 Strategies like this would have worked well *ex post* during the dollar's rise in 1989 and 1990 while at the same time protecting against a sudden drop in the dollar.

6 Constant volatility is a requirement of the Black–Scholes model inputs. Obviously, volatility is not constant, so this is simply a convenient simplification. There is much scope to consider alternative currency option pricing models here. Models such as those surveyed in Hull (1989) may be considered.

7 This is a binomial currency change model and it was convenient to utilize when analyzing broad currency trends as I wished to do. There are alternative theories of currency movements such as those based on the Fisher effect, purchasing power parity, and expectation theory that could be utilized and compared as well.

8 This is the scenario that is closest to what actually happened before the yen firmed again in mid 1990.

9 The yen did in fact fall to the 150–60 range and the no hedge strategy did the best *ex post* returning 170 versus 153 for strategy A1 and 176 versus 168 if the dollar rises past 170.

10 In May 1990, this differential against the US dollar was more on the order of 1% per year rather than the 3.75% it was in the fall of 1988 and in January 1992 it was minus 1%. Given this, one may wish to consider more risky strategies, such as those discussed above for Japanese investment in the US.

11 Thanks to Andy Turner for his help on this section. See also Gillies and Turner (1990).

12 This assumption, as discussed above, is not particularly valid for Japan.

13 In December 1990 the Cdn$/yen futures was at a discount of about 4.5% per year so that the investor could hedge currency and stock price level risks and still have a floor well above long-term NSA put warrants had costs much less than this per year. For US investors the currency difference was actually slightly negative. Hence the floor is slightly below A.

REFERENCES

M. Adler and J. Detemple (1988), "Hedging with futures in an intertemporal portfolio context," *Journal of Futures Markets*, 8: 249–69.

Edward J. Altman and Yoshiki Minowa (1989), "Analyzing risks and returns and potential interest in the US high yield corporate debt market for Japanese investors," *Japan and the World Economy*, 13: 163–86.

Warren Bailey and William T. Ziemba (1991), "An introduction to Japanese stock index options," in W. T. Ziemba, W. Bailey, and Y. Hamao (eds), *Japanese Financial Market Research*, North Holland, Amsterdam.

Darrell Duffie and M. Jackson (1986), "Optimal hedging and equilibrium in a dynamic futures market," Research Paper 814, Graduate School of Business, Stanford University.

Darrell Duffie and Henry R. Richardson (1989), "Mean-variance hedging in continuous-time," Working Paper, Graduate School of Business, Stanford University.

M. Eaker and D. Grant (1987), "Cross-hedging foreign currency risks," *Journal of International Money and Finance*, March, 6(1): 85–108.

John M. Gillies and Andrew L. Turner (1990), "On the relationship between currency hedged and local rates of return," *Russell Technical Notes*, March.

John Hull (1989), *Options, Futures and Other Derivative Securities*, Prentice Hall, New Jersey.

Philippe Jorion (1989), "Asset allocation with hedged and unhedged foreign stock and bonds," *Journal of Portfolio Management*, Summer: 49–54.

Andre F. Perold and Evan C. Schulman (1988), "The free lunch in currency hedging: implications for investment policy and performance standards," *Financial Analysts Journal*, May–June: 45–50.

Bruno Solnik (1989), "Optimal currency hedge ratios: the influence of the interest rate differential," in S. G. Rhee and R. P. Chang (eds), *Pacific-Basin Capital Markets Research*, North Holland, Amsterdam: 441–65.

Douglas Stone and Chris R. Hensel (1989), "Strategic currency hedging non-US investments for US-based investor," *Russell White Paper*, August.

Douglas Stone and William T. Ziemba (forthcoming), "Land and stock prices in Japan," *Journal of Economic Perspectives.*

L. E. O. Svensson (1988), "Portfolio choice and asset pricing with non-traded assets," Working Paper, Institute for International Economic Studies, University of Stockholm.

William T. Ziemba (forthcoming), *Strategies for Making and Keeping Excess Profits in the Stock Market*, William Morrow.

William T. Ziemba and Sandra L. Schwartz (1992), *Trading Japan: The Future, Options, Warrants and Derivative Securities Markets*, Probus Publishing Company.

Comment on "Currency hedging strategies for US investment in Japan and Japanese investment in the US"

YUKO BEPPU

1 Among world equity markets, the largest is the Tokyo Stock Exchange, the second is the New York Stock Exchange, the third is the US NASDAQ market, and the fourth will be the JASDAQ market which Japan is going to organize and start in 1992, exploring the possibility of connecting computers with NASDAQ. In recent years, therefore, Japanese investment in the US and American investment in Japan have been important activities in rapidly integrating world capital markets. Professor William T. Ziemba presented a timely and significant study, entitled "Currency Hedging Strategies" at the Conference on Financial Optimization held at The Wharton School of the University of Pennsylvania on 10 November 1989. His principal conclusion is as follows:

> For the American investing in Japan, the hedge provides a substantial bonus: an essentially risk free gain of about 3–5% per year due to the difference in interest rates between the two countries. . . . The situation is much more difficult and complicated for Japanese investment in the US. The forward/futures hedge will eliminate the currency risk but at a cost of about 3–5% per year.

In his chapter, he explored mainly the latter case, upon which I will comment.

2 He has dealt with the period 1985–9, when the yen/dollar rate fluctuated most. "The drop in the yen/dollar rate from the 260 range in the fall of 1985 to the 120 range at the end of 1987 and its sharp rise back to 150 in mid 1989." During this period, his currency hedging strategies worked well as explained in the paper.

It seems to me, though, that his strategies might not always be appropriate. He might have suggested alternative strategies for other periods with different trend rates or situations. The reason is that the efficiency of forward/futures hedges depends on the trend rate or average change of the yen/dollar exchange rate over time, and that the efficiency of our options-based hedging strategy depends on the volatility of rates.

We might find that the same strategies are not always appropriate *ex-post*. In order to arrive at a strong conclusion as to the appropriate nature of any of our strategies, it is important to look at long periods which encompass various market scenarios. Unfortunately this is difficult to do here – there is little data with futures and option prices. Nevertheless, Professor Ziemba's study is useful because it examines the alternatives for hedging international investment.

Note that currency swap, foreign exchange annuity swap, foreign currency borrowing, interest-rate caps and floors, investment or deposit in foreign currencies (e.g., dollar deposit with a reverse floating rate against option contract), hybrid capital-market instruments (e.g., index bond), etc., are also available for Japanese investors. Likewise, American investors have access to some of these strategies.

3 Professor Ziemba assumed the 3–5% hedge return or hedging cost per year because of the difference in interest rates between Japan and the US. However, the expiration time of future options is usually one month, three months at most for adequate liquidity in practice. In his example, it is necessary to roll over about 12–36 times, so that transaction costs should be dealt with more explicitly. (Market prices for forward options are more difficult to get than prices for future options and they may in fact be more expensive.) The transaction costs should reduce the "American bonus" and increase the "Japanese cost" significantly, so that they would also affect the efficiency of the hedging operations which certainly depends on the amount of basis risk equivalents.

4 Assuming that Japanese investment in the US portfolio itself always generates a certain amount of gains (capital gains and dividends), he proposes that the "strategies that do not lose the interest rate differential and in fact collect positive premiums are available by selling and buying yen put and call options." Interest-rate changes and price changes have been excluded from consideration by this assumption, so that the currency hedging strategies were extremely simplified. Since Japanese investors today are very sensitive to interest-rate changes under the rapid development of liberalization of interest rates, I think that an extension associated with interest-rate hedging strategies, etc., should also have been suggested.

5

Simulations using an investment in three year treasury bonds investigate this using recent actual currency movements and Black–Scholes estimated prices for plausible future scenarios to evaluate the risk-reward trade-offs and the worst possible outcome using the various strategies.

With regard to his simulations, however, I cannot agree with the assumption of volatility = 11.8% but double for the weeks with shocks, saying that "every three months there is a shock" and that "for one week the volatility doubles."

The efficiency of the use of options to hedge depends on volatility realized versus volatility forecast at the time of pricing options. The constancy assumed for volatility is, therefore, a crucial assumption in simulation. In this respect, his simulations are apparently different from a Black–Scholes simulation.

6 Professor Ziemba asks, "How can you protect against the chance that the yen will keep rising and possibly go to 100 against the dollar, as some suggest because of the US twin trade and budget deficits?" In other words, "this study was oriented towards the concern that the yen would strengthen to provide poor returns on the investment." Strictly speaking, only in the above situation and forecast can he conclude that "the added risk seems well worth taking for many investors to achieve substantial expected gains."

So, I had a somewhat empty feeling when I wrote this chapter on 5 January 1990, with the yen at 145.25 to the dollar. As of this period to date, from the end of his simulation period, Japanese insurance companies' strategies, which he portrayed as crazy in his presentation, might have been the best. That is, "no hedge strategies."

Furthermore, I understand that they have neglected the currency risk for the purposes of long-term investment in the higher interest rates in the US and international diversification of their portfolios. The main reason is that their capacity to shoulder large risks has come from the soaring prices of their accumulated real estate, stocks, and bonds in the domestic markets during the same period since 1985.

7 Finally, some noteworthy empirical evidence has to be introduced for reference. Jack D. Glen wrote an article, "Exchange rate uncertainty, forward contracts and the performance of global equity portfolios," Working Paper 37-1989, for the Rodney L. White Center for Financial Research at The Wharton School of the University of Pennsylvania. Glen performed empirical research on the effects of foreign exchange risk between the dollar and pound on internationally diversified equity portfolios.

His conclusion was that, when sufficiently long periods of time were considered, there was no improvement in the risk-adjusted performance of a hedged portfolio over an unhedged portfolio. For typical levels of foreign equity holdings, exchange risk was not found to be a substantial component of the overall risk of a portfolio and the value of hedging was found to be very period specific. He said that, surprisingly, this result came not through the use of forward contracts to hedge exchange risk, but rather through the use of their ability to reduce the overall variance of the portfolio, which often meant taking long positions in foreign currencies.

During periods of rapid currency appreciation, for example 1980–5, hedged portfolios did outperform unhedged portfolios. However, this result was reversed during periods of rapid currency depreciation. "Overall, while hedging does reduce the riskiness of a global portfolio, it also reduces returns so that no statistically significant improvement in portfolio performance is obtained."

10 Incorporating transaction costs in models for asset allocation

JOHN M. MULVEY

1 Introduction

An old adage states that "where you stand depends on where you sit." In the context of asset allocation, this translates into: sound investment advice must be based on the investor's unique situation. Some investors accept great risk for hopefully greater rewards. Others attempt to immunize their portfolios for fear of loss, however slight. Most investors fit somewhere between these two extremes.

A related issue involves the costs for making changes to an existing portfolio. Several asset categories require a substantial payment for either entry or exit, for example, real-estate or venture capital. Other investment categories generate small commissions, but trade in a relatively thin market. Therefore, institutions and other large investors may pay substantial market impact costs whenever a change is made in the makeup of their portfolios. Smaller capitalized US stocks display this feature – estimates range from 80 to over 400 basic points for each side of these transactions.

Despite the importance of turnover and transactions costs, most asset allocation programs treat the rebalancing issue in a simplistic fashion. Recommendations do not depend upon the investor's current portfolio; for example, "average" transaction costs are subtracted from expected returns. The rebalancing costs are often ignored.

This chapter develops a systematic approach for rebalancing a portfolio. Two optimization-based approaches are proposed. First, we describe a single-period myopic optimizer based on a von-Neumann–Morgenstern (V–M) family of utility functions, including the popular growth optimal curve. This work is an extension of R. Grauer and N. Hakansson.[1] Next, a multiple period generalization of the myopic (V–M) model is presented; empirical results show the benefits of this extension for taking into account transaction costs. The multiperiod model looks further ahead than the

Part III

Methodologies

single-period model and thereby profits. While the multiperiod model is more complex than the myopic approach, both possess a special network (graph) structure that can be exploited by the solution procedure as discussed in the appendix. Computational results are reviewed in the chapter.

The asset allocation decision involves the choice of proportions for various investment categories, such as S&P 500 stocks, cash, and long-term government bonds.[2] Much has been written concerning the importance of sound asset allocation. The approach fits naturally in the framework of a large institutional investor – e.g., a pension plan or university endowment administrator – who must select a group of managers and turn over some percentage of their funds for management. Asset allocation suggests target mixes, depending upon the plan administrator's degree of risk aversion and other issues. Each manager's objective would be then to meet or exceed the returns for their own category. Alternatively, index funds could be used in an effort to reduce administrative costs.

2 Single-period asset allocation

Traditional asset allocation procedures that are based on optimization have largely depended upon a single-period strategy. Do your best at each step and the long run will take care of itself. Grauer and Hakansson have formalized this notion. They showed that a myopic approach optimizes the investor's long-run wealth (or the expected utility of wealth) when transaction costs, market impact, costs, and liquidity available considerations are ignored. The general model fits within the context of maximizing the expected utility of wealth, as originally proposed by John von-Neumann and Oskar Morgenstern (V–M) as a method for making decisions under uncertainty.[3]

An important special case of the V–M approach as advocated by Grauer and Hakansson and others (Mulvey (1989)), is called the "growth optimal strategy."[4] Here, the investor possesses preferences that minimize the probability of a deficit, while providing the maximum expected geometric growth of the portfolio over the long run. To illustrate the performance of the growth optimal approach over eight recent years, we have employed the strategy on a quarterly basis – using thirty-six quarters worth of historical data for calibrations – during the first quarter 1979 to the second quarter 1988. Figure 10.1 shows a path of the predicted and the actual returns, assuming that initial wealth equals one and the recommendations are followed at each step. These results are consistent with previous studies which have ignored transaction and market impact costs (T-costs). An objective of this chapter is answering the question: what kind of performance would be forthcoming when T-costs are added to the process?

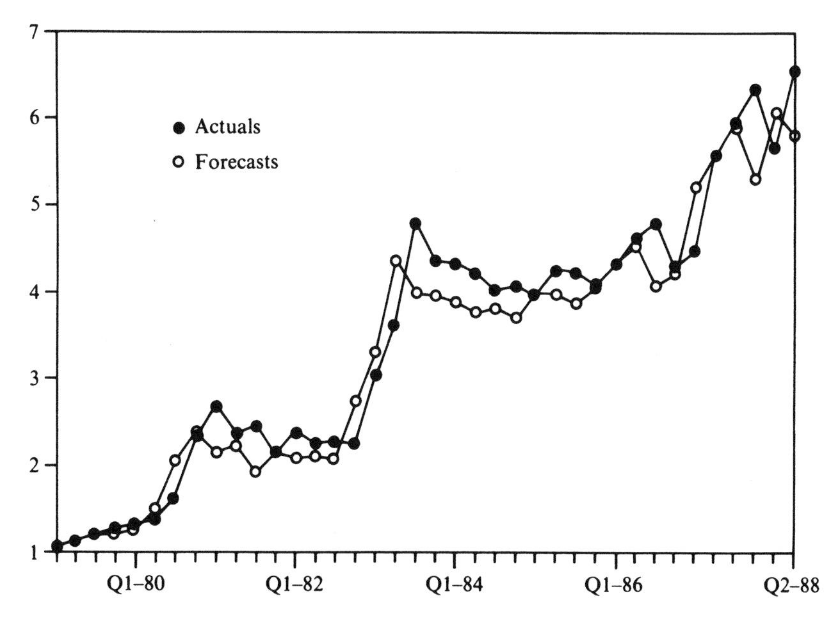

Figure 10.1 Cumulative value of assets using growth optimal strategy (initial value unit, 1979)

Traditional mean/variance (Markowitz) modeling also operates within a single-period framework. It is generally assumed that the expected returns are reduced by projected transaction costs and that the investor will periodically rebalance his asset mix to bring it in line with target percentages. Figure 10.2 depicts the impact of adding the following transaction costs when running the mean/variance model; S&P500 = 0.3%, SM-CAP = 1%, CASH = 0%, INT-GOVT = 0.2%, LT-GOVT = 0.2%, LT-CORP = 0.2%, HI-YIELD = 0.4%, EAFE = 2.0%, REIT = 6.0%, and VENT-CAP = 4.0%. Note that the effect of T-costs is more pronounced for the higher levels of risk in the mean/variance context.

Since investment dynamics are not considered explicitly within a quadratic mean/variance model and due to the problem of picking a consistent point on the efficient frontier at each step, it is difficult to evaluate the results of mean/variance models. Also a multiperiod generalization of mean/variance is limited. For these and other reasons, we focused on von-Neumann–Morgenstern expected utility approaches.

As discussed in the appendix, the resulting V–M asset allocation models are generally equivalent to single-period stochastic networks. There are several advantages to representing the portfolio problem as a network. First, most people understand the graphical context much better than the

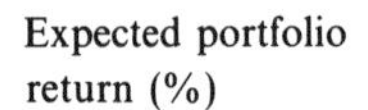

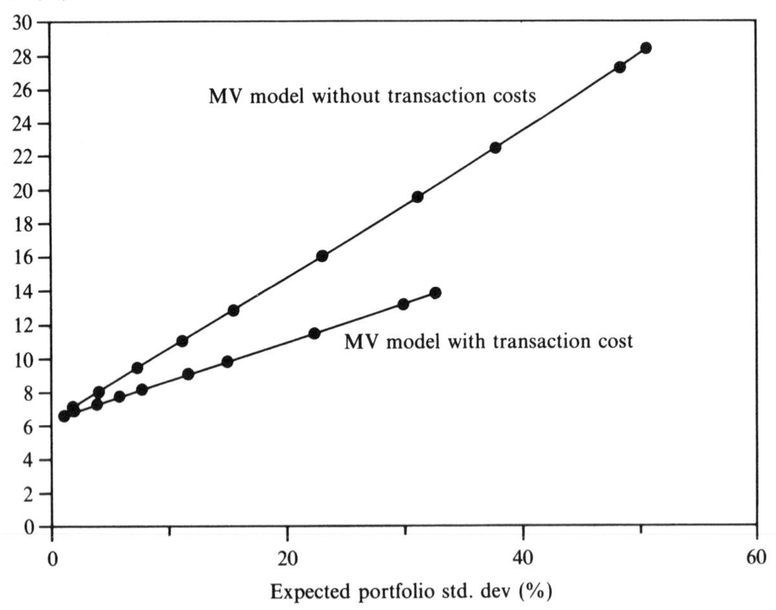

Figure 10.2 Mean-variance efficient portfolios

algebraic. Second, there are enormous computational efficiencies that grow out of exploiting the graph structure. See Dembo, Mulvey, and Zenios (1989). Third, the network model can be defined so that its parameters are random variables. A truly stochastic representation is possible.

The results shown in figure 10.1 and those of Grauer/Hakansson have demonstrated that the single-period V–M model can be effective for the asset allocation problem in the absence of transaction costs.[5] The next step is to test this approach under less favorable conditions by adding transaction costs to the model. Several strategies have been analyzed. First, we ignored T-costs in the decision-making portion, but then reduced the amount available for investing in the subsequent period by the projected T-costs.[6] Consider this strategy to be a naive implementation of the single-period V–M model. It is defined as strategy #1. The historical results of employing this strategy for various levels of the risk-aversion parameter (γ) are listed in table 10.1. Two points are worth noting. First, the impact of T-costs is more pronounced for the higher levels of risk, i.e., for risk-aversion parameters close to zero or positive values. The geometric

Table 10.1. *Performance of single-period V–M models*

Risk-aversion parameter	Ignore T-costs		Ignore T-costs for decisions, but include for measuring performance			
	GEO mean	Std. dev.	GEO mean	Std. dev.	Total T-costs	AVG turnover
1	23.04	23.95	24.0	24.0	0.396	0.08108
0	23.12	23.98	19.68	24.0	0.410	0.12212
−1	20.54	24.01	17.62	24.14	0.374	0.13705
−3	20.27	23.85	17.79	24.04	0.399	0.11320
−7	14.32	11.85	13.36	9.97	0.193	0.13514
−25	10.13	4.83	10.28	4.60	0.005	0.01814
−50	9.28	2.188	9.28	1.37	0.0	0.0

returns are reduced by well over 1% on an annual basis, due to average quarterly turnover between 8% and 13.7%. Under the high-risk strategies, the V–M model suggests frequent movements into and out of asset categories that possess relatively higher transaction costs. (Remember the naive strategy has no knowledge of T-costs in its decision-making phase.)

Second, the low-risk V–M strategies display almost no effect on performance when T-costs are added. Of course, the more conservative asset categories possess low entry/exit fees.

The next strategy to be analyzed consists of adding transaction costs directly to the V–M optimization model. Here the model maximizes expected utility at the end of its planning horizon – in this case one quarter – after T-costs have been paid for rebalancing the portfolio. In effect, the model decides if the T-costs are worth paying in the context of an optimal portfolio mix. Table 10.2 shows the results of applying this strategy across a range of risk aversion. Comparing Tables 10.1 and 10.2 we see that the incorporating of T-costs results in much reduced average transaction costs and turnover over the naive strategy. However, the variances in return are only moderately less than the naive approach. This approach is defined as strategy #2.

The third single-period strategy, one that is often used in practice, involves the imposition of limits on the proportion of total assets in any particular category.[7] Ideally, this approach favors putting upper limits on asset categories possessing relatively high transaction costs. To demonstrate the approach, we employed the following constraints: S&P

Table 10.2. *Performance of single-period V–M model with T-costs*

Parameter	GEO mean	Std. dev.	Total T-costs	AVG turnover
1	19.22	23.63	0.1647	0.05405
0	16.37	22.36	0.1373	0.0778
−1	16.32	19.49	0.175	0.07696
−3	17.94	20.50	0.1116	0.08108
−7	15.82	15.40	0.0909	0.0516
−20	11.18	4.04	0.0117	0.0248
−50	9.75	1.36	0.0	0.0

500 ≤ 60%, SM-CAP ≤ 20%, CASH ≤ 50%, INT-GOVT ≤ 50%, LT-GOVT ≤ 50%, LT-CORP ≤ 50%, HI-YIELD ≤ 20%, EAFE ≤ 20%, REIT ≤ 10%, and VENT-CAP ≤ 10%. The optimization problems were solved once again – for strategies 1 and 2, but including the predefined constraints. Results are provided in tables 10.3 and 10.4. Once again and as expected, the inclusion of T-costs directly in the model reduced total transaction costs and turnover. While there appears to be an advantage of the direct approach in terms of geometric means and variances over the naive strategy when limits are placed on the asset categories, the previous unconstrainted V–M model – strategy #2 – gave opportunity for greater gains in conjunction with greater risks, especially for low risk-aversion results (e.g., $\gamma = 1, 0, -1$).

A primary difficulty with the setting of limits on asset categories entails the arbitrariness of the process. Regardless of any potential gain, the model cannot place additional resources in an asset category beyond the *a priori* constraints. It seems reasonable, therefore, that the model's performance would suffer as a consequence of the asset limits, and our empirical results bear out this expectation.

In summary, for the range of transaction costs that were tested the myopic single-period model can be extended to handle T-costs. However, the three single-period approaches that were tested had limits. The naive strategy #1, in which T-costs were ignored in the decision process, resulted in substantial payment of transaction costs and high variance, especially for moderate and high-risk strategies. The second and third strategies, in which T-costs were directly included in the decision-making, resulted in much reduced volatility as compared with the naive strategy. But the model

Table 10.3. *Performance of single-period V–M model (with limits on asset categories, ignore T-costs for decision making, include T-costs when revising portfolio)*

Parameter	GEO mean	Std. dev.	Total T-costs	AVG turnover
1	13.28	15.8	0.23	0.1455
0	13.83	14.12	0.23	0.1866
−1	12.65	13.00	0.22	0.1989
−3	12.08	11.42	0.170	0.2197
−7	11.62	9.94	0.181	0.1707
−25	10.22	4.59	0.039	0.0325
−50	10.52	5.22	0.084	0.0740

Table 10.4. *Performance of single-period V–M model (with limits on asset categories, include T-costs in decision making and when revising portfolio)*

Parameter	GEO mean	Std. dev.	Total T-costs	AVG turnover
1	15.27	14.76	0.0563	0.09756
0	14.19	14.36	0.0751	0.12903
−1	13.41	13.67	0.0687	0.1294
−3	13.00	11.99	0.062	0.1366
−7	12.99	10.67	0.0721	0.1442
−25	10.27	4.53	0.0468	0.02065
−50	6.96	6.25	0.04238	0.0866

performance as measured by realized geometric means over the historical test period, suffered. Of the two, strategy #2 produced better returns for slightly less risk – compared with strategy #3 in which limits were placed on asset categories. Another conclusion is that including T-costs directly in the decision process appears to be a better idea than ignoring them.

The next section takes up the topic of developing a practical multiple-period model.

3 Multiperiod asset allocation

A complete representation of transaction costs cannot be conducted in a single-period context. A single-period model does not have the ability to anticipate far enough ahead in order to avoid costly rebalancing errors. The situation becomes worse when the amount of the transactional costs depends upon performance, for example, when costs are much larger during market downturns.

A second problem occurs when the investor's interest turns to dynamic strategies. These strategies are most effective when transaction and information processing costs are extremely low, e.g., large institutional investors who possess a research staff and an effective method for trading. Otherwise, the stated advantage of the dynamic strategy can be overwhelmed by the costs of conducting the approach.

In these cases, part of the problem stems from the model's inability to look ahead – and take advantage of the expected future course of events. Also, a look ahead feature prevents what we call "box canyon errors." These situations arise when an investor finds himself in a situation which has no obvious (or painless) exit. The undesirable investment is often kept for these circumstances under the illusion that the situation will eventually turn itself around, salvaging the manager's somewhat tarnished reputation. Misapplication of the concept of sunk costs sometimes leads to sunk reputations and wealth.

The implementation of multiperiod portfolio models has been restricted for several reasons. First and foremost, the multiperiod model results into extremely large non-linear optimization problems. A practical solution technology was unavailable. Second, the usefulness of a multiperiod context has not yet been demonstrated. It requires greater information requirements and expense. Yet Grauer and Hakansson (1982) showed that a myopic optimization is sufficient in the absence of T-costs. Is the extra effort worthwhile?

To address this issue, we designed a multiple-period portfolio model which includes transaction costs and tested this model with the identical asset categories and time span that was discussed in the previous section. A four-period (4 quarters) model was implemented via a network and solved by means of a new algorithm, called progressive hedging.[8] The results appear in table 10.5. For most cases, the multiperiod program gave superior recommendations over the single-period approach. Two procedures were used to assess the needed probability forecasts. First, a simple probability model employing historical data in a manner analogous to the study by Grauer and Hakansson was implemented. Second, the statistical method principal component analysis (PCA) transformed the historical

Table 10.5. *Results of multiple-period model (4-period look ahead, including T-costs, no limits on asset categories)*

Parameter	GEO mean	Std dev.	Total T-costs	Forecasting method
1	23.80	22.72	0.1236	PCA
0	21.60	20.82	0.1748	PCA
−1	20.47	20.26	0.1871	PCA
−3	19.16	18.41	0.1608	PCA
−5	17.73	14.13	0.1487	PCA
1	19.51	20.35	0.0984	simple
0	18.23	20.19	0.1810	simple
−1	18.01	19.47	0.1905	simple
−3	17.43	18.02	0.1879	simple
−5	16.34	15.74	0.1510	simple
−10	16.07	12.13	0.1350	simple
−25	15.83	12.03	0.0984	simple
−50	15.74	11.97	0.0983	simple

data and the forecast was derived from the uncorrelated transformed data. Both approaches produced superior performance as compared with the myopic strategies. Figure 10.3 depicts a chart of the realized geometric returns and the standard deviation of the portfolio returns during the test period (Q1-1979 to Q2-1988).

Last, these performance results should be evaluated with respect to a few standard benchmark portfolios. Table 10.6 lists the target proportions for six common benchmark portfolios. Since the benchmark portfolios consist of four asset categories that possess low transaction costs (S&P500, LT-GOVT, LT-CORP, and CASH), the impact of transaction costs are negligible on their performance. The results for maintaining the benchmark portfolios during the test period (Q1-1979 to Q1-1988) are displayed in figure 10.4. Note in particular the benchmark results as compared with the multiperiod models – even in the face of transaction costs.

Why does the multiperiod V–M model produce better results when employing much the same information as its single-period counterpart? The primary reason, in our opinion, lies with the issue of transaction costs. The multiperiod V–M model addresses, within a wider context than the myopic model, the tradeoff between paying an entry or exit fee and the potential gain for making a portfolio modification. This tradeoff can be a critical issue for making sound asset allocation decisions.

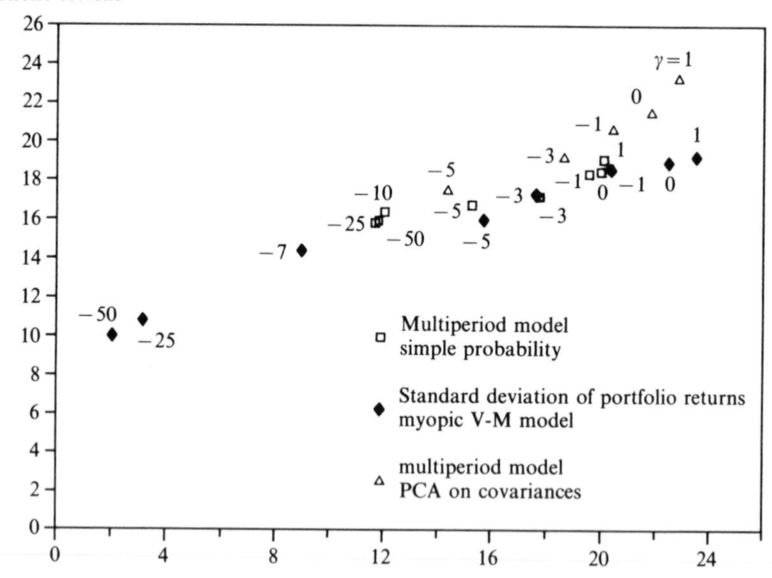

Figure 10.3

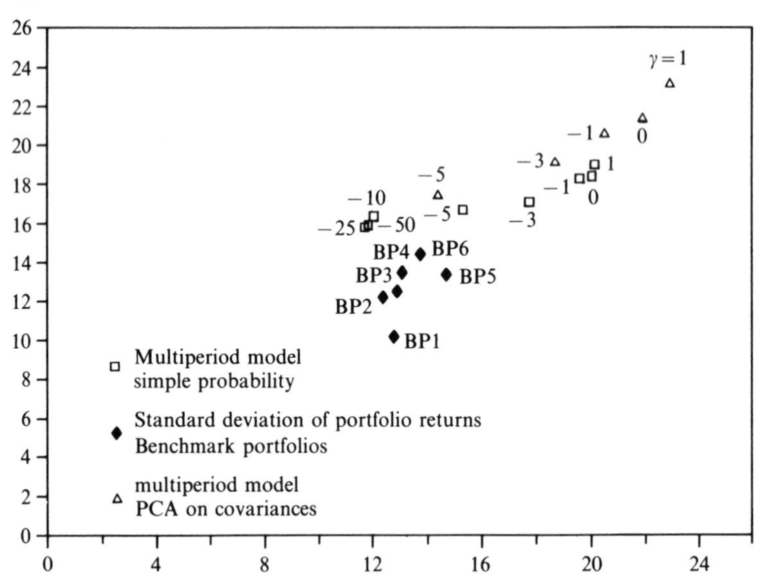

Figure 10.4

Table 10.6. *Compositions of fixed weight benchmark portfolios*

	Proportion in			
Portfolio	S&P500	LT-GOVT	LT-CORP	CASH
BP1		0.40	0.40	0.20
BP2	0.20	0.30	0.30	0.20
BP3	0.45	0.20	0.25	0.10
BP4	0.55	0.15	0.20	0.10
BP5	0.60	0.40		
BP6	0.65	0.10	0.15	0.10

4 Greater realism for asset allocation

Incorporating transaction and market impact costs complicates the asset allocation process. Not only must T-costs be estimated and added to the asset allocation model but also the investor's current portfolio must be considered. In this context it is difficult to provide general, widely applicable recommendations. Rather, asset allocation advice must be tailored to each investor's unique circumstances.

Specialization, of course, has always taken place – but typically outside of the mathematical model. However, the T-cost issue has gained significance especially in light of the superior performance that has occurred over the past decade for asset categories that possess relatively steep entry and/or exit fees – such as venture capital or smaller-capitalized foreign stocks. Asset allocation procedures should be able to address the two critical issues: (1) When is it wise to move into (or out of) high T-cost categories? (2) How much movement should be made?

This report has described two stochastic optimization models that handle T-costs as integral parts of the "portfolio revision" decision. First, the single-period V–M model (which extends the Grauer–Hakansson approach) compared favorably with the naive – ignore T-costs – strategy. Greater returns were received with commensurate risk, or alternatively less risk was possible with the same amount of return. Computational costs were not noticeably affected. Also, this model was able to keep T-costs under control without resorting to the somewhat arbitrary approach of putting constraints on asset categories.

Second, a four-period V–M portfolio model was defined and shown to produce superior recommendations over the single-period myopic optimizers, once T-costs were accounted for. In effect, its look ahead feature seems to be able to avoid dead-end asset categories. On the negative side, however, the multiperiod is more computationally intensive than the single-period approach, requiring a large mainframe computer for its solution. Further improvements in parallel computers will undoubtedly bring the cost of this technology down over the next few years.

The next step will be to increase the number of time periods and forecasted scenarios. Empirical testing will be needed to ascertain whether the increased model complexity will lead to improved performance, however. Also note that the V–M approach can be easily adapted to any forecasting system that produces discrete scenarios.

In summary, more realistic and practical asset allocation models are possible. Greater realism may come at the cost of the "generality of the recommendations," but we believe that at least in this case – more details are better than less.

Appendix: Financial network models

Substantial benefits are gained by modeling the asset allocation problem as a stochastic network. This discovery paves the way for a practical approach for multiperiod portfolio models, and as a consequence gives us a tool for handling transaction costs in a realistic fashion. The technology would be impossible, however, without the network interpretation. This appendix gives further details of network portfolio models.

Figures 10A and 10B depict the single-period and the multiperiod network portfolio models, respectively. Each node (circle) on these diagrams refers to an algebraic equation which enforces the conditions of resource conservation: flow in equals flow out. Each line (arc) in the graph represents a variable (x_{ij} – flow between node i and node j). The variable's value depends upon the amount of movement along the arc; these decision variables are determined by the optimization program. Instead of physical transfers such as autos or computer chips – common in logistics and transportation planning – we are interested in the flow of financial funds across time and space.

When flow takes place between time periods, a correction factor called a *multiplier* is needed. Forward pointing (left to right) arcs result in gains – representing increased flow due to interest, dividends, or price appreciation. An example is the arc (1,5) which indicates the amount of funds invested in asset category #1 – S&P500 for our test case. Backward pointing (right to left) arcs result in losses, i.e., they indicate the cost of borrowing through a reduction in funds.

It is important to note that in many cases the multiplier parameters are uncertain (stochastic), and must be treated as random variables in the optimization procedure. Otherwise, the model will base its recommendations on a certain future, whereas portfolio modeling is clearly fraught with uncertainty. Adding stochastic elements to an optimization model is a newly developing area, one which has received scant attention due to the

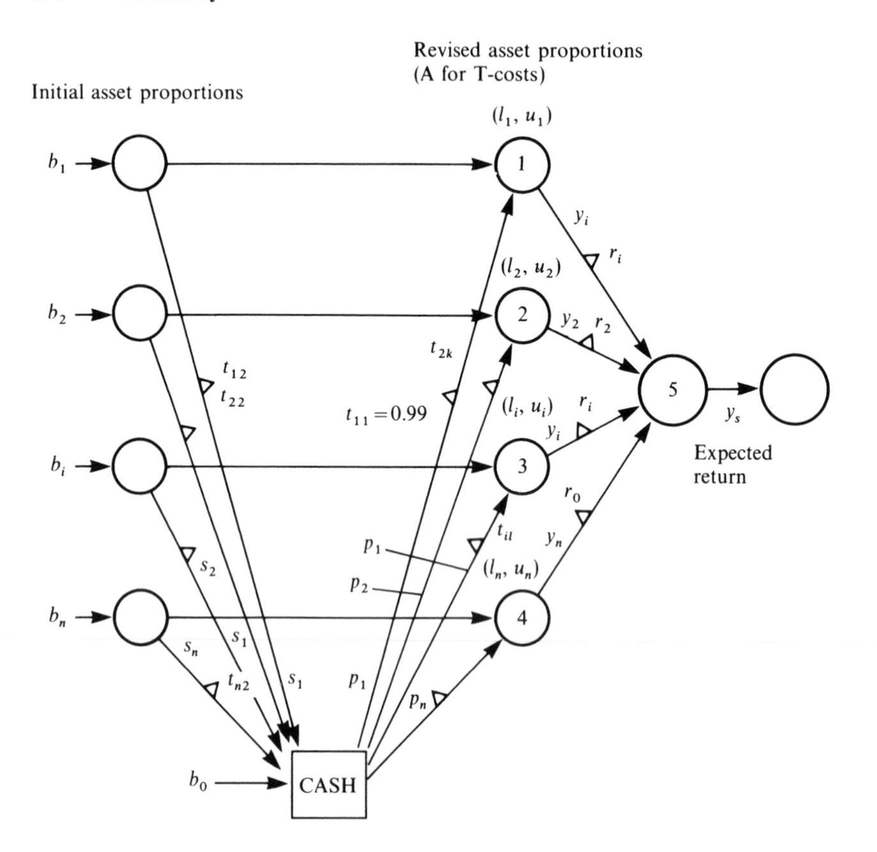

Figure 10A Network model for the single period asset allocation problem

lack of suitable software. However, much progress has been made overcoming the obstacles in the recent past.

Multipliers are also needed when revising a portfolio. Herein, the projected transaction costs are incorporated as loss factors on the appropriate arcs (variables). For example in figure 10A, the arc between cash and asset #1 possesses a multiplier equal to 0.99, indicating a 1% transaction cost.

The objective of von Neumann–Morgenstern (V–M) models is to maximize the expected utility of wealth at the end planning horizon – either one period or n periods in the future. Naturally, many functions are possible, including the isoelastic family which was tested in this report.

An algebraic description of the single-period portfolio problem follows:

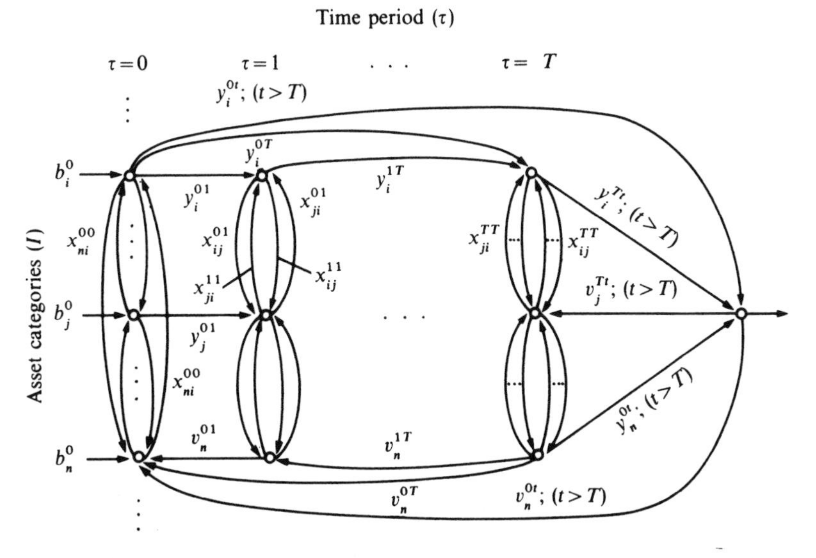

Figure 10B Basic structure of the network model for multiperiod financial planning problems

$$\text{Maximize} \sum_{i=1}^{k} p_i u(w_i) \tag{1}$$

$$\text{subject to } Ax = b \tag{2}$$

$$l \leq x \leq u \tag{3}$$

where:

w_i wealth, given scenario i,
$u(\cdot)$ V–M utility function,
x vector of decision variables,
b supply/demands on nodes,
$l(u)$ lower (upper) bounds on the arcs,
A node-arc incidence matrix for the network,
p_i probability of scenario i,
k number of scenarios.

The objective function (1) is generally non-linear due to risk-aversion behavior. Algebraic constraints (2) and (3) define the network graph as shown in the figures. For most individuals, the pictorial representation as a graph is easier to understand than the algebraic description.

NOTES

Partial support from Pacific Financial Companies and the National Science Foundation Grant No. DCR-861-4057 is gratefully acknowledged.

1 See Grauer and Hakansson (1982) for empirical results of employing this strategy.

2 Ten asset categories have been selected for empirical testing: S&P500, SM-CAP (smaller capitalized US stocks), CASH (three-month T-bills), INT-GOVT (intermediate-term governments bonds), LT-GOVT (long-term government bonds), LT-CORP (long-term corporate bonds), HI-YIELD (high-yield corporate bonds), EAFE (international stocks – Europe, Australia, and the Far East), REIT (real estate), and VENT-CAP (venture capital).

3 The von-Neumann–Morgenstern family of functions displaying this characteristic is called iso-elastic: $u(w)=\frac{1}{\gamma}w^{\gamma}$, where w is the investor's wealth and gamma (γ) is a risk-aversion parameter. Note that the growth optimal strategy results when $\gamma=0$, i.e., $u(w)=\ln(w)$.

4 The classic textbook is J. von Neumann and O. Morgenstern (1947).

5 Clearly, some asset categories provide T-costs low enough to justify the use of a myopic optimizer, such as the big three categories – SP500, US Government and corporate bonds, and CASH.

6 Every investor will have his or her own scale of T-costs. However, we attempted to find some common values by providing a median estimate from a variety of sources. We have found the overall conclusions of the study to be relatively insensitive to moderate changes in the T-cost estimates.

7 For an alternative approach, see Schreiner (1980).

8 The progressive hedging algorithm was recently invented by R. Wets and T. Rockafellar; see Mulvey and Vladimirou (1991) and Rockafellar and Wets (1990).

REFERENCES

R. S. Dembo, J. M. Mulvey, and S. A. Zenios (1989), "Large scale nonlinear network models and their application," *Operations Research*, 37, 353–72.

R. R. Grauer and N. H. Hakansson (1982), "Higher return, lower risk: historical returns on long run, actively managed portfolios of stocks, bonds and bills," *Financial Analysts Journal*, March–April: 1936–78.

J. M. Mulvey (1989), "A surplus optimization perspective," *Investment Management Review*, 3: 31–9.

J. M. Mulvey and H. Vladimirou (1991), "Applying the progressive hedging algorithm to stochastic generalized networks," *Annals of Operational Research*, 13: 399–424.

R. T. Rockafellar and R. J. B. Wets (1990), "Scenario and policy aggregation in optimization under uncertainty," *Mathematics of Operational Research*, 16: 119–47.

J. Schreiner (1980), "Portfolio revision: a turnover-constrained approach," *Financial Management Review*, Spring, 9: 67–75.

J. von Neumann and O. Morgenstern (1947), *Theory of Games and Economic Behavior*, Princeton University Press, Princeton, New Jersey.

11 Bond portfolio analysis using integer programming

ROBERT M. NAUSS

1 Introduction

The use of optimization techniques to design or modify bond portfolios is over twenty years old. The development of these techniques parallels the growth in computer power over the same time frame. Large-scale problems solved on the most powerful computers in the late 1960s may now be solved on run-of-the-mill mainframes and in some cases on personal computers. Bradley and Crane's paper (1972) provides a view of one of the first such models and reviews earlier published research. Hodges and Schaefer (1977) formulate a linear program (LP) that minimizes the cost of a portfolio while requiring that a series of cash outlays be met on time. Income taxes are assessed by multiplying coupons by the marginal tax rate. Alexander and Resnick (1985) formulate a linear goal program that maximizes the portfolio yield-to-maturity while immunizing the portfolio with a given duration and placing bounds on bond quality and on the mean absolute deviation of individual bond durations. More recently Ronn (1987) developed an LP approach which seeks to buy "underpriced" bonds and sell "overpriced" bonds while simultaneously considering taxes and bond portfolio cashflows. Leibowitz (1986a and b) presents a general overview of the issues involved in bond portfolio dedication and immunization procedures.

In this chapter we extend the analysis in a number of ways. First multiple objective functions in the before and after tax cases are considered. Included is a non-linear objective of maximizing the internal rate of return of the portfolio (Nauss, 1988). Second, the set of constraints that are considered are expanded to include cashflow, duration, bond quality, portfolio convexity, callable bond exposure, limits on realized capital gains or losses, and yield curve scenarios. Third, since bonds are sold in lots (generally $5,000 lots) and minimum purchase and sale amounts are restricted to say $100,000 or more (in order to obtain "good" prices) we introduce integer variables to account for lot sizes. Finally, we use an elastic constraint approach in order to assure feasibility by allowing selected constraints to be violated using user-specified penalty costs.

The chapter is organized as follows. First, various bond portfolio categories are defined. Then, the integer programming formulation is given. Next, various possible objective functions and constraint sets are presented. A section on solution considerations is presented, and finally, results are given from the analysis of four sample portfolios.

2 Categories of bond portfolios

We consider three overlapping categories of bond portfolios: dedicated, parallel shift immunization, and general investment. A bond portfolio is said to be *dedicated* if its principal and reinvestment cashflows are sufficient over time to pay off a set of liabilities. In a pension fund, for example, the set of liabilities is a list of cash pension benefits paid out monthly to current retirees. Since pension benefits cease upon death (or are reduced for a surviving spouse), such liabilities are not certain. Thus actuaries are commissioned with the task of generating the liability stream. The liability stream may be updated periodically to account for actual deaths and to include recent retirees. Recent retirees' cash benefits are sometimes paid out of a cash account until the annual actuarial revision takes place. At this time additional funds may need to be added to the dedicated portfolio. Alternatively a separate dedicated account could be set up for recent retirees.

A bond portfolio is said to be *immunized under parallel interest-rate shifts* when the market value of the portfolio is at least as great as the present value (PV) of the liabilities to be satisfied and the portfolio duration is matched with the liability duration. Both the duration and PV calculations for the liability stream depend upon the discount rate used. One approach for designating a discount rate is to use the internal rate of return (IRR) of the bond portfolio. A second approach is to use a discount rate as specified by the actuaries. Such a rate tends to be based on the historical levels of long-term interest rates and generally is not modified too often. The latter approach is somewhat easier to implement since the former requires a dynamic recalculation of the PV of the liabilities and the corresponding liability duration. It is also a more conservative approach when the specified rate is smaller than the IRR of the bond portfolio. Finally we reiterate that the duration matching used to achieve immunization is valid only when the yield curve moves up or down in a parallel fashion. We address this shortcoming once again in the next section.

We note that a combination of dedication and immunization may be applied to a bond portfolio. For example, a pension fund may require that the portfolio be dedicated for liabilities occurring during the next five years, but that cashflow requirements may be relaxed after that time so long as the duration of the portfolio and the liabilities are matched.

The third category of bond portfolios concerns *general investment*. In this case there may or may not be cashflow constraints. For example, a bank may need sufficient cashflow from its portfolio to pay quarterly taxes and to pay monthly overhead expenses. This would limit investment in long-term zero coupon bonds. In this case *some* portfolio cashflows are required, but the market value of the portfolio is much larger than the present value of the liabilities. General investment portfolios may be passively (a buy and hold strategy) or actively managed. The key difference, then, between dedicated and/or immunized portfolios and general investment portfolios is the absence of a set of liabilities whose present value is approximately equal to the market value of the portfolio. Certain cashflow requirements may still have to be met, but they are relatively small when compared to the size of the bond portfolio.

3 A model for bond portfolio optimization

In this section we describe a general purpose integer programming model that may be used to optimize a bond portfolio. Let E be the index set of existing securities. Let C be the index set of candidate securities, and let O be the index set of cashflow obligations that must be met. Finally let S be the index set of possible future yield curve scenarios as of a specified future date. The decision variables are defined as follows.

Let:

x_j = par dollar amount of existing security j to be sold, $j \in E$,

y_k = par dollar amount of candidate security k to be purchased $k \in C$,

Z_t = dollar amount of cashflow reinvested after obligation $t, t \in O$,

W_t = dollar amount to be borrowed to meet obligation $t, t \in O$,

P_k = logical 0-1 variable to require a minimum lot size purchase (if any) of security $k, k \in C$,

S_j = logical 0-1 variable to require a minimum lot size sale (if any) of security $j, j \in E$.

The term $\bar{x}_j$ is the par dollar amount of security j *currently* held in the portfolio. Technically speaking it is not a variable but a fixed value. Define the following data coefficients:

c_j = bid price plus accrued interest per par dollar sold of security $j, j \in E$,

c_k = asked price plus accrued interest per par dollar purchased of security $k, k \in C$,

a_{jt} = cashflow per par dollar due to security j at obligation $t, t \in O$,
a_{kt} = cashflow per par dollar due to security k at obligation $t, t \in O$,
m_j = minimum par amount of security j that may be sold, $j \in E$,
u_j = maximum par amount of security j that may be sold, $j \in E$,
d_j = denomination of par amounts sold, $j \in E$,
m_k = minimum par amount of security k that may be purchased, $k \in C$,
u_k = maximum par amount of security k that may be purchased, $k \in C$,
d_k = denomination of par amounts purchased, $k \in C$,
e_{ji} = bid price plus accrued interest per par dollar for security j, as of a specified future date under scenario $i, j \in E, i \in S$,
e_{ki} = bid price plus accrued interest per par dollar for security k as of a specified future date under scenario $i, k \in C, i \in S$,
cash = net cash position for existing portfolio as of a specified future date assuming the reinvestment and borrowing rates, r and b respectively,
D_j = duration of security $j, j \in E$,
D_k = duration of security $k, k \in C$,
r_t = annual reinvestment rate, r, multiplied by the number of days between obligation t and $t+1$ and divided by 365,
b_t = annual borrowing rate, b, multiplied by the number of days between obligation t and $t+1$ and divided by 365,
R_t = dollar amount to be removed from the portfolio at obligation $t, t \in O$,
s = maximum absolute dollar difference between amount sold and purchased,
p = maximum proportion of AA securities in the portfolio,
MXS = maximum dollar amount of the portfolio that may be sold,
MXP = maximum dollar amount that may be purchased (possibly including additional funds over and above the amount realized from the sale of portfolio securities),
MXD = maximum duration of the portfolio in years,
MND = minimum duration of the portfolio in years,
T = a specified index $t \in O$ where market value of modified portfolio must be greater than the present portfolio.

We note that the a_{jt} and a_{kt} coefficients are obtained by finding the future value of all security cashflows between obligation $t-1$ and t. The future value is calculated using the annual reinvestment rate. For example suppose a cashflow c occurs on 1 March, the next obligation date is 1 April, and the reinvestment rate is 6% per year. Then the future value is $c*(1+(0.06*31/365)) = c*1.005096$. It is easy to show that such an approach is equivalent to requiring an obligation (and hence a constraint) for every possible date of a cashflow. Unfortunately the same is not true when borrowing is required. In the case of borrowing the following situation creates an overestimate of the true cost. Suppose at some obligation date, borrowing is undertaken due to insufficient cashflows to meet an obligation. Now suppose a security cashflow occurs. In the former case we assume this cashflow is reinvested at the reinvestment rate until the next obligation. However it is more prudent to assume in the real world that the level of borrowing would be reduced *immediately* upon receipt of the cashflow. Thus, since the borrowing rate is generally greater than the reinvestment rate, the cost of borrowing will be higher in terms of the model. Such a modeling inaccuracy may be limited by introducing additional obligations (with value 0) so that immediate payback of borrowing is more precisely modeled.

The following integer linear programming formulation may be used for a non-taxable bond portfolio. Taxable portfolios have additional negative cashflows that occur (in the US) on four estimated tax payment dates to account for taxes due on an accrual basis for interest income and capital gain or loss amortization. These payments depend upon the individual security and thus must be included in the a_{jt} and a_{kt} coefficients. For the sake of brevity the taxable case is omitted. Depending on portfolio requirements, certain of the constraints given below may be omitted:

$$\text{Minimize} \sum_{j \in E} c_j(\bar{x}_j - x_j) + \sum_{k \in C} c_k y_k \tag{1}$$

subject to:

$$\sum_{j \in E} a_{jt}(x_j - \bar{x}_j) + \sum_{k \in C} a_{kt} y_k + (1 + r_{t-1})Z_{t-1} - Z_t - (1 + b_{t-1})W_{t-1} + W_t = R_t,\ t \in O \tag{2}$$

$$-s \le \sum_{k \in C} c_k y_k - \sum_{j \in E} c_j x_j \le s \tag{3}$$

$$\sum_{j \in E} c_j x_j \le MXS \tag{4}$$

$$\sum_{k \in C} c_k y_k \le MXP \tag{5}$$

$$\sum_{\substack{j \in E \\ j \text{ is AA rated}}} c_j(\bar{x}_j - x_j) + \sum_{\substack{k \in C \\ k \text{ is AA rated}}} c_k y_k \le p\left(\sum_{j \in E} c_j(\bar{x}_j - x_j) + \sum_{k \in C} c_k y_k\right) \tag{6}$$

$$\sum_{j \in E} e_{ji}(\bar{x}_j - x_j) + \sum_{k \in C} e_{ki} y_k + Z_T - W_T \ge \sum_{j \in E} e_{ji}\bar{x} + \text{cash}, i \in S \tag{7}$$

$$\sum_{j\in E} D_j c_j(\bar{x}_j - x_j) + \sum_{k\in C} D_k c_k y_k \leq MXD\left(\sum_{j\in E} c_j(\bar{x}_j - x_j) + \sum_{k\in C} c_k y_k\right) \quad (8)$$

$$\sum_{j\in E} D_j c_j(\bar{x}_j - x_j) + \sum_{k\in C} D_k c_k y_k \geq MND\left(\sum_{j\in E} c_j(\bar{x}_j - x_j) + \sum_{k\in C} c_k y_k\right) \quad (8a)$$

$$m_j S_j \leq x_j \leq u_j S_j,\ j\in E \quad (9)$$

$$m_k P_k \leq y_k \leq u_k P_k,\ k\in C \quad (10)$$

$$S_j, P_k = 0 \text{ or } 1\ j\in E,\ k\in C \quad (11)$$

$$x_j = 0, m_j, m_j + d_j, m_j + 2d_j, \ldots, u_j,\ j\in E \quad (12)$$

$$y_k = 0, m_k, m_k + d_k, m_k + 2d_k, \ldots, u_k,\ k\in C \quad (13)$$

$$W_t, Z_t \geq 0,\ t\in 0 \quad (14)$$

The objective function (1) minimizes the market value cost of the portfolio. Other objective functions may be substituted including maximizing IRR (cf. Nauss, 1988). Constraint (2) is the set of cashflow constraints allowing reinvestment and borrowing. The third constraint ensures that the total values of securities bought and of securities sold are approximately equal. Constraints (4) and (5) limit the maximum amount to be bought and sold. An upper limit of $p\%$ to be invested in AA securities is depicted in constraint (6). Other percentage limits as described in section 5 may also be included. Constraint (7) requires that the total value of any new portfolio must be greater than the total value of the original portfolio (without sales or purchases) as of a specified future date and under a specified yield curve scenario. Note that the cash balance for the new portfolio $(Z_t - W_t)$ may also reflect a change in the reinvestment rate r and borrowing rate b. This can be accomplished by adding cashflow constraints similar to (2) with modified r' and b' rates. In this case a cash value for the original portfolio would also be used in (7) and corresponding values for Z' and W'. Constraint (8) limits the duration of the portfolio to no more than a specified maximum duration, MXD, and (8a) limits duration to some minimum, MND. Constraints (9)–(13) are logical constraints that control the minimum and maximum par amount to be sold and the denominations.

It should be noted that constraints (2), (8), and (8a) may appear simultaneously. Consider the example of a general investment portfolio where certain cashflow requirements must be met on time, but where the portfolio manager also uses portfolio duration limits as a risk-management tool.

4 Bond portfolio objectives

Bond portfolio objectives vary according to the category in which a portfolio is placed as well as according to the trading strategy adopted. Probably the easiest example is a dedicated portfolio where the objective is simply to minimize the market value of the portfolio subject to meeting the cashflow requirements of the liability stream. Such a portfolio would generally be passively managed over time. Another example of a passive management style is a conservative bank's investment portfolio. The only trading that may take place is for reinvestment of after tax interest and matured principal. Investments are made with an eye toward managing the gap between the bank's asset and liability duration.

Just recently another objective fell out of favor due to governmental decree. Fixed-income mutual funds often advertised current income return on an annual basis. High coupon bonds of course provide high current income. However, over time the value of the bond drops until it reaches par at maturity. This advertising "loophole" allowed-fixed income funds to invest in high coupon bonds and to report high current income while principal value fell over time. The new government regulation now requires that results be stated on a total return basis and/or on an IRR basis.

Another objective is to maximize the total return of a portfolio. Total return is measured by comparing the market value of the portfolio one year ago, say, with the market value today. Today's market value includes interest income, reinvestment income, and the current market value of outstanding securities. The market value differences divided by the market value one year ago is total return. Generally this objective is used in an active trading strategy. Such a trading strategy might entail buying high coupon securities when their yields are high relative to comparable low coupon securities. Another approach is to buy Agency securities (and sell Treasury securities) when the Agency yield compared to Treasury yields is high. When the spread narrows the reverse trade is made.

An active trading strategy of maximizing the yield or IRR may also be pursued in more conservative portfolios. In this strategy duration and portfolio quality are held constant. The idea is to capitalize on yield curve distortions while keeping risk more or less constant. For example, at some point in time three-year Treasuries may yield measurably less than some combination of two- and five-year Treasuries bought in such proportion that the durations are the same.

Other approaches include maximizing the future value of a portfolio on some specified future date and maximizing the IRR utilizing a specified reinvestment rate. This differs from the standard maximization of IRR in that reinvestment is not assumed to occur at the IRR rate.

5 Portfolio constraints

Portfolios in each of the three categories defined in section 2 are often required to satisfy some collection of structural constraints. We enumerate a number of such constraints below. References to similar-type constraints in the ILP formulation are given in parentheses.

1 (cf. equation (6)) An upper limit on the maximum percentage of a portfolio that may be invested in any security type.
Example: IBM holdings must be limited to 5%

2 (cf. equation (6)) Lower and upper limits on the percentage of a portfolio invested in various security ratings.
Example: 50% to 100% in AAA
0% to 50% in AA
0% to 10% in A

3 (cf. equation (6)) Lower and upper limits on the percentage of a portfolio invested in various sectors.
Example: 50% to 100% in Treasuries
0% to 45% in Agencies
0% to 10% in Corporates

4 (cf. equation (6)) Lower and upper limits on the percentage of a portfolio invested in various maturities.
Example: 5% to 20% must mature in one year or less
25% to 40% must mature in one to two years
25% to 50% must mature in two to three years
20% to 40% must mature in three to seven years

5 (cf. equations (8,8a)) Lower and upper limits on the duration of the portfolio.
Example: A portfolio must have a duration between 4.9 years and 5.1 years.

6 (cf. equations (8, 8a)) Lower and upper limits on the modified duration of a portfolio.

7 (cf. equations (8, 8a)) Lower and upper limits on the average maturity life of the portfolio.

8 (cf. equations (8, 8a)) Lower and upper limits on the portfolio duration convexity.
This constraint aids in controlling price sensitivity of the portfolio.

9 (cf. equation (2)) Constraints on cashflows.
Must cashflows be of sufficient magnitude to satisfy liabilities as they occur? Can the cashflow constraint be relaxed after some specified date?

10 An upper limit on the par amount of callable bonds in a portfolio.

11 The weighted average of the yield minus coupon of callable bonds must be greater than some specified number of basis points. By requiring that

yields be larger than coupons on an overall weighted average basis, the risk of large-scale calls is reduced.

As mentioned in an earlier section, constraint type 5 allows a portfolio to be duration matched with a corresponding liability stream. It is well known that such a matching remains relatively stable as long as interest rates move in a parallel fashion. The idea is that, if interest rates rise, the market value of the portfolio will decline, but reinvested funds will now earn a higher market rate of return. If interest rates fall, the market value of the portfolio will rise, but reinvested funds will now earn a lower market rate of return. Thus the portfolio is said to be immunized against interest-rate risk with respect to the liability stream. When interest rates do not move in a parallel fashion the portfolio may no longer be immunized. In this case the portfolio is rebalanced (securities are bought and sold) in order to achieve the match on duration. Since the market value of the portfolio must also be no less than the present value of the liabilities it may also be necessary to add funds to the portfolio. Finally, we remark that rebalancing is also generally performed over time since the duration of assets and liabilities grows smaller but not necessarily at the same rate.

In order to lessen the rebalancing risk we introduce a scenario approach to interest-rate changes. When a portfolio is to be initially immunized or when rebalancing must be undertaken, a number of possible interest-rate scenarios are posited to occur at some future point in time. One method for describing a scenario is to specify the interest-rate change for each sector (Treasury, Agency, Corporate, etc.) over all possible maturities. The interest-rate change in each case could be a basis point shift in yield, or a percentage of yield change, or a combination of the two. For example we might specify that Treasuries with under two years to maturity will drop in yield by 50 basis points while Treasuries with over two years to maturity will rise in yield by 10% of their individual current market yields. Through the specification of these scenarios it is possible to match the duration of the portfolio with the (possibly modified) duration of the liability stream *and/or* to require the market value of the portfolio to be no less than the PV of the liabilities under each scenario. Such a method allows immunization over hypothesized non-parallel shifts in the yield curve.

Cashflow variations due to called securities may be handled by assuming that individual callable securities are called. Such an assumption would affect the $a_{.t}$ coefficients for the security by eliminating any security cashflows after the call date and adding the call price cashflow to the coefficient corresponding to the call date. A conservative approach assumes all calls occur while a more liberal approach would be to assume only that securities currently in jeopardy of being called are called.

Additional constraints may be used to control the set of securities sold and purchased for any of the three portfolio categories, utilizing any of the objectives presented in section 4 subject to any subset of the portfolio constraints given earlier in this section and in section 3. The additional constraints are enumerated below.

12 (cf. equations (12, 13)) Securities are generally sold in $5,000 lots. In addition prices may vary depending on the purchase amount. Thus a minimum purchase or sale amount (if either occurs) of say $100,000 may be specified for a security.

13 (cf. equations (12, 13)) Maximum purchase or sale amount for each security.

14 Taxes (if applicable) are accounted for on an accrual basis using a standard four times per year payment schedule. Both income and capital gains at the state and federal levels are computed. Tax payments then become a dynamic addition to the liability stream.

15 Additional dollar limits may be specified for realized capital gains and/or losses.

16 Bid prices may be used for securities to be sold, and asked prices may be used for securities to be purchased. Yields are then calculated for the bid or asked price.

17 A limit may be placed on the amount of new funds to be added to the portfolio (if any). A limit on the amount of funds to be withdrawn from the portfolio may also be specified.

18 A limit on the total amount of bonds to be sold and a limit on the total amount of bonds to be purchased may be specified.

19 Individual callable bonds may be assumed to be called as of a particular date or be assumed to mature. Automatic exercise of calls may be assumed if the difference between yield and coupon is sufficiently small or if the difference between market price and call price is sufficiently large.

20 (cf. equation (2)) Reinvestment of maturing principal and interest may be specified at some rate. A second rate effective after some specified date may also be specified.

21 (cf. equation (2)) Borrowing to meet liabilities may be allowed at some specified rate.

22 (cf. equation (7)) Yield curve scenarios may be posited where the portfolio remains immunized under the various scenarios and where the market value under each scenario is no less than the present value of the liabilities.

Table 11.1 *Characteristics of the test problems*

Problem	Existing securities in portfolio	Candidate securities that may be added to portfolio	Total general integer variables	Total variables (integer and continuous)	Number of constraints
1	20	465	485	673	140
2	16	465	481	610	95
3	41	376	417	585	102
4	18	179	197	293	80

6 Solution considerations

While modeling the mixed-integer linear program is helpful in understanding the nuances involved in analyzing a bond portfolio, current solution approaches fall short of generating optimal solutions in a reasonable amount of computer time. Table 11.1 depicts the size of four models analyzed in the next section.

As is evident, the number of *general* integer variables ranges from approximately 200 to 500. The number of constraints varies from 80 to 140. In light of the difficulty experienced in generating and proving an optimal solution, the analyst may nonetheless be able to generate good heuristic solutions. Of primary importance is the introduction of elastic constraints. It is often the case that constraints in the real world are "not hard." For example, a portfolio may be required to have a duration of no more than five years and have no more than 30% of the portfolio invested in AA securities. In most instances, however, a duration of 5.01 years (the equivalent of five years and three days) is acceptable. Likewise, an investment of 30.1% of the portfolio in AA securities may be permissible. Thus, an elastic constraint may be modeled using the hard constraint $\sum_j a_j x_j \leq b$ in the following way. Introduce the non-negative variable s with a penalty cost of p in the objective function:

$$\sum_j a_j x_j - s \leq b$$

Note that an upper bound may be placed on s to limit the constraint elasticity. Also, s_1 and s_2 may be used in the same constraint with s_1 being bounded from above and the penalty cost for s_2 being larger than the penalty cost for s_1. Note that the feasible region under such constraint elasticity is increased, and hence the possibility of no feasible solution is lessened. We obtain feasible integer solutions by first solving the associated linear programming relaxation. Candidate securities with highly unattractive reduced costs are then locked at 0. The remaining variables including both existing and candidate securities are locked at a neighboring integer value one by one. After a variable is locked at an integer value, the linear program is reoptimized over all free (not locked) variables. This process continues until an integer feasible solution is found. If an infeasible solution is detected at some point, backtracking occurs using the next closest integer value until feasibility is restored.

7 Bond portfolio examples

In this section we present results for four bond portfolios. Two of the portfolios are taxable and two are non-taxable. Two of the portfolios also had cashflow requirements on a monthly or quarterly basis. The summary table (table 11.2) presents the improvement in dollar and percentage return terms from a buy/sell recommendation date to the evaluation date of 7 July 1989. In order to compare "apples and apples" various reinvestment rates for cashflows occurring between the buy/sell recommendation date and 7 July are used and dollar differences are calculated. The remaining tables (11.3 through 11.10) are comprised of a buy/sell recommendation page for the buy/sell date and a page comparing the old and new portfolios on 7 July 1989. Note that only securities bought and sold are printed. Existing securities that are not sold are omitted for brevity.

REFERENCES

G. Alexander and B. Resnick (1985), "Using linear and goal programming to immunize bond portfolios," *Journal of Banking and Finance*, 9: 35–54.

S. Bradley, and D. Crane (1972), "A dynamic model for bond portfolio management," *Management Science*, 19: 139–51.

M. Granito (1983), *Bond Portfolio Immunization*, Heath, Lexington, MA.

S. D. Hodges and S. M. Schaefer (1977), "A model for bond portfolio improvement," *Journal of Financial and Quantitative Analysis*, June: 243–60.

Martin L. Leibowitz (1986a), "The dedicated bond portfolio in pension funds – Part I: motivations and basics," *Financial Analysts Journal*, January–February: 68–75.

(1986b), "The dedicated bond portfolio in pension funds – Part II: immunization, horizon matching and contingent procedures," *Financial Analysts Journal*, March–April: 47–57.

R. M. Nauss (1988), "On the use of internal rate of return in linear and integer programming," *Operations Research Letters*, 7: 285–9.

E. Ronn (1987), "A new linear programming approach to bond portfolio management," *Journal of Financial and Quantitative Analysis*, 22: 439–66.

S. M. Schaefer (1982), "Tax induced clientele effects in the market for British government securities," *Journal of Financial Economics*, 10: 121–59.

D. Stock and D. Simonson (1989), "Tax-adjusted duration for amortizing debt instruments," *Journal of Financial and Quantitative Analysis*, 24: 313–27.

J. Yawitz and W. Marshall (1981), "The shortcomings of duration as a risk measure for bonds," *Journal of Financial Research*, 4: 91–101.

Table 11.2. *Summary of model buy/sell recommendations evaluated as of 7 July 1989*

Portfolio description as of 7/7/89	Buy/sell date	Recommended sale amount	Assumed reinv. %	Incr. in total market value as of 7/7/89	Basis point incr. (annualized)	Assumed reinv. %	Incr. in total market value as of 7/7/89	Basis point incr. (annualized)
1 $15 million charitable foundation	03/06/89	$4,832,529	8.60%	$7,400	15 (45)	10.60%	$7,590	16 (47)
2 $540 million no-load mutual fund	03/06/89	$93,873,228	8.30%	$91,372	10 (29)	10.30%	$95,686	10 (30)
3 $49 million S&L investment	10/31/88	$16,295,537	7.70%	$66,329	41 (60)	9.70%	$134,210	82 (121)
4 $79 million tot. return treas. fund	03/06/89	$20,458,782	8.20%	$116,920	57 (170)	10.20%	$117,399	57 (170)

The "increase in total market value" columns reflect the difference in market value as of 7 July 1989 for outstanding securities, matured principal, and income of the recommended securities to be purchased versus the securities recommended to be sold. The "assumed reinv. %" columns give the assumed annual reinvestment rate for cashflows occurring between the buy/sell date and 7 July 1989. Taxes (including capital gains) for the taxable portfolio 3 have been accounted for in full. Note that in each portfolio the recommended trades were duration neutral and that the mix of Treasury and agency securities remained the same.

Table 11.3. *Portfolio A*

Index	Security	Coupon	Mat. date	Par amt. sold	Orig. par amt.	Price	Yield to mat.	Duration	Sale price	Accrued interest
Securities sold (*denotes amount sold is limited by user supplied upper bound or entire holdings sold)										
EX1–8	*TNOT AAA	13.875	8/15/89	1,000,000.00	1,000,000.00	101.812	9.581	0.444	1,081,125.00	7,282.46
EX1–9	*TNOT AAA	8.875	2/15/96	2,200,000.00	2,200,000.00	97.562	9.358	5.285	2,146,375.00	10,247.93
EX1–10	*TBON AAA	8.125	2/15/98	750,000.00	750,000.00	92.687	9.348	6.394	695,156.25	3,198.38
EX1–14	*FHLB AAA	7.375	12/27/93	1,000,000.00	1,000,000.00	91.469	9.628	4.041	914,687.50	14,135.42
EX1–18	CASH AAA	0.000	3/7/89	23,321.48	877,148.00	100.000	0.000	0.003	23,321.48	
Totals				4,973,321.48	5,827,148.00				4,797,665.23	34,864.18

Index	Security	Coupon	Mat. date	Par amt. purchased	Price	Yield to mat.	Duration	Purchase price	Accrued interest
Securities purchased (*denotes amount purchased is limited by user supplied upper bound)									
NW1–70	TNOT AAA	11.625	1/15/92	280,000.00	105.156	9.514	2.491	294,437.50	4,495.86
NW1–97	*TNOT AAA	11.875	8/15/93	2,000,000.00	108.594	9.460	3.600	2,171,875.00	12,465.47
NW1–117	TNOT AAA	11.250	5/15/95	1,255,000.00	108.469	9.408	4.553	1,361,282.81	43,292.30
NW3–148	FHLB AAA	7.750	4/25/96	1,015,000.00	90.250	9.669	5.357	916,037.50	28,624.41
Totals				4,550,000.00				4,743,632.81	88,878.04

For securities sold	
Price plus accrued interest	$4,832,529.41
Duration in years	4.153
Modified duration in years	3.983
Portfolio duration convexity	0.706%
Internal rate of return	9.40902%
For securities purchased	
Price plus accrued interest	$4,832,510.85
Duration in years	4.152
Modified duration in years	3.943
Portfolio duration convexity	0.706%
Internal rate of return	9.49615%

Table 11.4. *Portfolio A*

Name	Coupon	Mat. date	Par amt.	Actual price in 32nd	Actual price decimal	Ytm on 7/7/89	Sale price	Accd. int.	Duration	Accd. fac.
Securities sold on 3/6/89 and priced as of 7/7/89										
TNOT	13.75	8/15/89	$1,000,000	100.14	100.438	9.269	$1,004,375	$54,425	0.1069	0.7845
TNOT	8.875	2/15/96	$2,200,000	104.05	104.156	8.049	$2,291,437	$76,587	5.0102	0.7845
TNOT	8.125	2/15/98	$750,000	100.07	100.219	8.087	$751,641	$23,903	6.167	0.7845
FHLB	7.375	12/27/93	$1,000,000	97.13	97.406	8.077	$974,062	$2,050	3.876	0.0556
Totals			$4,950,000			8.070	$5,021.516	$156,964	3.967	
							Total market value	$5,178,480		
Secur. bought on 3/6/89 and priced as of 7/7/89										
TNOT	11.625	1/15/92	$280,000	108.08	108.250	7.949	$303,100	$15,556	2.161	0.9558
TNOT	11.875	8/15/93	$2,000,000	113.12	113.375	7.986	$2,267,500	$93,159	3.287	0.7845
TNOT	11.250	5/15/95	$1,255,000	114.25	114.781	8.032	$1,440,505	$20,331	4.49	0.288
FHLB	7.750	4/25/96	$1,015,000	96.16	96.500	8.404	$979,475	$15,733	5.303	0.4
Totals			$4,550,000			8.108	$4,990,580	$144,779	3.950	
							Total market value	$5,135,358		

Sensitivity analysis			
Reinvestment rate for income and maturing principal	8.6%	9.6%	10.6%
For securities sold			
Value of income plus maturing principal for entire portfolio			
From 3/6/89 to 7/7/89	$1,257,981	$1,261,457	$1,264,939
Market value of sold securities on 7/7/89	$5,178,480	$5,178,480	$5,178,480
Total market value as of 7/7/89	$6,436,461	$6,439,937	$6,443,419

For securities purchased			
Value of income plus maturing principal for entire portfolio			
From 3/6/89 to 7/7/89	$1,308,503	$1,312,077	$1,315,651
Market value of purchased securities on 7/7/89	$5,135,358	$5,135,358	$5,135,358
Total market value as of 7/7/89	$6,443,861	$6,447,435	$6,451,009
Difference between recommended and original portfolios	$7,400	$7,498	$7,590

Notes: Refer to table 11.3 for buy/sell data from 3/6/89.
Summary of bond portfolio analysis program trade recommendations of 6 March 1989.
Priced as of 7 July 1989 for a charitable foundation investment portfolio with market value of $15 million.
This spreadsheet analyzes a recommended buy/sell program from 6 March 1989.
Prices were obtained from Bloomberg Financial Services as of late afternoon on 7 July 1989.

Table 11.5. *Portfolio B*

Index	Security	Coupon	Mat. date	Par amt. sold	Orig. par amt.	Price	Yield to mat.	Duration	Sale price	Accrued interest
Securities sold (*denotes amount sold is limited by user supplied upper bound or entire holdings sold)										
EX1–5	*TNOT AAA	14.625	2/15/92	$20,000,000.00	$25,000,000.00	113.281	9.359	2.512	$22,656,250.00	$153,522.10
EX1–8	TNOT AAA	13.125	5/15/94	$15,350,000.00	$53,000,000.00	114.875	9.428	3.887	$17,633,312.50	$617,763.29
EX1–11	TNOT AAA	8.875	7/15/95	$16,375,000.00	$17,700,000.00	97.562	9.388	4.917	$15,975,859.37	$200,729.45
EX1–14	TNOT AAA	8.875	11/15/98	$11,050,000.00	$39,575,000.00	97.250	9.308	6.474	$10,746,125.00	$300,708.05
EX1–15	FHLB AAA	12.000	2/25/94	$8,000,000.00	$8,000,000.00	109.406	9.575	3.929	$8,752,500.00	$29,333.33
EX1–16	FHLB AAA	12.150	12/27/93	$15,000,000.00	$15,000,000.00	109.719	9.572	3.759	$16,457,812.50	$349,312.50
Totals				$85,775,000.00	$158,275,000.00				$92,221,859.37	$1,651,368.73

Index	Security	Coupon	Mat. date	Par amt. purchased	Price	Yield to mat.	Duration	Purchase price	Accrued interest
Securities purchased (*denotes amount purchased is limited by user supplied upper bound)									
NW1–1	CASH AAA	0.000	3/7/89	$9,657.19	100.000	0.000	0.003	$9,657.19	
NW1–44	TNOT AAA	10.750	8/15/90	$2,625,000.00	101.594	9.536	1.370	$2,666,835.94	$14,810.95
NW1–70	*TNOT AAA	11.625	1/15/92	$20,000,000.00	105.156	9.514	2.491	$21,031,250.00	$321,132.60
NW1–97	*TNOT AAA	11.875	8/15/93	$20,000,000.00	108.594	9.460	3.600	$21,718,750.00	$124,654.70
NW1–117	*TNOT AAA	11.250	5/15/95	$20,000,000.00	108.469	9.408	4.553	$21,693,750.00	$689,917.13
NW3–148	FHLB AAA	7.750	4/25/96	$6,825,000.00	90.250	9.669	5.357	$6,159,562.50	$192,474.48
NW3–156	*FHLB AAA	8.250	9/25/96	$20,000,000.00	92.562	9.658	5.472	$18,512,500.00	$237,916.67
Totals				$89,459,657.19				$91,792,305.62	$2,080,906.51

For securities sold	
Price plus accrued interest	$93,873,228.10
Duration in year	4.016
Modified duration in years	3.800
Portfolio duration convexity	0.718%
Internal rate of return	9.41893%
For securities purchased	
Price plus accrued interest	$93,873,212.13
Duration in years	4.014
Modified duration in years	3.818
Portfolio duration convexity	0.710%
Internal rate of return	9.52466%

Table 11.6. *Portfolio B*

Name	Coupon	Mat. date	Par amt.	Actual price in 32nd	Actual price decimal	Ytm. on 7/7/89	Sale price	Accd. int.	Duration	Accd. fac.
Securities sold on 3/6/89 and priced as of 7/7/89										
TNOT	14.625	2/15/92	$20,000,000	116.01	116.031	7.711	$23,206,250	$1,147,331	2.185	0.7845
TNOT	13.125	5/15/94	$15,350,000	120.05	120.156	8.025	$18,443,984	$290,115	3.794	0.288
TNOT	8.875	7/15/95	$16,375,000	104.01	104.031	8.017	$17,035,117	$630,288	4.651	0.8674
TNOT	8.875	11/15/98	$11,050,000	105.02	105.062	8.090	$11,609,406	$141,219	6.547	0.288
FHLB	12.000	2/25/94	$8,000,000	113.31	113.969	8.300	$9,117,500	$351,984	3.621	0.7333
FHLB	12.150	12/27/93	$15,000,000	114.13	114.406	8.233	$17,160,937	$50,666	3.637	0.0556
Totals			$85,775,000			8.052	$96,573,195	$2,611,603	3.834	
							Total market val.	$99,184,798		
Securities bought on 3/6/89 and priced as of 7/7/89										
TNOT	10.750	8/15/90	$2,625,000	102.20	102.625	8.206	$2,693,906	$110,688	1.033	0.7845
TNOT	11.625	1/15/92	$20,000,000	108.08	108.250	7.949	$21,650,000	$1,111,118	2.161	0.9558
TNOT	11.875	8/15/93	$20,000,000	113.12	113.375	7.986	$22,675,000	$931,594	3.287	0.7845
TNOT	11.250	5/15/95	$20,000,000	114.25	114.781	8.031	$22,956,250	$324,000	4.49	0.288
FHLB	7.750	4/25/96	$6,825,000	96.16	96.500	8.433	$6,586,125	$105,788	5.303	0.4
FHLB	8.250	9/25/96	$20,000,000	99.05	99.156	8.404	$19,831,250	$467,528	5.449	0.5667
Totals			$89,450,000			8.159	$96,392,531	$3,050,714	3.824	
							Total market val.	$99,443,246		
Sensitivity analysis										
Reinvestment rate for income and maturing principal							8.3%	9.3%	10.3%	

For securities sold			
Value of income plus maturing principal for entire portfolio			
From 3/6/89 to 7/7/89	$11,959,326	$11,974,807	$11,990,289
Market value of sold securities on 7/7/89	$99,184,798	$99,184,798	$99,184,798
Total market value as of 7/7/89	$111,144,124	$111,159,606	$111,175,087
For securities purchased			
Value of income plus maturing principal for entire portfolio			
From 3/6/89 to 7/7/89	$11,792,251	$11,809,887	$11,827,527
Market value of purchased securities on 7/7/89	$99,443,246	$99,443,246	$99,443,246
Total market value as of 7/7/89	$111,235,496	$111,253,133	$111,270,773
Difference between recommended and original portfolios	$91,372	$93,527	$95,686

Notes: Refer to table 11.5 for buy/sell data from 3/6/89.
Summary of bond portfolio analysis program trade recommendations of 6 March 1989.
Priced as of 7 July 1989 for a no-load mutual fund with market value of $540 million.
This spreadsheet analyzes a recommended buy/sell program from 6 March 1989.
Prices were obtained from Bloomberg Financial Services as of late afternoon on 7 July 1989.

Table 11.7. *Portfolio C*

Index	Security	Coupon	Mat. date	Par amt. sold	Orig. par amt.	Price	Yield to mat.	Duration	Sale price	Accrued interest
Securities sold (*denotes amount sold is limited by user supplied upper bound or entire holdings sold)										
EX1–14	*TNOT AAA	7.375	1/31/90	$1,000,000.00	$1,000,000.00	98.813	8.379	1.999	$988,130.00	18,437.50
EX1–15	*TNOT AAA	6.500	2/15/90	$1,000,000.00	$1,000,000.00	97.719	8.388	1.245	$977,190.00	13,600.54
EX1–16	TNOT AAA	7.125	2/28/90	$650,000.00	$1,000,000.00	98.407	8.401	1.278	$639,645.50	7,804.04
EX1–17	*TNOT AAA	7.375	3/31/90	$1,000,000.00	$1,000,000.00	98.719	8.346	1.361	$987,190.00	6,280.91
EX1–18	*TNOT AAA	7.625	4/30/90	$1,000,000.00	$1,000,000.00	98.969	8.371	1.442	$989,690.00	
EX1–19	*TNOT AAA	7.875	5/15/90	$1,000,000.00	$1,000,000.00	99.282	8.377	1.427	$992,820.00	36,165.08
EX1–20	*TNOT AAA	7.875	8/15/90	$1,000,000.00	$1,000,000.00	99.157	8.380	1.678	$991,570.00	16,477.58
EX1–21	*TNOT AAA	8.000	11/15/90	$1,000,000.00	$1,000,000.00	99.313	8.370	1.855	$993,130.00	36,739.13
EX1–22	*TNOT AAA	7.375	2/15/91	$1,000,000.00	$1,000,000.00	97.938	8.374	2.119	$979,380.00	15,431.39
EX1–23	*TNOT AAA	8.125	5/15/91	$1,000,000.00	$1,000,000.00	99.407	8.386	2.259	$994,070.00	37,313.18
EX1–24	*TBON AAA	6.750	2/15/93	$300,000.00	$300,000.00	94.219	8.377	3.728	$282,657.00	4,237.09
EX1–26	*FHLB AAA	8.125	11/27/89	$1,000,000.00	$1,000,000.00	99.438	8.673	1.016	$994,380.00	34,531.25
EX1–27	*FHLB AAA	8.250	12/26/89	$1,000,000.00	$1,000,000.00	99.594	8.609	1.095	$995,940.00	28,416.67
EX1–28	FHLB AAA	7.300	3/26/90	$750,000.00	$1,000,000.00	98.094	8.762	1.348	$735,705.00	5,170.83
EX1–29	*FHLB AAA	7.700	4/25/90	$1,000,000.00	$1,000,000.00	98.657	8.681	1.427	$986,570.00	1,069.44
EX1–30	*FHLB AAA	7.700	4/25/90	$500,000.00	$500,000.00	98.681	1.427	1.427	$493,285.00	534.72
EX1–31	*FHLB AAA	8.300	1/25/91	$1,000,000.00	$1,000,000.00	99.407	8.586	2.043	$994,070.00	21,902.78
EX1–32	*FHLB AAA	7.650	2/25/91	$1,000,000.00	$1,000,000.00	98.219	8.503	2.141	$982,190.00	13,812.50
Totals				$16,200,000.00	$16,800,000.00				$15,997,612.50	297,924.64

Index	Security	Coupon	Mat. date	Par amt. purchased	Price	Yield to mat.	Duration	Purchase price	Accrued interest
Securities purchase (*denotes amount purchased is limited by user supplied upper bound)									
NW1–1	*TNOT AAA	10.625	12/31/88	$2,000,000.00	100.532	7.125	0.167	$2,010,640.00	$71,025.82
NW1–2	*TNOT AAA	6.250	12/31/88	$2,000,000.00	99.876	6.863	0.167	$1,997,520.00	$41,779.89
NW1–3	*TNOT AAA	6.125	1/31/89	$2,000,000.00	99.751	7.031	0.252	$1,995,020.00	$30,625.00

NW1–4	TNOT AAA	14.625	1/15/89	$1,550,000.00	101.595	6.518	0.208	$1,574,722.50	$66,527.85
NW1–5	*TNOT AAA	8.000	2/15/89	$2,000,000.00	100.189	7.216	0.293	$2,003,780.00	$33,478.26
NW1–88	CASH AAA	0.000	11/1/88	$8,342.87	100.000	0.000	0.003	$8,342.87	$0.00
NW3–66	*FHLB AAA	10.750	5/25/93	$2,000,000.00	107.220	8.794	3.605	$2,144,400.00	$92,569.44
NW3–69	*FHLB AAA	11.700	7/26/93	$2,000,000.00	111.032	8.795	3.725	$2,220,640.00	$61,100.00
NW3–71	FHLB AAA	11.950	8/25/93	$1,700,000.00	112.157	8.793	3.795	$1,906,669.00	$36,679.86
Totals				$15,258,342.87				$15,861,734.37	$433,786.13

Index	Security	Coupon	Mat. date	Tax basis	Sale price	Sale Tax basis	Sale proceeds	Cap. gain or loss	Cap. gains Tax paid (+)
Capital gains tax report for securities sold:									
Assumes a marginal effective capital gains tax rate of 31.000%									
EX1–14	TNOT	7.375	1/31/90	99.969	98.813	$999,690.00	$988,130.00	–$11,560.00	–$3,583.60
EX1–15	TNOT	6.500	2/15/90	99.955	97.719	$999,550.00	$977,190.00	–$22,360.00	–$6,931.60
EX1–16	TNOT	7.125	2/28/90	99.874	98.407	$649,181.00	$639,645.50	–$9,535.50	–$2,956.00
EX1–17	TNOT	7.375	3/31/90	99.981	98.719	$999,810.00	$987,190.00	–$12,620.00	–$3,912.20
EX1–18	TNOT	7.625	4/30/90	99.980	98.969	$999,800.00	$989,690.00	–$10,110.00	–$3,134.10
EX1–19	TNOT	7.875	5/15/90	99.954	99.282	$999,540.00	$992,820.00	–$6,720.00	–$2,083.20
EX1–20	TNOT	7.875	8/15/90	99.900	99.157	$999,000.00	$991,570.00	–$7,430.00	–$2,303.30
EX1–21	TNOT	8.000	11/15/90	100.000	99.313	$1,000,000.00	$993,130.00	–$6,870.00	–$2,129.70
EX1–22	TNOT	7.375	2/15/91	99.691	97.938	$996,910.00	$979,380.00	–$17,530.00	–$5,434.30
EX1–23	TNOT	8.125	5/15/91	99.723	99.407	$997,230.00	$994,070.00	–$3,160.00	–$979.60
EX1–24	TBON	6.750	2/15/93	99.799	94.219	$299,397.00	$282,657.00	–$16,740.00	–$5,189.40
EX1–26	FHLB	8.125	11/27/89	100.000	99.438	$1,000,000.00	$994,380.00	–$5,620.00	–$1,742.20
EX1–27	FHLB	8.250	12/26/89	100.000	99.594	$1,000,000.00	$995,940.00	–$4,060.00	–$1,258.60
EX1–28	FHLB	7.300	3/26/90	100.000	98.094	$750,000.00	$735,705.00	–$14,295.00	–$4,431.45
EX1–29	FHLB	7.700	4/25/90	100.000	98.657	$1,000,000.00	$986,570.00	–$13,430.00	–$4,163.30
EX1–30	FHLB	7.700	4/25/90	100.000	98.657	$500,000.00	$493,285.00	–$6,715.00	–$2,081.65
EX1–31	FHLB	8.300	1/25/91	100.000	99.407	$1,000,000.00	$994,070.00	–$5,930.00	–$1,838.30
EX1–32	FHLB	7.650	2/25/91	100.000	98.219	$1,000,000.00	$982,190.00	–$17,810.00	–$5,521.10
Totals								–$192,495.50	–$59,673.61

Table 11.7 (*cont.*)

Index	Security	Coupon	Mat. date	Tax basis	Sale price	Sale Tax basis	Sale proceeds	Cap. gain or loss	Cap. gains Tax paid (+)
Thus a tax credit of $59,673.61 is received									
For securities sold									
Price plus accrued interest					$16,295,537.14				
Duration in years					1.602				
Modified duration in years					1.543				
Portfolio duration convexity					0.073%				
Internal rate of return					8.47178%				
For securities purchased									
Price plus accrued interest					$16,295,520.50				
Duration in years					1.600				
Modified duration in years					1.515				
Portfolio duration convexity					0.232%				
Internal rate of return					8.65008%				

Table 11.8. *Portfolio C*

Name	Coupon	Mat. date	Par amt.	Actual price in 32nd	Actual price decimal	YTM on 7/7/89	Sale price	Accd. int.	Duration	Accd. fac.
Securities sold on 10/31/88 and priced as of 7/7/89										
TNOT	7.375	1/31/90	$1,000,000	99.18	99.563	8.166	$995,625	$31,985	0.552	0.8674
TNOT	6.500	2/15/90	$1,000,000	98.31	98.969	8.256	$989,688	$25,496	0.595	0.7845
TNOT	7.125	2/28/90	$650,000	99.11	99.344	8.159	$645,734	$16,235	0.63	0.7011
TNOT	7.375	3/31/90	$1,000,000	99.10	99.312	8.335	$993,125	$19,747	0.714	0.5355
TNOT	7.625	4/30/90	$1,000,000	99.19	99.594	8.127	$995,938	$14,091	0.796	0.3696
TNOT	7.875	5/15/90	$1,000,000	99.25	99.781	8.125	$997,813	$11,340	0.836	0.2880
TNOT	7.875	8/15/90	$1,000,000	99.25	99.781	8.073	$997,813	$30,890	1.051	0.7845
TNOT	8.000	11/15/90	$1,000,000	100.00	100.000	8.000	$1,000,000	$11,520	1.301	0.2880
TNOT	7.375	2/15/91	$1,000,000	99.05	99.156	7.936	$991,562	$28,928	1.506	0.7845
TNOT	8.125	5/15/91	$1,000,000	100.10	100.312	7.931	$1,003,125	$11,700	1.741	0.2880
TNOT	6.750	2/15/93	$300,000	96.15	96.469	7.890	$289,406	$7,943	3.173	0.7845
FHLB	8.125	11/27/89	$1,000,000	99.24	99.750	8.711	$997,500	$9,019	0.392	0.2220
FHLB	8.250	12/26/89	$1,000,000	99.27	99.844	8.575	$998,437	$2,520	0.472	0.0611
FHLB	7.300	3/26/90	$750,000	99.05	99.156	8.506	$743,672	$15,360	0.7	0.5611
FHLB	7.700	4/25/90	$1,000,000	99.14	99.438	8.417	$994,375	$15,400	0.781	0.4000
FHLB	7.700	4/25/90	$500,000	99.14	99.438	8.417	$497,188	$7,700	0.781	0.4000
FHLB	8.300	1/25/91	$1,000,000	100.01	100.031	8.273	$1,000,313	$37,350	1.436	0.9000
FHLB	7.650	2/25/91	$1,000,000	98.29	98.906	8.370	$989,063	$28,049	1.529	0.7333
Totals			$16,200,000			8.174	$16,120,375	$325,273	0.987	
							Total market val.	$16,445,648		

Table 11.8. (*cont.*)

Name	Coupon	Mat. date	Par amt.	Actual price in 32nd	Actual price decimal	YTM on 7/7/89	Sale price	Accd. int.	Duration	Accd. fac.
Secur. bought on 10/31/88 and priced as of 7/7/89										
FHLB	10.750	5/25/93	$2,000,000	108.15	108.469	8.156	$2,169,375	$25,080	3.269	0.2333
FHLB	11.700	7/26/93	$2,000,000	111.11	111.344	8.341	$2,226,875	$104,645	3.233	0.8944
FHLB	11.950	8/25/93	$1,700,000	112.17	112.531	8.300	$1,913,031	$74,485	3.306	0.7333
Totals			$5,700,000			8.266	$6,309,281	$204,209	3.267	
Total market val.								$6,513,491		
Sensitivity analysis										
Reinvestment rate for income and maturing principal							7.7%	8.7%	9.7%	
For securities sold										
Value of income plus maturing principal for entire portfolio										
From 10/31/88 to 7/7/89							$20,851,163	$20,926,098	$21,001,215	
Market value of sold securities on 7/7/89							$16,445,648	$16,445,648	$16,445,648	
Total market value as of 7/7/89							$37,296,811	$37,371,746	$37,446,863	
For securities purchased										
Value of income plus maturing principal for entire portfolio										
From 3/6/89 to 7/7/89							$30,849,649	$30,958,487	$31,067,583	
Market value of purchased securities on 7/7/89							$6,513,491	$6,513,491	$6,513,491	
Total market value as of 7/7/89							$37,363,140	$37,471,978	$37,581,074	
Difference between recommended and original portfolios							$66,329	$100,232	$134,210	

Notes: Refer to table 11.7 for buy/sell data from 10/31/88.
Summary of bond portfolio analysis program trade recommendations of 31 October 1988.
Priced as of 7 July 1989 for a saving and loan investment portfolio with market value of $49 million.
This spreadsheet analyzes a recommended buy/sell program from 31 October 1988.
Prices were obtained from Bloomberg Financial Services as of late afternoon on 7 July 1989.

Table 11.9. *Portfolio D*

Index	Security	Coupon	Mat. date	Par amt. sold	Orig. par amt.	Price	Yield to mat.	Duration	Sale price	Accrued interest
Securities sold (*denotes amount sold is limited by user supplied upper bound or entire holdings sold)										
EX1-3	*TNOT AAA	9.000	1/31/91	$6,500,000.00	$6,500,000.00	99.062	9.541	1.780	$6,439,030.00	$54,944.75
EX1-5	*TNOT AAA	8.875	2/15/96	$4,500,000.00	$4,500,000.00	97.593	9.352	5.285	$4,391,685.00	$20,961.67
EX1-9	*TNOT AAA	9.000	5/15/98	$5,500,000.00	$5,500,000.00	98.187	9.294	6.244	$5,400,285.00	$151,781.77
EX1-11	*TNOT AAA	8.875	11/15/98	$4,000,000.00	$4,000,000.00	97.281	9.303	6.475	$3,891,240.00	$108,853.59
Totals				$20,500,000.00	$20,500,000.00				$20,122,240.00	$336,541.78

Index	Security	Coupon	Mat. date	Par amt. purchased	Price	Yield to mat.	Duration	Purchase price	Accrued interest
Securities purchased (*denotes amount purchased is limited by user supplier upper bound)									
NW1-2	CASH AAA	0.000	3/7/89	$508.57	100.000	0.000	0.003	$508.57	0.00
NW2-73	*TNOT AAA	6.625	5/15/92	$10,000,000.00	92.250	9.488	2.858	$9,225,000.00	203,142.27
NW2-89	TNOT AAA	7.625	5/15/93	$1,395,000.00	93.812	9.440	3.562	$1,308,684.37	32,615.83
NW2-124	TNOT AAA	7.250	11/15/96	$4,200,000.00	88.687	9.341	5.770	$3,724,875.00	93,368.78
NW2-128	TNOT AAA	8.125	2/15/98	$5,050,000.00	92.906	9.309	6.397	$4,691,765.62	21,535.74
NW2-171	TBON AAA	9.875	11/15/15	$1,050,000.00	107.187	9.148	9.990	$1,125,468.75	31,793.68
Totals				$21,695,508.57				$20,076,302.32	382,456.30

For securities sold	
Price plus accrued interest	$20,458,781.78
Duration in years	4.665
Modified duration in years	4.466
Portfolio duration convexity	1.020%
Internal rate of return	9.33577%
For securities purchased	
Price plus accrued interest	$20,458,758.62
Duration in years	4.666
Modified duration in years	4.494
Portfolio duration convexity	1.016%
Internal rate of return	9.34911%

Table 11.9. *Portfolio D*

Index	Security	Coupon	Mat. date	Par amt. sold	Orig. par amt.	Price	Yield to mat.	Duration	Sale price	Accrued interest
Securities sold (*denotes amount sold is limited by user supplied upper bound or entire holdings sold)										
EX1–3	*TNOT AAA	9.000	1/31/91	$6,500,000.00	$6,500,000.00	99.062	9.541	1.780	$6,439,030.00	$54,944.75
EX1–5	*TNOT AAA	8.875	2/15/96	$4,500,000.00	$4,500,000.00	97.593	9.352	5.285	$4,391,685.00	$20,961.67
EX1–9	*TNOT AAA	9.000	5/15/98	$5,500,000.00	$5,500,000.00	98.187	9.294	6.244	$5,400,285.00	$151,781.77
EX1–11	*TNOT AAA	8.875	11/15/98	$4,000,000.00	$4,000,000.00	97.281	9.303	6.475	$3,891,240.00	$108,853.59
Totals				$20,500,000.00	$20,500,000.00				$20,122,240.00	$336,541.78

Index	Security	Coupon	Mat. date	Par amt. purchased	Price	Yield to mat.	Duration	Purchase price	Accrued interest
Securities purchased (*denotes amount purchased is limited by user supplier upper bound)									
NW1–2	CASH AAA	0.000	3/7/89	$508.57	100.000	0.000	0.003	$508.57	0.00
NW2–73	*TNOT AAA	6.625	5/15/92	$10,000,000.00	92.250	9.488	2.858	$9,225,000.00	203,142.27
NW2–89	TNOT AAA	7.625	5/15/93	$1,395,000.00	93.812	9.440	3.562	$1,308,684.37	32,615.83
NW2–124	TNOT AAA	7.250	11/15/96	$4,200,000.00	88.687	9.341	5.770	$3,724,875.00	93,368.78
NW2–128	TNOT AAA	8.125	2/15/98	$5,050,000.00	92.906	9.309	6.397	$4,691,765.62	21,535.74
NW2–171	TBON AAA	9.875	11/15/15	$1,050,000.00	107.187	9.148	9.990	$1,125,468.75	31,793.68
Totals				$21,695,508.57				$20,076,302.32	382,456.30

For securities sold	
Price plus accrued interest	$20,458,781.78
Duration in years	4.665
Modified duration in years	4.466
Portfolio duration convexity	1.020%
Internal rate of return	9.33577%
For securities purchased	
Price plus accrued interest	$20,458,758.62
Duration in years	4.666
Modified duration in years	4.494
Portfolio duration convexity	1.016%
Internal rate of return	9.34911%

For securities purchased			
Value of income plus maturing principal for entire portfolio from 3/6/89 to 7/7/89	$1,496,380	$1,498,528	$1,500,677
Market value of purchased securities on 7/7/89	$21,715,057	$21,715,057	$21,715,057
Total market value as of 7/7/89	$23,211,437	$23,213,585	$23,215,733
Difference between recommended and original portfolios	$116,920	$117,159	$117,399

Notes: Refer to table 11.9 for buy/sell data from 3/6/89.
Summary of bond portfolio analysis program trade recommendations of 6 March 1989.
Priced as of 7 July 1989 for a total return US Treasury fund with market value of $79 million.
This spreadsheet analyzes a recommended buy/sell program from 6 March 1989. (No high coupons purchased).
Prices were obtained from Bloomberg Financial Services as of late afternoon on 7 July 1989.

12 Scenario immunization

RON S. DEMBO

1 Introduction

Traditionally portfolio immunization has referred to the problem of finding a set of bonds whose present value matches that of a predefined set of liabilities. This problem arises, for example, in the context of pension fund management where one seeks a way of investing a portion of a fund in a manner that will protect its value relative to the fund's projected liabilities. In this way, regardless of external factors such as interest-rate changes, the fund's assets and liabilities will have similar values. Any surplus funds may then be used to ensure capital growth.

In the early 1980s an extremely volatile interest-rate environment and high levels of interest rates gave prominence to models for portfolio immunization. In particular, pension fund managers found that they were valuing assets and liabilities inconsistently. The effect was dramatic when short-term rates approached 20% whereas internal actuarial discount rates were conservatively set to approximately 5%.

Since that time the use of "proper evaluation techniques" has been legislated in the United States and, more recently, in certain European countries as well. Optimization models for portfolio immunization are now used routinely for managing fixed-income portfolios.

It is curious to note that immunization models are almost identical across investment banking environments. The same models that were developed in the late 1970s are still used today with little or no variation. Typically they are formulated as linear programming models that maximize duration-weighted yield or minimize cost subject to present value, duration, and convexity constraints. Examples of such formulations are given in Dahl *et al.* and Nauss (chapters 1 and 11 in this volume). There has been almost no modeling innovation in this area since its inception.

This chapter is devoted to an analysis of the models that are currently used. Our conclusion is that, because of their deterministic nature, they are inherently limited and require extensive massaging to give results that conform to accepted intuition.

As an alternative we describe stochastic models and an approach which we refer to as *Scenario Immunization.* These models, first introduced in Dembo (1991), more closely address the problem of immunization and provide more acceptable results without massaging. On the downside, they are slightly more complicated and may require more information than those currently in use.

Immunization is also a far richer problem area than is implied by the classical statement of the problem. The definition of portfolio immunization that we use here is a more general one. In this chapter an immunized portfolio refers to one whose behavior is the same as that of some given target portfolio under a predefined set of risk measures and criteria, for the foreseeable future.

As such, the classical definition of immunization may be viewed as a special case concerned with a target portfolio made up of a set of liabilities and the risk measure being solely interest-rate risk. In the definition we prefer to use, a portfolio is said to be immunized against certain risk measures, for example, interest-rate, volatility, market sector exposure, etc. The definition is also meant to cover the entire spectrum of synthetic securities, that is portfolios that behave "identically" to some target portfolio.

For example, in our terminology, the securitization of fixed-income assets may be expressed as an immunization problem. In this case cashflows from a pool of fixed-income securities are used to fund the expected flows generated by the new assets that are created. An example of this is the construction of Collateralized Mortgage Obligations (CMOs) from a pool of mortgage-backed securities.

Another example of an immunized portfolio might be one that tracks a prespecified index. In such a case the risks with respect to which the portfolio is immunized are all the market forces that might affect its value.

2 A model for structuring immunized portfolios

The immunization model, stated in English, is to find a portfolio, referred to as the desired portfolio, whose (present) value matches the (present) value of a given target portfolio, for the foreseeable future. Usually, in addition and depending on the context, other requirements might be imposed such as:

the desired portfolio should have as high a yield as possible; or
the total return of the desired portfolio should be as high as possible over a given horizon; or
the cost of the desired portfolio should be minimal.

We can now write down a mathematical description that approximates the specifications of an immunization model.

Let x_j denote the amount of instrument j in the desired portfolio and V_j denote the present value of the future coupon streams of instrument j (one may find it easier to follow this presentation by assuming that the instruments are all simple bonds); V_T denotes the present value of the target portfolio. The "English" statement of the model may now be translated into an equivalent mathematical one.

Find x_j that satisfy	(Find how many instruments of type j to include in the portfolio so that. . .)
$V(x) \equiv \sum_j V_j x_j = V_T$	(the present value of the desired portfolio equals the present value of the target portfolio)

. . . for the foreseeable future and under context dependent criteria.

We have deliberately neglected to translate the "foreseeable future" and "context dependent criteria" requirements into a mathematical format. It is precisely here where our approach differs from the standard models. The translation of these requirements into mathematics is where we think existing models are deficient.

Just how do we translate "find a portfolio whose (present) value matches the (present) value of a target portfolio, for the forseeable future"?

As time passes, factors in the economy that will affect fixed–income instruments, amongst others, are changes in interest rates and credit risk. Yield curves shift in a complex stochastic manner as do exchange rates and other factors that are perhaps difficult to describe in mathematical terms. Yet these movements cannot be ignored in an immunization since they could have a significant effect on the desired and target portfolios. (For a discussion of the types of risks that one might need to consider refer to Dahl, Meeraus, and Zenios, chapter 1 in this volume.)

For the most part, existing models implicitly or explicitly focus on interest-rate yield-curve shifts as the primary cause of uncertainty. A portfolio whose present value matches that of a target portfolio today might find a significant imbalance tomorrow if interest rates change. This is true for changes in any of the factors influencing the portfolio's value. Such changes pose a serious problem for pension plans, for example. Each pension plan consists of a target portfolio made up of a projected stream of future payments (liabilities) and a portfolio that has been invested so as to be able to cover these future payments. Interest-rate changes occur for many reasons, political as well as economic, but in all cases these changes are environmental, i.e., out of the control of the pension plan manager. So that, without protection (immunization) a plan may lose its value relative to the projected liabilities.

In this context, therefore, if we were able to *protect value under changing interest-rate scenarios* then this could be viewed as a surrogate for the "*foreseeable future*" requirement.

One way of approximately achieving this is to match not only the present values but the rate of change of present value with respect to average yield. The duration of a portfolio measures its sensitivity to small changes in the average yield of the portfolio. Thus if the asset and liability portfolios have the same duration then they will exhibit similar same interest-rate sensitivity. A (small) change in interest rates will result in approximately the same change in assets and liabilities if the durations of the two are equal. Therefore, a matched portfolio will remain so for small changes in the term structure of interest rates.

Let $d(x)$ denote the duration of the desired portfolio; d_j denote the duration of the jth asset in this portfolio, and d_T the duration of the target portfolio. Then, in mathematical terms, the statement that the asset and liability portfolios have the same durations may be expressed as:

$$d(x) = \sum_j d_j x_j = d_T$$

This constraint is often referred to as duration matching.

Unfortunately, it is precisely for hedging against large and non-parallel shifts in the term structure of interest rates that one needs to immunize. There is no guarantee that duration matching will accomplish this.

If we were to use cost as a criterion for selecting the desired portfolio then one possible formulation of the problem stated so far might be:

Find x_j that (unknown variables)

minimize: $\sum_j c_j x_j$ (minimum cost objective)

subject to: $\sum_j V_j x_j = V_T$ (present-value matching)

$\sum_j d_j x_j = d_T$ (duration matching)

$l_j \leq x_j \leq u_j$ (position limits)

This is the basic core of portfolio immunization models that are currently used to manage billions of dollars worldwide. In practise, minor modifications such as additional constraints on convexity are usually added. The primary reason for adding such restrictions is to avoid frequent rebalancing in volatile markets. A serious difficulty with such models is the fact that duration is often difficult to compute or define for many fixed-income instruments, thereby limiting the utility of these models.

When the data of this model, namely the present value and duration of instrument $j(V_j,d_j)$ are known with certainty, very efficient mathematical techniques for obtaining a solution, x_j, exist.

The model we have described would be adequate (as it has been for 99% of the sellers and users of portfolio immunization) were it not for the fact that a more technical analysis shows a number of serious deficiencies.

In the above model, for reasonable choices of the position limits, a solution would always contain only two bonds, regardless of how complicated the liabilities might be! This result violates our well-founded intuition that a properly chosen, diverse portfolio would be our best hedge against interest-rate movements.

The most glaring deficiency is the assumption that present values and durations are constant over the period for which immunization is sought. It is precisely when there are radical changes in interest rates that one needs to ensure that one's assets do not devalue radically with respect to ones liabilities. Assuming that present value and durations remain constant will definitely not provide protection in such circumstances. A fundamental observation is therefore that the data of this model are uncertain (random).

A more in-depth view

One way to force the model to generate a diverse and hence more believable solution is to impose unrealistically narrow position limits or other constraints which restrict the quantity of a given bond or collection of bonds that may be selected in a solution. This is precisely what is done in practise to obtain solutions that money managers will believe, since their experience would dictate the choice of a diverse portfolio. The practice of imposing constraints that force a given outcome, however common, is extremely poor from the mathematical modeling point of view. In effect, the model is massaged to give the behavior that we expect or are prepared to believe rather than providing such a solution naturally. It is the model that is inherently deficient and not our intuition. No amount of massaging will remove this deficiency.

Unfortunately it takes a very technical money manager to recognize that the diverse portfolio suggested by the output of an immunization model is the result of an appropriate model and not because the analyst has fudged the position limits in order to force diversity.

Getting back to our immunization model, the primary problem with the mathematical formulation above is that the data of the model, namely the present value and duration of instrument $j(V_j,d_j)$, are values that will be affected by changes in interest rates (the yield curve). Thus for the foreseeable future we cannot assume that these are constant values unless interest rates are likely to remain constant over the horizon being considered. Thus, precisely when we need protection, the above model does not give it. It will provide some protection due to the duration constraint, however, only in environments where interest rates are not very volatile.

This does not mean that such a model is useless. Rather, it means that the portfolio it suggests is suboptimal and must be monitored closely and that the immunization policy might have to be altered frequently (rebalanced) during the period in which protection is sought. This could therefore be costly in a volatile environment. Unfortunately, most people selling or purchasing immunization do not take into consideration the potential high expected trading costs associated with managing and rebalancing an immunized portfolio.

3 A stochastic immunization model

A primary problem with the immunization model as formulated here is that the data of the model vary over time. The present value of a stream of cashflows depends on the discount rates that are used and these will vary as the term structure of interest rates varies. In particular, each change in the term structure could cause the present value and durations of assets and liabilities to move in different and unpredictable ways. For each realization of the yield curve, the coefficients of the model change and so does the optimal solution, sometimes radically. The model does not hedge against uncertainty in interest rates. It does not calculate a tradeoff between the cost of an immunized portfolio and its quality, measured in terms of how well it hedges against possible interest-rate scenarios or alternatively, how frequently it will have to be rebalanced under various volatility and tracking assumptions.

Instead of attempting to hedge against uncertainty in future interest rates by matching the durations of the target and desired portfolios it is best to take a step back to our original statement of the immunization problem and rephrase it as a stochastic system.

Find x_j that satisfy:

$$V(x) \equiv \sum_j V_j x_j = V_T$$

. . . under context dependent criteria.

The present value V_j of instrument j's expected future cash flows is an uncertain quantity which takes on values consistent with the interest rates expected in the foreseeable future.

This brings us one step closer to a model that will produce the desired tradeoffs. However, it is still necessary to elaborate on the precise manner in which uncertainty in the coefficients may be modeled.

Although this more accurately reflects the situation, it moves us from the realm of deterministic systems to stochastic (uncertain) systems. No longer do we necessarily have efficient ways of computing a solution. The situation is further complicated because there are many ways in which the stochastic aspects of the problem may be modeled, each of which has its own associated benefits/drawbacks from a computational point of view. One thing is certain, however, a stochastic model will produce a diverse portfolio naturally, that is, when diversity is appropriate.

An example

To understand the differences between a deterministic immunization model and a stochastic one let's look at a simple example.

Assume our liability stream was a large single payment of $3,898,766 that was due exactly 4.87 years from now.

Assume that the only assets available to us were:

one-year and thirty-year zero coupon bonds

a certificate of deposit that we could purchase today that would pay exactly the amount of the liability at exactly year 4.87.

These could be represented schematically as follows.

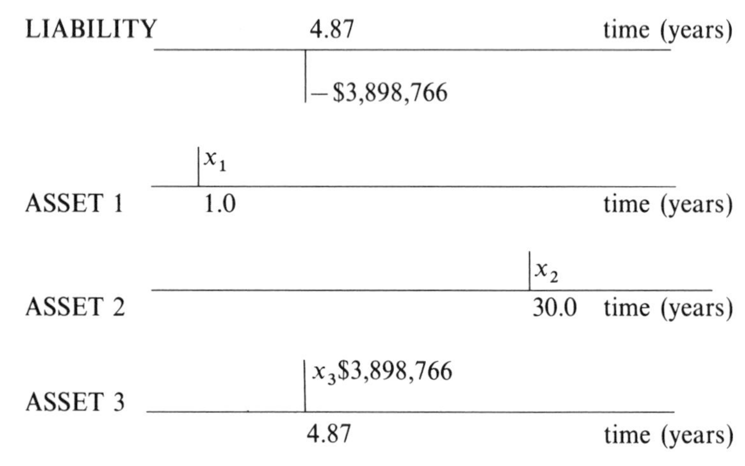

The duration of our liability is 4.87 years and of our assets 1, 30 and 4.87 years respectively.

The immunization problem is to find an optimal (cheapest) combination of assets 1, 2, and 3 (x_1,x_2,x_3) that will have the same present value and duration as the liability, regardless of changes in interest rates. Since ASSET 3 is a customized certificate of deposit that we must purchase from an insurance company or a bank, it is likely to be expensive relative to the other assets which are available on the open market. A portfolio consisting of one unit of ASSET 3 and zero of the other assets does, however, meet the requirements of our problem definition perfectly. In this case the present value and duration of our assets and liabilities will be exactly the same regardless of what happens to interest rates.

This is where optimization enters into the picture. It is quite likely that one would get cheaper protection against interest-rate fluctuations and still meet the requirements by choosing some combination of assets 1 and 2 to immunize the liability. An optimization model will always pick such a solution if it were cheaper. The deterministic model above will pick such a solution regardless of how volatile interest rates are over the period under consideration. A well-formulated stochastic model would pick a combination of ASSETS 1, 2, and 3 in a manner that will depend on how volatile we assume interest rates to be over the immunization period. This is because a stochastic model attaches a cost to the expected degree of mismatching over the immunization horizon. For low volatility, the solution would look much like a deterministic one, that is, use the appropriate combination of ASSETS 1 and 2 because they are cheap and the likelihood that a matched portfolio will become unmatched is low. The higher the volatility, the more of ASSET 3 one would see in the solution. This is because ASSET 3 provides a perfect, albeit costly, hedge. Thus the stochastic model is a hedge against uncertainty. The more the uncertainty the costlier the hedge. It explicitly trades off the risk of mismatching versus the cost of a better hedge. A deterministic model completely ignores this tradeoff which is the essence of immunization.

Traditionally, stochastic optimization models have been very complicated, requiring inordinate amounts of computing power for their solution. This probably explains why 99% of immunization is still done with the deterministic model presented here. However, there have been recent developments which show that there are simple stochastic models which are able to capture the tradeoffs discussed in this chapter and require only marginally more computational time than a deterministic model. One such development is discussed below.

4 A scenario optimization model for portfolio immunization

Scenario optimization, Dembo (1991), is a simple approach to optimization problems under uncertainty in which the uncertain component may be represented by discrete scenarios. In this approach, the problem is solved for each of the scenarios as a deterministic problem. The scenario solutions are then combined into a single policy using a "tracking" or "coordination" model. The approach is computationally simple and easy to understand. Because of its generality, it can handle multiple competing objectives, complex stochastic constraints, and may be applied in contexts other than optimization such as in the solution of stochastic systems of equations.

A scenario, in this context, refers to one single realization of uncertain future events. For example, one scenario might be some particular yield curve, since this is equivalent to specifying the coefficients of the stochastic linear system we have defined. Scenarios might also be far more complicated. An example is the tuple (yield-curve time). Use of scenarios to express uncertainty, however, is a very powerful tool since it enables us to deal with situations which are very difficult to quantify. An example of this is the case of immunization using instruments whose cashflows are interest-rate path dependent.

The essential idea behind scenario optimization is to solve the stochastic model for each possible scenario and to combine the solutions into a single feasible policy using a "tracking" or "coordination" model. For details refer to Dembo (1991). The underlying optimization model is deterministic under the assumption that a particular scenario occurs with probability 100% and may therefore be solved using known methods. The innovation is the method for combining scenario solutions.

A deterministic scenario formulation of the portfolio immunization problem, assuming a cost criterion, would have the general form:

$$\begin{aligned} &v^s \equiv \text{Minimize: } (c^s)^{\mathrm{T}} x \\ &\text{subject to: } l \leq x \leq u \\ &V^s(x) = V^s_T \end{aligned}$$

where:

$V^s_j \equiv$ present value of instrument j under discount scenario s,

$V^s(x) \equiv \sum_{j \in J} V^s_j x_j$; present value of the unknown portfolio x under discount scenario s;

$V^s_T \equiv$ present value of target portfolio under discount scenario s,

$x_j \equiv$ amount of instrument j in the optimal immunizing portfolio,

$c^s \equiv$ the vector of instrument market prices under scenario s,
$u_j \equiv$ maximum units of instrument j allowed,
$l_j \equiv$ minimum units of instrument j allowed,
$J \equiv$ set of instruments j available for immunizing.

The uncertainty in this problem stems from the present-value coefficients, V_j^s. These fluctuate as interest rates (and hence discount rates) change over time. The deterministic model above, which assumes a single scenario s, is very easy to compute. However, notice that, since it is a knapsack problem, if the bounds (u_j and l_j) are large enough, the solution will contain only one instrument, regardless of the scenario chosen. Moreover, the optimal immunizing instrument will typically be different for different scenarios. As mentioned previously, this solution may be diversified to two instruments if a duration constraint is added and up to three if both duration and convexity constraints are present. However, in all cases they are included only as an approximate hedge against non-parallel yield-curve shifts.

It is clear that a solution involving one or two instruments is not satisfactory since we know that this is likely to track poorly if the assumed discount scenario does not occur. We expect an "optimal" solution to contain a diverse portfolio in order to hedge against the uncertain future long- and short-term interest rates. Notice also that, whereas the solution to any single-scenario subproblem does not appear to offer a good solution to the immunization problem, one may be able to solve many such problems extremely cheaply. This observation appears to make immunization a good candidate for scenario optimization.

In current practise, in order to achieve diversity, additional constraints are added to the deterministic immunization model and arbitrary bounds are placed on the variables. Prescribing a solution, however, cannot be considered to be good modeling practise.

In contrast, the stochastic scenario immunization model developed below is quite simple and naturally produces a diverse portfolio. The resulting solution is likely to track the present value of the liabilities (or any other analytical function of the cashflows for that matter) over time, without need for significant rebalancing, under many possible realizations of the future yield-curve discount scenarios.

The Scenario Optimization approach calls for the solution of the deterministic subproblems under each possible scenario. Since the scenario subproblems are knapsack problems, for reasonable choices of the bounds l_j and u_j the scenario subproblem solution, x^s, will always satisfy:

$$V^s = V_T^s$$

The generic coordination model prescribed by Scenario Optimization (Dembo (1991)) for combining the scenario solutions x^s into a single implementable policy is:

$$\underset{x}{\text{Minimize:}} \sum_{s \in S} p_s\{\|(c^s)^{\mathrm{T}}x - v^s)\| + \|(V^s(x) - V^s_T)\|\}$$

$$\text{subject to: } l \leq x \leq u$$

where $\|\cdot\|$ is any convenient norm and v^s is the optimal portfolio cost under scenario s and p_s is the probability of scenario s occurring.

Many alternative coordination models exist (see the discussion in Dembo (1991)), some of which are more convenient computationally than others. The exact choice of tracking model will be context dependent. For example, another possible tracking or coordinating model could be:

$$\underset{x}{\text{Minimize:}} \sum_{s \in S} p_s[((c^s)^{\mathrm{T}}x - v^s)^2 + (V^s(x) - V^s_T)^2]$$

$$\text{subject to: } l \leq x \leq u$$

where V_s is the optimal portfolio cost under scenario s and p_s is the probability of scenario s occurring.

This model is incomplete, since if the universe of securities available for immunizing actually contained the target portfolio it would almost surely be selected because it would track the target perfectly (the minimum objective value in the above problem would be zero) regardless of the cost of the target instrument. Thus it is important to bring in other relevant criteria such as cost or total return.

Let C be the total budget available for constructing the immunizing portfolio. An improved tracking model would then be the following parametric quadratic programming problem:

$$Q(C) \equiv \underset{x}{\text{Minimize:}} \sum_{s \in S} p_s[((c^s)^{\mathrm{T}}x - v^s)^2 + (V^s(x) - V^s_T)^2]$$

$$\text{subject to: } l \leq x \leq u$$

$$\sum_{j \in J} c_j x_j \leq C$$

In this model one can explicitly examine the tradeoff between the cost of the immunization and its "quality," $Q(C)$, as measured by the error in tracking the scenario solutions. This cost versus "risk exposure" tradeoff is extremely useful in commercial applications of portfolio immunization.

An alternative tracking model could be:

$$Q(C) \equiv \underset{x}{\text{Minimize:}} \sum_{s \in S} p_s[|(c^s)^{\mathrm{T}}x - v^s| + |V^s(x) - V^s_T|]$$

subject to: $l \leq x \leq u$

$$\sum_{j \in J} c_j x_j \leq C$$

This corresponds to the following linear programming problem:

$$\underset{w^+, w^-, y^+, y^-, x}{\text{Minimize:}} \quad \Sigma_s p_s[(y_s^+ + y_s^-) + (w_s^+ + w_s^-)]$$

subject to: $V^s(x) - (y_s^+ - y_s^-) = V_T^s$; $\forall s \in S$

$(c^s)^{\mathrm{T}}x - (w_s^+ - w_s^-) = v^s$; $\forall s \in S$

$l \leq x \leq u$

$$\sum_{j \in J} c_j x_j \leq C$$

By dropping the terms involving y_s^+ and w_s^+ in the objective function the model only minimizes the downside deviations from the target. Here, y_s^+ and y_s^- may be interpreted as the positive and negative absolute deviations from tracking the present value of the target portfolio under the sth scenario.

The precise choice of norm is important and depends on the application. For example, in some cases one would be interested in minimizing the downside errors only. Such is the situation if one were managing a portfolio that is immunized against a set of liabilities. On the other hand, one would minimize both up and downside errors in models designed to track an index. There is no absolute rule we can give for choice of norm. It depends solely on the context in which the model is being used. One advantage of using the absolute-value function is that the solution will be less affected by extreme scenarios.

The number of possibilities for the tracking model are virtually limitless. Its form may be molded to suit the application context. It is flexible and can deal with many diverse requirements and context dependent constraints. Most important, however, is the consequence that the solutions generated by these models will be naturally diverse, the diversity depending on the scenarios with respect to which immunization is carried out.

An example of a scenario immunization model for synthesizing mortgage-backed securities may be found in chapter 6. It shows how additional constraints might be incorporated in the tracking model.

The above tracking models could also be extended to handle multiple criteria as follows. Assume that, in addition to present value, we were interested in the total return of the portfolio over a given horizon under scenario s, $TR_s(x)$, where:

$$TR_s(x) \equiv \sum_{j \in J} V_{sj} Y_{sj} x_j$$

and Y_{sj} is the average yield of bond j under scenario s.

Let TR_{sT} be the total return of the target portfolio over the same horizon under the sth scenario. A reasonable tracking model might be:

$$Q(C) \equiv \underset{x}{\text{Minimize:}} \sum_{s \in S} p_s[c_s^{\mathrm{T}} x - v_s]^2 +$$

$$\sum_{s \in S} p_s[V_s(x) - V_{sT}]^2 +$$

$$\sum_{s \in S} p_s[TR_s(x) - TR_{sT}]^2$$

$$\text{subject to: } l \leq x \leq u$$

$$\sum_{j \in J} c_j x_j \leq C$$

A natural consequence of a scenario immunization model is that in situations of extremely high interest-rate volatility (i.e., widely disparate yield-curve scenarios), the model will tend to seek a cash match naturally especially for maturities that exhibit high volatility). In situations where there is little or no variation in expected yields the tracking-model solution will approximate a duration match. Most importantly, however, the type of solution is a consequence of the yield-curve scenario assumptions and not forced as is customary in current practise. A common type of model used at present is one which combines a forced cash match in early periods with an overall duration match to provide protection in later periods.

5 Scenario dedication (cashflow matching)

Instead of matching some function of cashflows, as is done in the above scenario immunization models, one might attempt to track cashflows directly. This results in a class of models we refer to as scenario dedication or alternatively as stochastic cashflow matching. Note that if the cashflows of both the target and dedicated portfolios are similar under all possible scenarios then various functions of these flows (e.g., present value, duration, convexity, etc.) will also be similar for all possible scenarios. This leads to the following model.

$$D(C) \equiv \underset{l \leq x \leq u}{\text{Minimize}}: \sum_{s} p_s \sum_{t} \frac{1}{(1+y_t^s)^t} \| m_t^s(x) - k_t^s \|$$

$$\text{subject to: } \sum_{j \in J} c_j x_j \leq C$$

where:

$m_t^s(x)$ = aggregate cashflow from unknown portfolio x at time period t under scenario s,

y_t^s = yield at maturity t under scenario s,

k_t^s = aggregate cashflow at time t from target portfolio under scenario s.

This model is analogous to the standard portfolio dedication models currently in use. It has many significant advantages over such models however. For one, it is always feasible. Also, notice that the cashflow matching terms are all measured in terms of current dollars so as to obtain a consistent objective function. This is not the case in portfolio dedication models where cashflow matching is done using constraints, thereby implicitly giving equal weights to every match. In addition, we allow for portfolios containing instruments whose cashflows might be scenario dependent.

If a cash matched solution is feasible within a budget of C, the model will compute one. This follows since the optimal solution will have an objective value of zero. Thus the model permits a range of solutions going from as cash matched as possible to a solution resembling classical dedication, depending on the budget. It is this fact that makes it so powerful. It is able to compute the tradeoff between quality of dedication and cost. The closer to a perfect cash match the less rebalancing will be required. Such a solution would be suitable for a volatile market. In contrast, in a flat market, a cheaper (less cash matched) solution would be more appropriate.

This model attempts a cash match over time weighted by the discount factor for the particular maturity. For a flat yield curve heavier weights will therefore be given to cashflows occurring in the near future and progressively lower weights given to matching flows in the distant future. In effect, the model in this case would attempt to do cashflow matching on the near-term flows and if possible, but with lower priority, on the far-term flows as well. When dedication is done with respect to a variety of yield-curve scenarios, each occurring with a different probability, it becomes more difficult to know which cashflows will be given priority for matching. Still, the model behaves in a manner consistent with our intuition. Cashflows corresponding to maturities with small yields, in scenarios with high probability, will be given the highest priority since they affect the value of the portfolio most.

Under certain conditions the above model is closely related to a standard portfolio dedication model. Consider the case in which we use the one-sided norm:

$$\|z\| = |z|_- = -z \quad \text{if } z \leq 0$$
$$0 \text{ otherwise}$$

The objective function, becomes equivalent to:

$$D(C) \equiv \underset{l \leq x \leq u}{\text{Minimize:}} \sum_s p_s \sum_t \frac{1}{(1+y_t^s)^t}(w_t^s)^-$$
$$\text{subject to: } m_t^s(x) - (w_t^s)^+ + (w_t^s)^- = k_t^s;\ t = 1, \ldots, T$$
$$(w_t^s)^+ \geq 0;\ (w_t^s)^- \geq 0$$

Since in a dedicated portfolio any excess cashflows from one period are always reinvested to meet the liabilities in the next period, the quantities $(w_t^s)^+$ and $(w_t^s)^-$ are related by:

$$(w_t^s)^+ = (w_{t+1}^s)^-/(1+f_t^s)^{\delta_t};\ \forall s,t$$

where:

$f_t^s \equiv$ the forward reinvestment rate for time t under scenario s,
$\delta_t \equiv$ the amount of time between the (t)th and $(t+1)$th liability.
This leads to the scenario dedication model:

$$D(C) \equiv \underset{l \leq x \leq u}{\text{Minimize:}} \sum_s p_s \sum_t \frac{1}{(1+y_t^s)^t}(w_t^s)^-$$
$$\text{subject to: } m_t^s(x) - (w_{t+1}^s)^-/(1+f_t^s)^{\delta_t} + (w_t^s)^- = k_t^s\ \forall s,t$$
$$(w_t^s)^- \geq 0;\ \forall s,t$$

$$\sum_{j \in J} c_j x_j \leq C$$

These constraints are the same as those found in portfolio dedication models (see chapter 1 for example). In this context w_t^- may be interpreted as being the cash infusion from period $t-1$ needed to supplement the portfolio flows so as to fund the target liabilities at time t. Our model minimizes the probability-weighted discounted value of the sum of these flows under different interest-rate scenarios and, parametrically, for different budgets.

An extremely important feature of the model is that it always has a feasible solution, regardless of the budget C. To see this note that for $C=0$ we can construct a feasible solution by setting:

$$(w_1^s)^- = \sum_t k_t^s \text{ and } (w_{t+1}^s)^- = (1+f_t^s)^{\delta_t}[(w_t^s)^- - k_t^s] \geq 0.$$

For a sufficiently large budget, if feasible, the solution determined by our model will be cash matched. Otherwise, in the sense of the norm used, the solution will be as close as possible to a cash matched one.

6 Relationship to scenario immunization

Observe that:

$$V_s(x) - V_{sT} = \Sigma_t \left[\frac{m_t^s(x)}{(1+y_t)^t} - \frac{k_t^s}{(1+y_t)^t} \right]$$

The basic scenario immunization model has the form:

$$Q(C) \equiv \text{Minimize: } \Sigma_s p_s \| V_s(x) - V_{sT} \|$$

$$\text{subject to: } \sum_{j \in J} c_j x_j \leq C.$$

Equivalently, the model may be written as:

$$Q(C) \equiv \text{Minimize: } \Sigma_s p_s \| \Sigma_t \frac{1}{(1+y_t^s)^t} (m_t^s(x) - k_t^s) \|$$

$$\text{subject to: } \sum_{j \in J} c_j x_j \leq C$$

It is not difficult to show that the scenario immunization model is a relaxation of the scenario dedication model, i.e., $Q(C) \leq D(C)$. Thus scenario dedication is more restrictive. This is also true for portfolio-immunization and dedication models (see Zipkin (1989) for example).

For the same choice of scenarios, the portfolio generated by a scenario dedication is likely to track the performance of the target portfolio far more closely than one obtained from a scenario immunization. It will attempt to match cashflows directly, especially those occurring in the near term. When the budget is sufficient, if possible, an exact cashflow match will be obtained. When it is insufficient, the yield-curve scenarios that are assumed will determine which cashflows will have priority in the matching.

One may also mix scenario immunization and dedication in a natural manner much in the same way as multiple objectives are treated. It is also often desirable to include additional constraints that force an exact cash match for certain maturities. Once again the exact formulation in such cases will be context dependent.

A hybrid model, in which scenario dedication is done for the first period followed by scenario immunization for subsequent periods appears to make the most sense in cases where the target portfolio contains cashflows with maturities far into the future. This is because attempting to match cashflows far into the future, especially when they are usually only actuarial estimates, does not make good modeling sense.

The basic hybrid tracking model has the form:

$$\underset{l \leq x \leq u}{\text{Minimize:}}\ \Sigma_s p_s \left\{ \sum_{t=1}^{t^1-1} \frac{1}{(1+y_t^s)^t} \| (m_t^s(x) - k_t^s \| \right.$$

$$\left. + \left\| \sum_{t=t^1}^{T} \frac{1}{(1+y_t^s)^t} (m_t^s(x) - k_t^s) \right\| \right\}$$

$$\text{subject to: } \sum_{j \in J} c_j x_j \leq C.$$

7 Scenario generation for portfolio immunization

An extremely important aspect of scenario optimization is the scenarios used to capture the stochastic behavior of the model. The immunized portfolio that is obtained will depend on the scenarios chosen and the probabilities that are assigned to them.

We have actually found this to offer a significant advantage in the commercial application of these models. Portfolio managers welcome the opportunity to provide subjective input and to measure the effect of different yield-curve scenario assumptions. The model provides them with a feel for the risk/reward tradeoffs.

For portfolio immunization, the constraint coefficients, V_{sj}, are obtained by discounting cashflows according to the sth discount function scenario. The discount function would typically be obtained from a yield curve (bond yield versus maturity) for the bond under consideration. Thus scenarios may be constructed from yield curves.

There are a number of existing models for generating yield-curve scenarios (see Black *et al.* (1990) for example) and this is currently an active area of research in finance. Almost all models are based on a "no arbitrage" condition.

One of these theoretical models could be used to generate scenarios. However, often in practise yield-curve scenarios may be generated purely from intuition and a knowledge of historical yield variance at different maturities. Scenario optimization enables a user to hedge against the chosen scenarios and consequently any scenarios that may be "close to" the chosen ones. In our view, whether the scenarios thus chosen are realistic or not does not matter. They permit an expert to quantify the effect of hedging against a personal subjective viewpoint. Often in the world of finance, this can be as or more valuable than using an artificially generated set of scenarios based on some imperfect analytical model.

A more interesting and richer view of scenarios in practise would be to make the tuple (time, yield curve) a scenario. In this way one could select different points in time and different expected yield curves associated with these points together with their associated probabilities of occurrence. This would result in a tracking model that would be able to hedge selectively at different points in time. These points in time could, for example, be chosen as the natural points at which the portfolio is to be reviewed. The scenarios could reflect hunches regarding the expected evolution over time of the yield curve.

8 Conclusions

Scenario immunization and dedication are not "black-box" models. They do not specify a single formulation that should be solved to get a solution to the underlying stochastic problem, as do most other methods. Instead they provide a framework in which much room is still left to the modeler. The art of choosing scenarios and the flexibility permitted by the tracking model leave much room for expert judgment.

ACKNOWLEDGMENTS

Motivation for this work came from the author's consulting experience on the design and implementation of immunization and dedication models. In particular we wish to thank Alan Shuch and Steve Friedman for their support and encouragement.

This research was supported in part by a grant from the National Science and Engineering Research Council under grant number OGPIN007 and by an equipment grant from the Digital Equipment Corporation.

REFERENCES

E. Adamidou, Y. Ben-Dov, L. Pendergast and V. Pica, "Optimal horizon-return portfolios under varying interest-rate scenarios," this volume, chapter 6.

F. Black, E. Derman, and W. Toy (1990), "A one factor model of interest rates and its application to Treasury bonds," *Financial Analysts Journal*, January–February, 33–39.

R. S. Dembo (1989), "Scenario optimization," Algorithms Inc. Working Paper 89.01, February. To appear in "Stochastic optimization," J. Birge and R. J-B. Wets (eds), *Annals of Operations Research* (1991).

H. Dahl, A. Meeraus, and S. A. Zenios, "Some financial optimization models: I, Risk management," this volume, chapter 2.

R. M. Nauss, "Bond portfolio optimization," this volume, chapter 11.

P. Zipkin (1989), "The structure of structured bond portfolios," Technical Report, Columbia University Graduate School of Business.

13 Mortgages and Markov chains: a simplified evaluation model

PAUL ZIPKIN

1 Introduction

The market for fixed-income securities has changed dramatically over the last decade. Two of the most significant changes have been the unprecedented volatility of interest rates and the introduction of mortgage-backed securities. The mortgage market has grown in the last decade from virtually nothing to become a substantial segment of the fixed-income market. There are nearly $1 trillion of these securities outstanding, or about half the total of US Treasury and Agency securities. See Person (1989), for example.

These developments have led to substantial innovations in the techniques used to analyze the values of fixed-income securities. Specifically, a collection of techniques known as *option-pricing methods* has come to be widely applied. This approach was developed originally to evaluate stock options, hence the name, but in fact it can be applied to any security whose value is affected by fluctuating economic variables.

The starting point is a stochastic-process model of the evolution of these variables over time. In fixed-income analysis the key variable is the short-term interest rate, though many models include other variables as well. Based on this model and the characteristics of the particular security being analyzed, one then derives a *valuation equation*. The solution of this equation provides an estimate of the value of the security under various conditions. A full exposition of this approach can be found in Ingersoll (1987), for example.

The form of the valuation equation and the methods required to solve it depend on the type of stochastic process used to model the interest rate and the other variables. Often, a diffusion process is employed, and in this case the valuation equation becomes a partial differential equation. This approach has a number of advantages, but to follow it requires a high level of mathematical sophistication.

One purpose of this chapter is to present the central concepts of stochastic interest-rate models and security-evaluation methods in a simpler mathematical setting. Specifically, we represent all stochastically changing variables by a discrete-time, finite-state Markov chain. In this case the valuation equation takes the form of a straightforward linear recursion, at least for relatively simple securities, including bonds and bond options.

Unfortunately, mortgage-backed securities are not "simple" in this sense. The other purpose of the chapter is to extend the Markov-chain, option-pricing framework to the evaluation of mortgage-backed securities. There are two factors which complicate the application of option-pricing methods to such securities:

1 These securities are affected by (somewhat) unpredictable *prepayments* on the underlying mortgages. Prepayments are in turn affected by interest rates, but the nature of this dependence seems to be quite complex, and interest rates themselves are stochastic. In particular, prepayments may be path-dependent, that is, they may depend in some fashion on the history of interest rates, not just on current rates.
2 There is another element of path dependence inherent in mortgages, in addition to the relation between interest rates and prepayments: Mortgages generate *path-dependent cashflows*. This is due to the simple fact that, at any point in time, future cashflows depend on the prepayment experience to date.

See Fabozzi (1985), for example; also, for a recent review of the evidence on prepayments, see Becketti and Morris (1990). Because of these complications, evaluation of such securities is usually very demanding computationally. Typically, an elaborate stochastic simulation is required, consuming a great deal of time on a very advanced computer. For examples of this kind of application see Brazil (1988), Carron and Hogan (1988), Davidson, Herskovitz and van Drunen (1988), and Schwartz and Torous (1989); also, see the chapter by Zenios elsewhere in this volume.

Recently, Zipkin (1989) has shown that, in a very broad class of stochastic models, the difficulties resulting from the second factor above, path-dependent cashflows, can be eliminated. Here, we specialize this result to Markov chains. We also obtain some additional simplifications and approximations in this case.

The complications in the first category above remain problematic. It is not clear, however, how much detail is truly needed to obtain reasonable value estimates. Here, we face the standard modeling tradeoff between simplicity and realism. The model reductions described below are useful, even when a very complex model is employed to describe interest rates and prepayments. In addition, our approach opens at least the possibility of using quite simple models, which can be solved quickly and cheaply.

We have learned (through confidential discussions) that the basic idea underlying our approach has been discovered independently and implemented by a couple of practitioners. On the other hand, we have spoken to several other practitioners and researchers who had been unaware of the result. Thus, while not entirely novel, our approach seems not to be widely known.

The chapter is organized as follows: Section 2 presents a basic model, in which the short-term interest rate is a function of some underlying Markov chain. We show how to use such a model to evaluate relatively simple securities, namely, those whose cashflows are not path dependent.

Section 3 describes the basic characteristics of mortgage-backed securities, and in section 4 we discuss two standard approaches for evaluating them: the first method recovers path independence of the cashflows by expanding the state space, while the second employs simulation to estimate security values. Both approaches are quite demanding computationally.

Section 5 presents a simplified approach to the evaluation of mortgage-backed securities. This approach is based on a certain homogeneity property of the values of these securities. Because of this property, the values can be computed using a linear recursion, as in the case of simple securities. In fact, the solution of this recursion can be expressed in closed form. (This is a slight exaggeration: while the formula is short, some of its terms may require substantial computation.) Also, we present an even simpler approximate formula.

Finally, in section 6, we illustrate the model with a simple numerical example based on a twenty-state Markov chain. No serious effort was made to specify the model's parameters, yet the values it produces are qualitatively plausible. Solutions are obtained in a few seconds using a personal computer.

2 Markov chain models of interest rates

Constant interest rates – review

Before turning to stochastic models, we briefly review the theory of cashflow valuation in discrete time, assuming a constant interest rate. These ideas are familiar to most readers, no doubt; we cover them here to provide a basis of comparison with the more complex models introduced later. See Brearly and Myers (1981), for example, for a fuller treatment.

We shall work with fixed time periods (say, months). Define:

r = one-period interet rate,

α = one-period discount factor $= 1/(1+r)$,

v = amount of some anticipated future cashflow,

τ = time (periods) until the cashflow occurs,

$v(\tau)$ = present value of cashflow v at time τ.

Then, according to the theory:

$$v(\tau) = \alpha^{\tau} v \tag{2.1}$$

We can also express this relationship in the form of a simple linear recursion:

$$v(\tau + 1) = \alpha v(\tau),\ v(0) = v \tag{2.2}$$

In particular, a discount (0-coupon) bond has $v = 1$, so

$$b(\tau) = \text{value of a discount bond} = \alpha^{\tau} \tag{2.3}$$

The rationale for these formulas is an equilibrium argument. Briefly, under certain assumptions, any departure from these values would open an opportunity for *arbitrage*: that is, it would be possible to combine borrowing and lending in such a way as to realize unlimited riskless profits. Such activities would tend to drive prices to their equilibrium values, namely, the values above.

A stochastic interest-rate model

Interest rates do fluctuate over time in somewhat unpredictable ways, and several *stochastic* interest-rate models have been developed. Many such models are based on diffusion processes. Most of the essential ideas, however, can be expressed using simpler processes. We now present a fairly general stochastic interest-rate model using the framework of Markov chains.

This approach is by no means new. The basic ideas here were developed by Pye (1966), and several similar models have been used subsequently. See Narula (1989) for a review of the literature.

The elements of the model are as follows:

$\boldsymbol{x}$ = an underlying finite-state, time-homogeneous Markov chain,
$= \{x(t): t = 0, 1, \ldots\}$,
$P = (p_{ij})$ = transition-probability matrix of $\boldsymbol{x}$,
r_i = one-period interest rate when $x(t) = i$,
$\alpha_i = 1/(1 + r_i)$.

Observe, the one-period interest rate itself need not be a Markov chain; rather, it is a function of such a process. Thus, $\boldsymbol{x}$ may include other variables as well.

Given this conception of the behavior of the one-period rate, how should future cash flows be evaluated? We now describe a *plausible* answer to this question. This approach covers known future cashflows, and more generally cashflows whose timing is certian, but whose amount depends on the prevailing state of $\boldsymbol{x}$. Define:

v_j = amount of future cashflow if $x(t)=j$ when the cashflow occurs,
$v_i(\tau)$ = present value of cashflow (v_j) at time τ, given $x(0)=i$:

$$\boldsymbol{v}=(v_j),\ \boldsymbol{v}(\tau)=[v_i(\tau)]$$

We have $\boldsymbol{v}(0)=\boldsymbol{v}$. Suppose we known $\boldsymbol{v}(\tau)$. To compute $\boldsymbol{v}(\tau+1)$,

1 Given we're in state i, compute the expected value one period ahead, using P and $\boldsymbol{v}(\tau)$;

2 Discount the result, using the factor α_i.

That is, defining $A=(\alpha_i p_{ij})$, we have:

$$\boldsymbol{v}(\tau+1)=A\boldsymbol{v}(\tau),\ \boldsymbol{v}(0)=\boldsymbol{v} \tag{2.4}$$

The solution to this recursion can be written simply as:

$$\boldsymbol{v}(\tau)=A^{\tau}\boldsymbol{v} \tag{2.5}$$

In particular, for a discount bond, letting **1** denote a vector of 1's:

$$\boldsymbol{b}(\tau)=A^{\tau}\mathbf{1} \tag{2.6}$$

Remarks:

1 Notice, these formulas (2.4)–(2.6) are the matrix analogues of those we saw before in the case of a constant interest rate (2.1)–(2.3). (Think of A as capital α.)

2 This approach may be plausible, but is it right? The simple answer is *yes*: This evaluation method precludes arbitrage, under certain simple assumptions.

3 However, other methods work too. The most general arbitrage-free evaluation method is of the same form, but A may be defined differently. Specifically, A is computed as above, but using some possibly different transition matrix P^* in place of P. (This result was proven in a much more general context by Harrison and Kreps (1979); see also Harrison and Pliska (1981) and Ingersoll (1987). For this relatively simple model, the result can be obtained as a simple application of linear-programming duality.)

4 The specification of P^* is essentially an empirical issue. (In this respect the diffusion-based models have the advantage that the relationship between the analogues of P and P^* is relatively simple; see Cox, Ingersoll, and Ross (1985), for example.) Fortunately, there is a huge body of empirical research to support the choices of P and P^*; see van Horne (1984) or Narula (1989) for a review.

5 For simplicity we shall continue to use $P^*=P$ below. This is in fact a common choice in practice, and it has an interesting economic interpretation; see Cox, Ingersoll, and Ross (1981).

6 The value of a security with cashflows at *several* future times can be computed simply as the sum of the values of the individual cashflows. Coupon-bearing bonds, for example, can be evaluated by this means. Also, by choosing $\mathbf{v}$ appropriately, we can evaluate *any* security with stochastic cashflows, *provided* each cashflow depends only on the current value of $x(t)$. Unfortunately, mortgage-backed securities do *not* satisfy this condition, as we shall see below.

3 Mortgage-backed securities

Basic characteristics

We shall be concerned here with the simplest type of mortgage-backed security, the *pass-through*, which is basically a collection of similar home mortgages. See Fabozzi (1985), for example, for a description of such instruments. To simplify somewhat, we shall assume the underlying mortgages are precisely identical; also, we shall treat only standard, fixed-rate mortgages. Such a security is often called a "mortgage" for short.

To describe a mortgage, we use the following notation:

c = interest rate on the mortgage,

$\beta = 1/(1+c)$,

T = original term of the mortgage,

$\tau = T - t$,

$y(t)$ = remaining principal at the start of period t.

A standard mortgage is structured so that the total scheduled payment is a constant in every period. For now, assume the actual payments are precisely those scheduled. Using standard calculations, one can show that:

$$y(t) = [(1-\beta^{\tau})/(1-\beta^{T})]y(0)$$

Also, the dynamics of $y(t)$ can be expressed through the recursion:

$$y(t+1) = [(1+c) - c/(1-\beta^{\tau})]y(t) \tag{3.1}$$

Modeling prepayments

In a standard mortgage the homeowner has the option to *prepay* part or all of the remaining principal at any time. (Also, defaults appear as prepayments to security holders.) For now, suppose we know what the prepayments will be in the future; specifically, define:

$$\pi(t) = \text{prepayment fraction at the end of period } t$$

Now, the dynamics of $y(t)$ in (3.1) must be modified:

$$y(t+1) = [1-\pi(t)][(1+c) - c/(1-\beta^{\tau})]y(t)$$

We can solve this recursion: Letting:

$$z(t)=\{\Pi_{s<t}[1-\pi(s)]\}y(0)/(1-\beta^{T})$$

one can show:

$$y(t)=(1-\beta^{\tau})z(t)$$

This derivation assumes a particular time-path of prepayments. In reality they are stochastic; specifically, they are influenced by interest rates. For our model we assume $\pi(t)$ depends on $x(t)$. Thus, we specify:

$$\pi_i=\text{prepayment rate in period } t \text{ when } x(t)=i.$$

Thus, $\boldsymbol{x}$ now includes an interest-rate model, plus any additional factors which influence prepayments. Depending on our ambitions towards realism, $\boldsymbol{x}$ may be simple or quite complex. (In practice a variety of such additional factors are sometimes used; some of these are described in Richard and Roll (1989), Schwartz and Torous (1989) and Kang and Zenios (1992).)

Cashflows

We are now prepared to describe the cashflows of a mortgage. Assume all cashflows occur at the ends of periods. Define:

$F_i^P(z,\tau)$ = principal cashflow in period $t=T-\tau$, with $z(t)=z$ and $x(t)=i$,
$F_i^I(z,\tau)$ = interest cashflow in period $t=T-\tau$, with $z(t)=z$ and $x(t)=i$,
$F_i(z,\tau)$ = total cashflow $=F_i^P(z,\tau)+F_i^I(z,\tau)$.

Let s denote the service cost charged by the issuing agency. One can show:

$$F_i^P(z,\tau)=[\pi_i(1-\beta^{\tau-1})+c\beta^{\tau}]z$$
$$F_i^I(z,\tau)=(c-s)(1-\beta^{\tau})z$$
$$F_i(z,\tau)=[\pi_i(1-\beta^{\tau-1})+c-s(1-\beta^{\tau})]z$$

So, the cashflow in period t depends on i and t, but also on $z=z(t)$, which is a function of the entire sample path of $\boldsymbol{x}$ up to time t. Because of this path-dependence, we cannot use the relatively simple methods of section 2 to evaluate a mortgage.

Traditional approach to evaluation

In section 4 we shall see how to adapt option-pricing methods to the evaluation of mortgages. First, however, we mention a much simpler approach: *Ignore stochastic interest rates and prepayments*. Specifically, the method is as follows:

1 Assume a fixed time path of prepayment rates $\pi(t)$. Now, the cashflows are certain, like those of a bond. (There are conventions for doing this.)

2 Proceed as if the mortgage *were* a bond: discount the cashflows using a fixed interest rate, the *yield* of the security. Or, given the security's price, solve for its yield.

This approach has the virtue of being simple and free of model assumptions. It remains the standard method in the market. Also, it is still widely used for portfolio selection, whether model-driven or not.

Of course, this approach obscures some of the risks of mortgages, and hence may produce distorted values.

4 Evaluation: direct approaches

Augmented state space

One way to adapt the option-pricing approach to mortgages is the classic method of supplementary variables. We regard $\mathbf{z}=\{z(t):t=0,1,\ldots\}$ as a stochastic process and consider the joint process $(\mathbf{x},\mathbf{z})$. The dynamics of $\mathbf{z}$ are given by:

$$\{z(t+1)|z(t)=z,x(t)=i\}=(1-\pi_i)z$$

So, the joint process is a Markov process, and the cashflows depend on the current state of the process.

Now, define:

$$M_i(z,\tau)=\text{value of the mortgage at the beginning of period } t=T-\tau,$$

when $x(t)=i,z(t)=z$. These quantities can be computed using the following recursion:

$$\begin{aligned} M_i(z,\tau)&=\alpha_i\{F_i(z,\tau)+\Sigma_j p_{ij}M_j[(1-\pi_i)z,\tau-1]\} \\ M_i(z,0)&=0 \end{aligned} \tag{4.1}$$

This recursion is completely analogous to (1.4). It looks more complex, but only because there are cashflows in all periods.

The problem here, of course, is that the state space of $\mathbf{z}$ is essentially continuous. To solve (4.1) directly thus requires a discrete approximation. Such calculations are usually quite time consuming.

Simulation

We can *estimate* $M_i(z,\tau)$ in the following way:

1 Generate a sample path of $\mathbf{x}$. From this generate the corresponding path of $\mathbf{z}$. Now, compute all the cashflows, and discount them using the interest rates generated by the sample path of $\mathbf{x}$. Call the result the *present value of the mortgage, conditional on the sample path.*

2 Repeat (1) for many sample paths, generated by some means consistent with the probability law governing $\boldsymbol{x}$. Average the conditional present values. The result is an estimate of $M_i(z,\tau)$.

This is the approach used most often in practice. Indeed, the same approach may be used to solve (2.4) when the state space of $\boldsymbol{x}$ is large.

5 Evaluation: simplified approach

Homogeneity

We now show how the additional complexities induced by path-dependent cashflows can be eliminated. This simplification is based on the following fact: The function $M_i(z,\tau)$ is homogeneous of degree 1 in z. This fact can be verified by a simple inductive argument using (4.1). It is proven in a much more general context by Zipkin (1989).

Using this fact we can obtain a reduced formulation: Define:
$f_i(\tau)=\alpha_i F_i(1,\tau)$, $m_i(\tau)=M_i(1,\tau)$,
$\boldsymbol{f}(\tau)=[f_i(\tau)]$, $\boldsymbol{m}(\tau)=[m_i(\tau)]$,
$G=[(1-\pi_i)\alpha_i p_{ij}]$.

Then, specializing (3.1) to the case of $z=1$, we obtain:

$$\boldsymbol{m}(\tau)=\boldsymbol{f}(\tau)+G\boldsymbol{m}(\tau-1),\ \boldsymbol{m}(0)=\boldsymbol{0} \tag{5.1}$$

Thus, we have eliminated the extra state variable z. This recursion can be solved directly, or, when $\boldsymbol{x}$ is complex, through simulation. Then, M can be recovered easily, using $M_i(z,\tau)=zm_i(\tau)$.

Closed-form solution

In fact, we can express the solution to (5.1) in closed form. For simplicity we shall treat only the case of zero service cost, that is, $s=0$. Denote:

$$\boldsymbol{\alpha}=(\alpha_i),\ \boldsymbol{\delta}=(\pi_i\alpha_i),\ H=\beta^{-1}G$$

Note, $I-G$ is invertible; assume $I-H$ is too. (This is not essential.) Then, it is easy to verify that:

$$\begin{aligned}\boldsymbol{m}(\tau)=&[(I-G^{\tau-1})(I-G)^{-1}-\beta^{\tau-1}(I-H^{\tau-1})(I-H)^{-1}]\boldsymbol{\delta}\\&+c(I-G^{\tau})(I-G)^{-1}\boldsymbol{\alpha},\ \tau>0\end{aligned} \tag{5.2}$$

This formula is reasonably short, but it is not that easy to evaluate. In addition to the solution of several linear systems (the terms involving matrix inverses), the formula requires the computation of powers of G. To obtain such powers directly is, of course, no easier than solving (5.1) itself. However, if we are willing to compute the spectral decomposition of G, then powers of G are easy to work with. Clearly, this approach is tractable only when G is small, that is, when our interest-rate and prepayment models are simple.

Notice, the limiting value $\boldsymbol{m}(\infty)$ (that is, the limit as $\tau \rightarrow \infty$) is quite simple:

$$\boldsymbol{m}(\infty) = (I - G)^{-1}(\boldsymbol{\delta} + c\boldsymbol{\alpha}) \tag{4.3}$$

This formula requires only the solution of one linear system. As shown in the next section, $\boldsymbol{m}(\infty)$ may work well as an approximation of $\boldsymbol{m}(\tau)$ for large τ.

6 Example

We now illustrate the results above using a simple numerical example. To specify $\boldsymbol{x}$, we discretize the continuous-time, continuous-state, single-factor model of Cox, Ingersoll, and Ross (1985). In that model the instantaneous risk-free interest rate r evolves according to the stochastic differential equation:

$$dr = \kappa(\theta - r)dt + \sigma r^{\frac{1}{2}} dw$$

This model has three parameters, κ, θ and σ. With no particular claim to realism, we specify the following values:

$$\kappa = 1,\ \theta = 0.07,\ \sigma = 0.1$$

(These are annual values. That is, the time units for the diffusion model are years.)

For the discrete model we use twenty states with (annualized) short-term rates ranging from 0.5% to 19.5% in increments of 1%. The time period is one month, so the r_i are monthly equivalents of these annual rates.

We used a standard though rather crude method to obtain the discrete approximation. First, we discretized the state space, leaving time continuous. The result of this step is a continuous-time Markov chain, specifically, a birth–death process; the transition rates were chosen to match the actual instantaneous mean and variance of changes in r. Second, we discretized time, using a few terms of the power-series expansion of the one-month transition matrix. The result is a *bona fide* Markov chain. More sophisticated procedures are discussed in Narula (1989) and the references therein.

Figure 13.1 displays the implications of this model for discount bonds. Prices are computed using (2.6). The figure shows the corresponding yields for four of the twenty one-period rates.

We use this model to evaluate a mortgage with an 8% annual interest rate. (The monthly rate is thus $c = 0.08/12$.) Also, we assume the service cost $s = 0$.

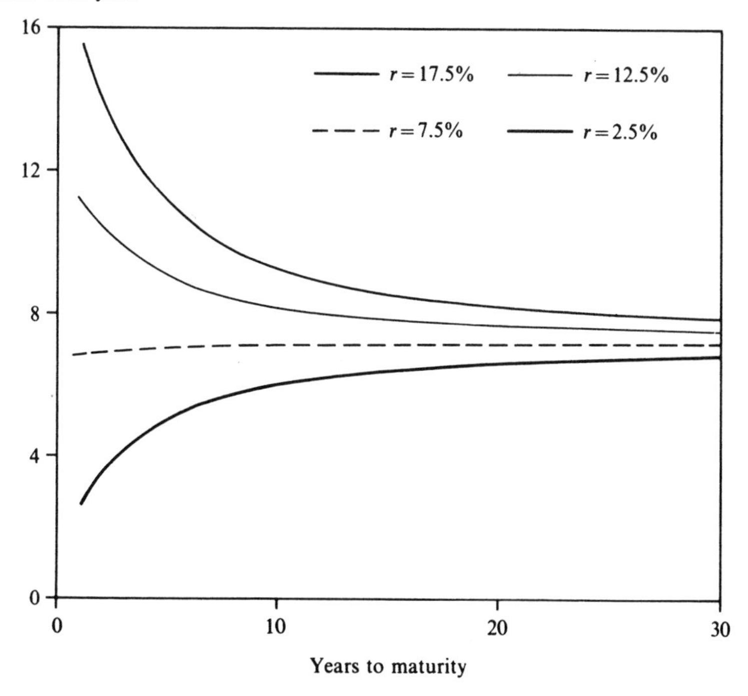

Figure 13.1 Yield curves for several short-term rates

Figure 13.2 shows the prepayment model. The figure displays the annual prepayment rate as a function of the one-period interest rate. Note, the prepayment rate is non-increasing in the interest rate, since homeowners have a greater incentive to refinance their mortgages when interest rates are low. Otherwise, the specific numbers were chosen arbitrarily. The monthly rates π_i are obtained from these annual rates.

Figure 13.3 displays the value of the mortgage as a function of the one-period rate for several different times to maturity τ. The value is computed as a fraction of the remaining principal $y(t)$, and the τs are converted to years to maturity (YTM). Also shown is the limiting value $\boldsymbol{m}(\infty)$, computed as in (5.3). Note, most new mortgages are originated with thirty years to term.

Qualitatively, these curves display the sort of behavior we expect of mortgage prices: Specifically, the price is decreasing in the interest rate; also, the curves are (slightly) convex for high interest rates, but concave for low rates, due to the effects of prepayments. See Fabozzi (1985), for example.

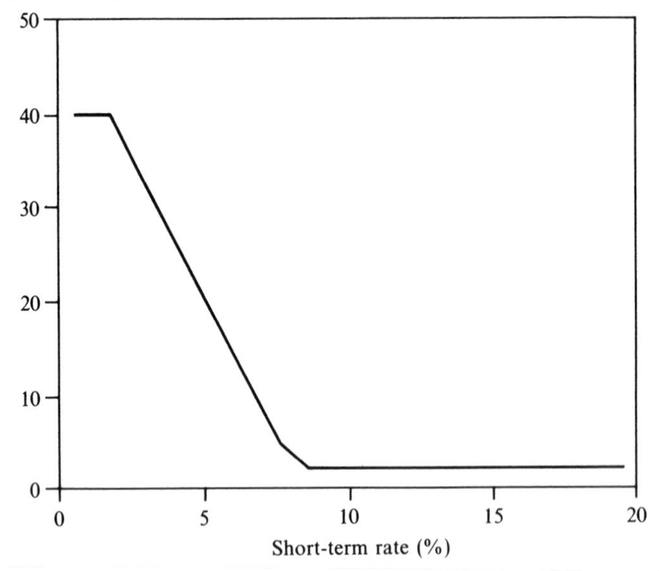

Figure 13.2 Prepayment model

Evidently, in this case, $\boldsymbol{m}(\infty)$ is a quite accurate approximation of the value of a new mortgage (YTM = 30). Even for older mortgages (YTM = 21 and 12) the approximation is reasonably close. Only when maturity is very near (YTM = 3) does the pattern of values change substantially.

Figure 13.4 displays the effects of changing the prepayment model. In each case the basic pattern of figure 13.2 is used, but all the prepayment rates are multiplied by a parameter called the *speed*. Thus, for speed 0 there are no prepayments, and speed 1 implies the specific rates shown in figure 13.2. The other curves demonstrate higher prepayment rates. In every case we evaluate a new, thirty-year mortgage.

As indicated in the figure, faster prepayments result in lower security values, and the effect is greatest for low interest rates. Again, this is roughly the pattern we would expect.

Finally, in figure 13.5 we display the values of strips. Here, the mortgage is divided into two securities. One, the "PO piece," receives the principal cashflows from the mortgage, while the other, the "IO piece" receives the interest payments. Such securities can be evaluated as above, using the cashflow functions F^P and F^I in place of F. Again, we use YTM = 30.

These curves correspond with the qualitative behavior we expect of

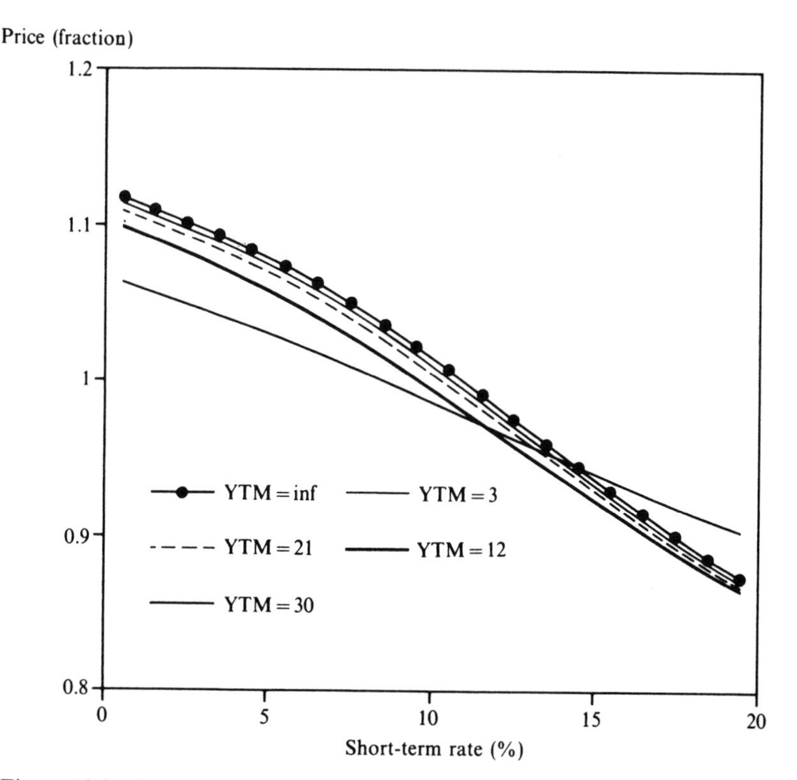

Figure 13.3 Maturity effects

strips: Briefly, the IO piece increases in value as rates rise, at least until rates become very high, and the PO's price decreases with interest rates much faster than that of the mortgage itself. See Asay and Sears (1988), for example. In sum, even this very casually specified model produces roughly plausible mortgage values.

We should mention that all the computations done to produce these figures were performed on a personal computer with an Intel 80286 processor and an 80287 numeric coprocessor. Programs were written in the language GAUSS. The recursive equations (5.1) were solved directly, not the closed-form solution (5.2). To obtain the values for a thirty-year mortgage ($\tau = 360$) required about 10 seconds in each case. The limiting approximation (5.3) required less than one second.

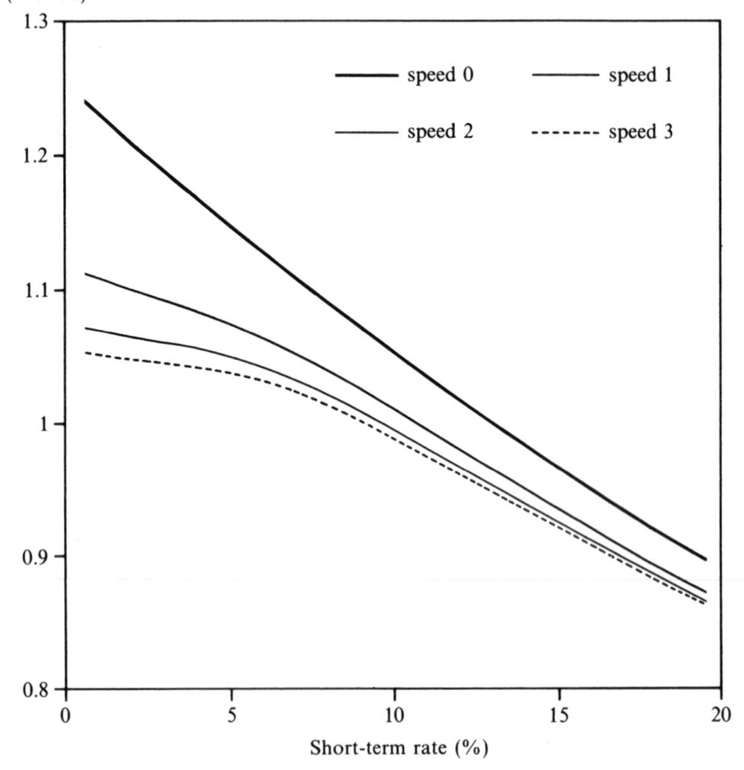

Figure 13.4 Prepayment speed effects

REFERENCES

M. Asay and T. Sears (1988), "Stripped mortgage-backed securities," Parts I and II. Research Report, Goldman, Sachs & Co., New York.

S. Becketti and C. Morris (1990), "The prepayment experience of FNMA mortgage-backed securities," Working Paper, Federal Reserve Bank of Kansas City, Kansas City, MO.

A. Brazil (1988), "Citicorp's mortgage valuation model: option-adjusted spreads and option-based duration," *Journal of Real Estate Finance and Economics*, 1: 151–62.

R. Brearly and S. Myers (1981), *Principles of Corporate Finance*, McGraw-Hill, New York.

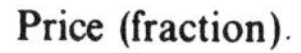

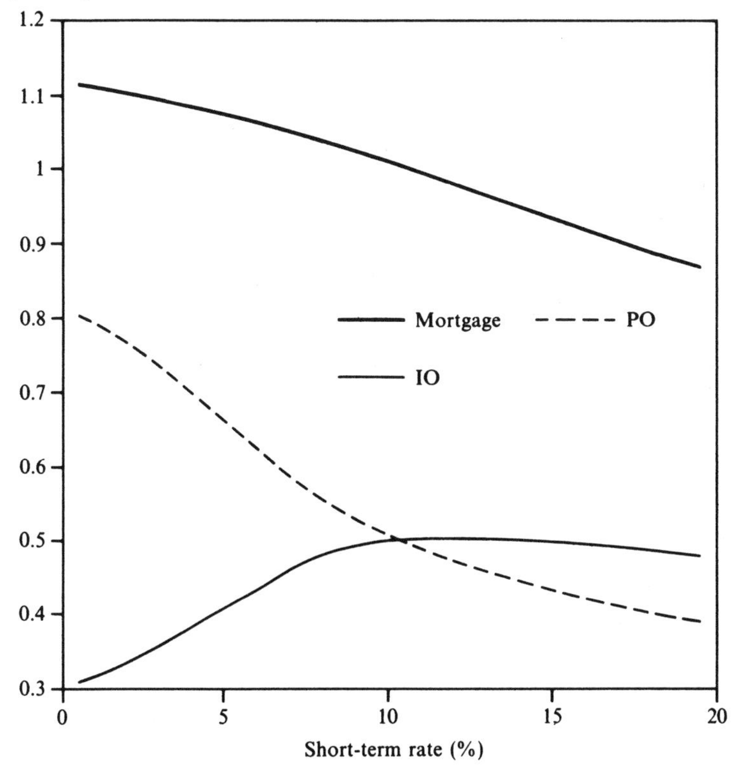

Figure 13.5 Strips

A. Carron and M. Hogan (1988), "The option valuation approach to mortgage pricing," *Journal of Real Estate Finance and Economics*, 1: 131–49.

J. Cox, J. Ingersoll, and S. Ross (1981), "A re-examination of traditional hypotheses about the term structure of interest rates," *Journal of Finance*, 36: 769–99.

(1985), "A theory of the term structure of interest rates," *Econometrica*, 53: 385–407.

A. Davidson, M. Herskovitz, and L. van Drunen (1988), "The refinancing threshold pricing model: an economic approach to valuing MBS," *Journal of Real Estate Finance and Economics*, 1: 117–30.

F. Fabozzi (ed.) (1985), *The Handbook of Mortgage-Backed Securities*, Probus, Chicago.

J. Harrison and D. Kreps (1979), "Martingales and arbitrage in multiperiod securities markets," *Journal of Economic Theory*, 20: 381–408.

J. Harrison and S. Pliska (1981), "Martingales and stochastic integrals in the theory of continuous trading," *Stochastic Procedures and Applications*, 11: 215–60.

J. Ingersoll (1987), *Theory of Financial Decision Making*, Rowman & Littlefield, Totowa, NJ.

P. Kang and S. A. Zenios, (1992), "Complete Prepayment Models for Mortgage Backed Securities", *Management Science*, Nov.

D. Narula (1989), "Inventory control with stochastic interest rates," Ph.D. Dissertation, Columbia University, New York.

K. Person (1989), "Introduction to mortgages and mortgage-backed securities: 1989 update," Research Report, Salomon Brothers Inc., New York.

G. Pye (1966), "A Markov model of the term structure," *Quarterly Journal of Economics*, 25: 60–72.

S. Richard and R. Roll (1989), "Prepayments on fixed-rate mortgage-backed securities," *Journal of Portfolio Management*, 15(3), Spring: 73–82.

E. Schwartz and W. Torous (1989), "Prepayment and the valuation of mortgage-backed securities," *Journal of Finance*, 44: 375–92.

J. van Horne (1984), *Financial Market Rates and Flows* (2nd edition), Prentice-Hall, Englewood Cliffs, NJ.

P. Zipkin (1989), "A simplifying principle in the evaluation of mortgage-backed securities," Working Paper, Graduate School of Business, Columbia University, New York.

14 Parallel Monte Carlo simulation of mortgage-backed securities

STAVROS A. ZENIOS

1 Introduction

Mortgage-backed securities – MBS for short – have emerged in the 1980s as an important class of securities. The total size of outstanding mortgage debt, the potential for MBS market growth, and the complexity of mortgage securities present unique opportunities and challenges for financial analysts. At the end of 1985 outstanding mortgage debt in the United States was approximately $2.2 trillion, of which nearly 70% was in residential mortgages. This sum dwarfs the more established corporate and government debt markets. To date only some 20% to 30% of the outstanding residential debt has been securitized via the issuance of mortgage-backed securities, but MBS represent the fastest growing segment of the debt markets. Growth in outstanding MBS was dramatic into the mid 1980s – from $3 billion in MBS outstanding in 1979 to $500 billion by the end of 1986. Trading in MBS after issuance has also increased significantly in the last few years – from $243 billion in 1981 to $1.2 trillion in 1985. Interest in MBS is not restricted to the US alone; for example, 90% of all residential debt in Denmark has been securitized.

MBS facilitate the flow of funds from the ultimate lenders in the capital markets to the mortgage borrower. Lending institutions can escape the mismatch of a short-term cost of funds (deposits) and the long-term return from a portfolio of mortgage loans by converting their portfolio of illiquid long-term loans into holdings of liquid MBS, while reaping additional income from servicing fees. For other investors, MBS are an interesting alternative instrument given their potential for high return with little credit risk, together with a wide range of available derivative securities.

Mortgage-backed securities are complex and difficult to value as they embody features of both bonds and options. The homeowner's ability to prepay outstanding principal represents a call option on the underlying mortgage. For any specific mortgage within a pool whether this call option will be exercised, and if so when, is uncertain. Many factors outside the characteristics of the pool may affect the option's value. Some of these factors include the level, structure, and history of interest rates, the market perception of future interest rates, total and disposable consumer income, and others. Adding to the complexity of early MBS has been the constant stream of innovative new derivative securities whose risk and return characteristics can bear little resemblance to the original MBS.

Several references discuss the emergence of mortgage-backed securities and their characteristics. For general interest we cite Pavel (1986), Pinkus *et al.* (1987), and Fabozzi (1985, 1987). Pavel discusses securitization, the process by which the assets (loans or other receivables) of financial intermediaries are converted into securities. The economic rationale for securitization is presented. Pinkus *et al.* present a comprehensive introduction to the primary and secondary mortgage markets, discuss various mortgage types and the variety of securities that are available to investors. They review the problems inherent to analyzing the prepayment of principal and outline several different approaches to MBS valuation. Fabozzi's volumes are compilations of published articles and research findings by some of the leading participants in the MBS market.

Pricing MBS is a complex process that relates the possible future paths of interest rates to the cashflows generated by MBS. Such cashflows should take into account both payment of principal and interest, as well as prepayment of the mortgage (i.e., exercise of the underlying call option). The analysis is carried out using Monte Carlo simulation to generate paths of interest rates, usually in monthly intervals for a period of thirty years. Such simulations are quite complex and time consuming. Not only do we have to generate several interest-rate scenarios – usually 1,000 – but a non-linear equation solver uses the result of the simulation to calculate an *option adjusted spread* – OAS for short. Hence, analyzing a single security may take several minutes of computer time on a large mainframe. Examined in the context of portfolio management, when several securities have to be analyzed for their relative attractiveness, the task may indeed become formidable.

In this chapter we show how the simulation model can be implemented in a parallel computing environment. Parallel computers – whereby multiple processors coordinate to solve a single problem – are a technological reality. The simulation of MBS is ideally suited for parallel computing. We discuss the implementation of such a model on an Alliant FX/4 and a distributed network of workstations. Section 2 describes our simulation model and illustrates its use. Section 3 describes the parallel implementation on two diverse paradigms of parallel computing: a shared memory multiprocessor, the Alliant FX/4, and a distributed memory configuration, represented here by a local area network of workstations. Computational results that evaluate the performance of the parallel implementation are given in the same section. Concluding remarks are given in section 4.

2 Monte Carlo simulation of mortgage-backed securities

The overall design of our model is illustrated in figure 14.1. The primary components of the model are:
1 Estimation of the treasury curve,
2 Monte Carlo simulation for the generation of short-term interest rates,
3 Generation of cashflows along each path of short-term rates, based on prepayment figures for the security of interest,
4 Non-linear equation solver for the calculation of option adjusted spread.
We analyze now each of those components, giving details and references for a complete description.

The Monte Carlo simulation model generates interest rates from a discretized diffusion process. The one-year forward rates implied from the current treasury curve over a period of thirty years are used to calibrate the diffusion process. Hence, the first step of our model is to build a curve that approximates observed data on treasury bonds. Our approach is based on the exponential spline fitting technique of Vasicek and Fong (1982). An exponential spline is fitted – using non-linear regression – to current market data on the spot rates of treasuries with different maturities and coupons.

2.1 Interest-rate simulation

2.1.1 Computing the rates

The interest-rate simulation model accepts as input the current treasury curve as estimated above. This curve represents spot rates. The prepayment model we use estimates prepayment rates based on monthly one-year forward rates. Hence, we must translate first the curve of spot rates into an equivalent curve of monthly one-year forward rates. This is a relatively simple process:

Step 1: Calculate the annual, one-year forward rates. These are the rates implied by the yield curve which will prevent arbitrage between spot rates for year t and year $t+1$.

Step 2: Calculate the monthly spot rates. These can be calculated once we know the annual spot rates and one year forward rates.

Step 3: Calculate the monthly one-year forward rates. Given the monthly spot rates this is straightforward.

Let sa_t denote the spot rate for year t, sm_m denote the monthly spot rate

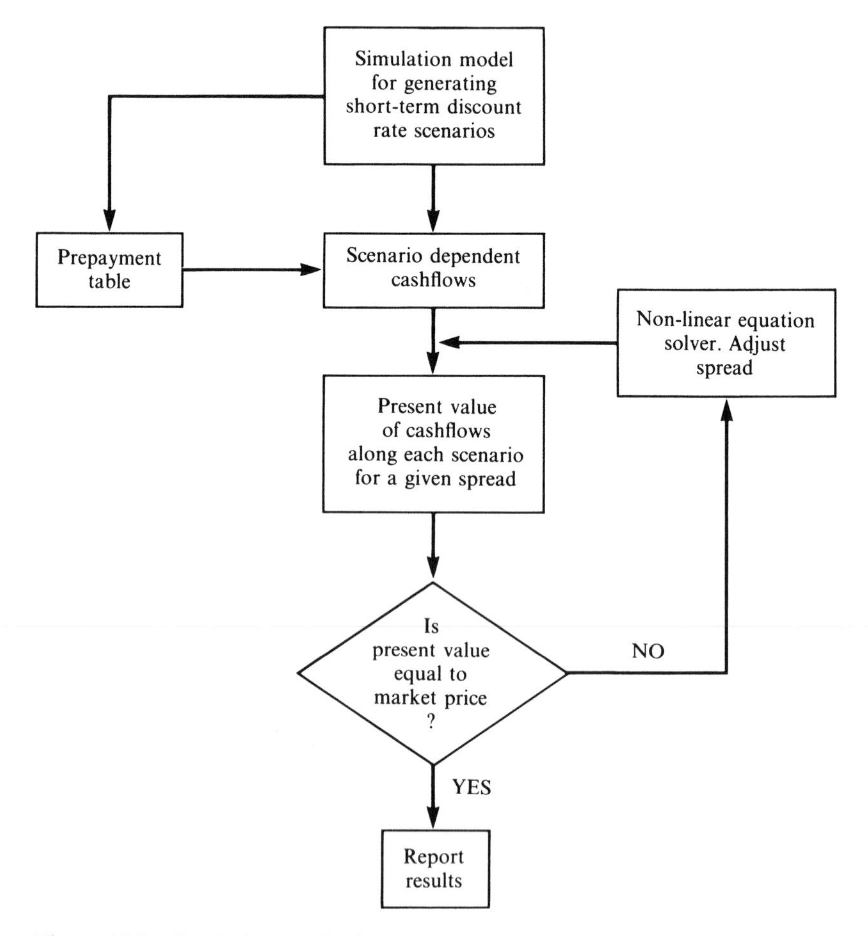

Figure 14.1 Simulation model for valuation of mortgage-backed securities

for month m, and fa_t denote the annual one-year forward rates for year t (i.e., the rate from $t-1$ to t). The one-year forward rate fm_m for month m (i.e., the rate from m to $m+12$) is calculated as follows:

At Step 1 compute the annual one-year forward rates for $t \in \{1,2,3, \ldots,30\}$, where $fa_1 = sa_1$ by definition. In general:

$$(1+fa_t)(1+sa_{t-1})^{t-1} = (1+sa_t)^t$$

Since sa_1 is known we can solve for all fa_t.

At Step 2 compute the monthly spot rates sm_m. For the first year $m \in \{1,2,3, \ldots,12\}$:

$$(1+sm_m)^{12}=(1+sa_1)$$

For subsequent years, $t \in \{2,3,\ldots,30\}$, the monthly spot rate is equivalent to the annual spot rate compounded for the year preceding the month, compounded again by the annual forward rate from this preceding anniversary for the fraction of the year the month is into the following year.

Similarly, for all months $m \in \{13,14,\ldots,360\}$ we calculate the monthly spot rate as follows. For $n \in \{2,3,\ldots,30\}$ and $m \in \{1,2,3,\ldots,12\}$, let mc be the month counter:

$$mc=(n-1)*12+m$$

For any month, determine the year number n for the preceding anniversary, and the month number m within the year $n+1$. Calculate mc and then calculate sm_{mc} using the following formula:

$$(1+sm_{mc})^{mc}=(1+sa_n)^n(1+fa_{n+1})^{\frac{m}{12}}$$

Once the monthly spot rates are known it is relatively straightforward to calculate the monthly, one-year forward rates. This rate, fm_m, is the rate from month m to month $m+12$, defined by:

$$(1+sm_m)^m(1+fm_m)=(1+sm_{m+12})^{m+12}$$

For each month $m \in \{1,2,3,\ldots,360\}$ we can calculate the one-year forward rate fm_m. This rate will be used to calibrate the rates generated by the Monte Carlo simulation and then used in the prepayment model to generate cashflows.

2.1.2 Monte Carlo simulation of the rates

Based on the treasury curve computed earlier we carry out a Monte Carlo simulation to obtain several paths of interest rates. The simulation takes into account both the current shape of interest rates and preserves arbitrage-free properties. Our simulation model follows standard industry practices, see for example Bartlett (1989). It is possible to use other models to generate interest-rate scenarios consistent with the term structure. Some recently developed models include the one-factor model of Black, Derman, and Toy (1990), the binomial lattice of Ho and Lee (1986), or the model of Heath *et al.* (1988) that is based on martingale theory. The Monte Carlo simulation model has a disadvantage that the arbitrage-free property is not inherent in the model but instead a calibration factor (drift) is used to enforce this property. The models cited above preserve the arbitrage-free property inherently, and in that sense they are more accurate. The generation of interest-rate paths based on the Black, Derman, and Toy (1990) model is the subject of the next section.

We use a lognormal distribution to generate forward-rate scenarios. Let $f = fm_m$ be the one-year forward rate (see section 2.1.1). The one-year forward rate f_t used in the prepayment model is generated as follows:

$$\log \frac{f_t}{f_{t-1}} = z * \sigma_t + R(f_t) + \mu_t$$

where:

z is a normally distributed random variable, $z \sim N(0,1)$,
σ_t is the volatility of interest rates per unit time, at instance t,
$R(f_t)$ is a mean-reversion term that forces the simulated rates to stay within historically acceptable limits (in our model in the range l = 6%, u = 16%).
It is defined by:

$$R(f_t) = \begin{cases} 0 & \text{if } l \leq f_t \leq u \\ -\gamma_0 (f_t - u)^2 & \text{if } f_t > u \\ \gamma_1 [(l - f_t)/f_t]^2 & \text{if } f_t < l \end{cases}$$

where γ_0 and γ_1 are constants estimated based on empirical observations. μ_t are drift factors estimated so that the model rates are consistent with the term structure. It is estimated by requiring that the present value of on-the-run treasuries, computed using the model rates, is in agreement with the current market prices.

2.2 *Generating interest-rate paths from a binomial lattice*

An alternative approach to the Monte Carlo simulation is to assume that term structure movements can be approximated by a discrete binomial process, represented by a lattice as shown in figure 14.2. Discrete points in time are marked on the horizontal axis, and nodes of the lattice represent possible states of interest rates at every point in time. The term structure can move to one of two possible states between successive points in time – conveniently called the "up" and "down" states. The lattice is connected in the sense that an "up, down" and a "down, up" path starting from the same state will lead to the same state after two periods. After t time periods from the origin the lattice has t possible states. Each one of these states can be reached through 2^t possible paths. For example, a binomial lattice of interest rates over a thirty-year time horizon in monthly intervals has 360 possible states at the end of the planning horizon, and a formidable 2^{360} possible paths of interest rates that lead to these states.

Short-term forward rates at the nodes of the lattice can be computed based on market data, in such a way that the arbitrage free property is satisfied. Alternative techniques for fitting a binomial lattice to market data

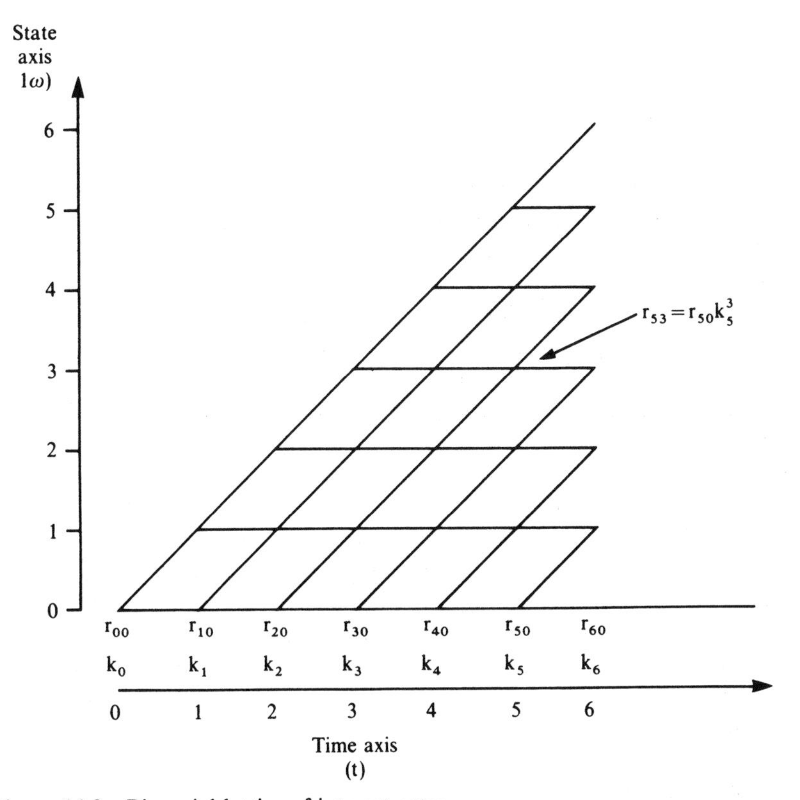

Figure 14.2 Binomial lattice of interest rates

are beyond our present task. Readers are referred to Ho and Lee (1986) or Sharpe (1985) for comprehensive discussions, and to Bookstaber, Jacob, and Langsam (1986) for critique. In our work we use the single-factor model of Black, Derman, and Toy (1990).

Once the binomial lattice has been fit to the current-term structure – in itself a difficult and computer intensive process – we can represent the short-term rate at time period t and at state ω by the relation:

$$r_{t\omega}=r_{t0}k_t^{\omega} \tag{1}$$

The quantities r_{t0}, k_t for $t=\{1,2,3,\ldots,T\}$ represent the 0-th (i.e., ground) state, and the volatility of short-term rates at period t; they are parameters estimated by the binomial lattice model of choice. A large number of interest-rate scenarios are computed by sampling paths from the binomial lattice using equation (1).

2.3 Cashflow model

The cashflow model determines the cashflow in each month for the security under study for each interest-rate scenario. The first step is to determine the monthly prepayment rate. We use data provided by Bear Stearns in the forms of tables of monthly survivorship (the complement of prepayment). Each table has as its dimensions the range of possible interest rates – 6% to 16% – and the age of the security. That is, if the one-year forward rate from the simulation model is f' in month m, we simply look up the monthly survivorship in the Bear Stearns table. These prepayment tables are large datasets: the range of interest rates (6% to 16%) is divided into 0.01% increments, and the maximum number of months under consideration for the security is 360, thus each table has 360,000 entries. An alternative approach would be to use a prepayment model for the estimation of survivorship factors at each run of the simulation program, such as the model of Kang and Zenios (1992).

Once the monthly survivorship is read from the Bear Stearns table it is straightforward to calculate the cashflows; see for example Fabozzi (1985, chapter 15). The interest rate on the underlying mortgages, i, and the servicing rate, s, are known. Also known is the remaining term, n, and the current balance b_0. For each month we must calculate the monthly cashflow that would occur without any prepayment of principal or servicing fees, and then add prepayment of principal and net the servicing fee.

The net cashflow to investors in this pass-through security in month t, cf_t, is the sum of net interest (ni), scheduled principal payments (sp), and principal prepayments (pr):

$$cf_t = ni + sp + pr$$

Let the monthly cashflow, interest, and principal, expected in month t without prepayment or servicing fees be denoted mp_t, and the remaining balance at the end of month t be mb_t. Then:

$$mp_t = mb_{t-1} * \frac{i(1+i)^{n-t+1}}{(1+i)^{n-t+1} - 1}$$

The dollar amount of interest paid by the homeowner, in month t, is:

$$I = mb_{t-1} * i$$

The net interest ni received by security holders in month t is this interest less the servicing fee, or:

$$ni = mb_{t-1} * (i - s)$$

The scheduled principal payment, *sp*, is the monthly payment less the mortgage interest, or:

$$sp = mp_t - I$$

The amount of principal prepaid, *pr*, is determined by the monthly survivorship factor, call it x. It is calculated as follows:

$$pr = (1 - x) * (mb_{t-1} - sp)$$

For each iteration of the simulation, we can determine the survivorship factor x and hence compute the cashflow by month for the security under analysis. These cashflows can then be used in one of two ways: Either to compute a theoretical price for the security, which can then be compared to the market price, or to determine the spread over the treasury curve (OAS) that will make the theoretical price equal to the market price.

2.4 Pricing model

For each iteration of the simulation, the pricing model accepts as input the simulated structure of interest rates, the calculated cashflows, and a market expectation about the incremental yield spread the security should provide over intermediate treasury rates. If these intermediate rates, the monthly one-month forward rates, are denoted r_m then we determine a theoretical price using the following formula:

$$\text{Theoretical price} = \sum_{m=1}^{n} \frac{cf_m}{(1 + r_1 + oas)(1 + r_2 + oas) \ldots (1 + r_m + oas)} \tag{2}$$

Since the cashflows (cf_m) were determined above, and *oas* is a supplied parameter, the only necessary work is to calculate the monthly one-month forward rates. For each iteration of the simulation, since we know the term structure of rates as given by the monthly one-year forward rates we can determine the monthly one-month forward rates. Again it is a multistep process: First, determine the monthly spot rates sm_m and then determine the monthly one-month forward rates.

For simplicity of notation, call the monthly one-year forward rate for month m, fy_m and the monthly one-month forward rate fm_m. Again, call the monthly spot rates sm_m but recognize that these are not the rates computed from the treasury curve, but instead reflect the results of the simulation model.

Step 1: (*Monthly spot rates*) As part of the simulation model above we also simulated the first year's spot rate. Use this sa_1 to derive monthly spot rates sm_m for $m \in \{1,2,3, \ldots,12\}$ according to:

$$(1+sm_m)^{12}=(1+sa_1)$$

Once the first year's spot rates are known, we can determine all the other spot rates iteratively since:

$$(1+sm_m)^m(1+fy_m)=(1+sm_{m+12})^{m+12}$$

We can use this to solve for sm_{m+12} for all remaining m.

Step 2: (*One-month forward rates*) Once the monthly spot rates are determined it is relatively trivial to determine the monthly one-month forward rates. That is, the one-month forward rate between month $m-1$ and month m is given by:

$$(1+sm_{m-1})^{m-1}(1+fm_m)=(1+sm_m)^m$$

Once the monthly one-month forward rates are determined, substitute these together with the *oas* parameter and calculated cashflows into the formula for theoretical price. Each iteration of the simulation will then produce an estimate of the theoretical price of the security, dependent upon the interest-rate structure modeled and the subsequent projected cashflows.

The pricing model can be used to detect relative under- or overpricing of securities given historical OAS levels. In addition, the distribution of prices across simulation iterations contains information that is useful for optimization models that embody more than the point estimate.

2.5 OAS model

The OAS model is nearly identical to the pricing model of the previous section except that the theoretical price is known as a parameter (i.e., it is assumed to be equal to the market price) and the *oas* must be calculated. The cashflows and the intermediate interest rates are the same. Instead of simply substituting an *oas* value in equation (2), however, we solve for this value against a given price. This is a problem of solving a non-linear equation in OAS and we utilize a numerical solver to determine its value. For a description of the non-linear equation solver used in our model see Kahaner *et al.* (1989).

3 Parallel and distributed computing

The simulation model presented in the preceding sections is quite complex and time consuming. The step of estimating the treasury curve is executed only once and the same curve is used to carry out all the simulations, and is also used for multiple securities. However the generation of interest-rate paths, and the associated cashflow calculations on each path, is time consuming. Furthermore, the simulation model has to be applied for multiple securities. It should be evident by now, however, that the simulation model is well suited for parallel computing. Each interest-rate scenario can be generated by a different processor. Multiple processors have to coordinate in calculating statistics from the simulations (i.e., average theoretical price, or average OAS, and the standard deviations of these quantities). In this section we discuss the implementation of our model on two diverse paradigms of parallel computing: a shared memory multiprocessor, the Alliant FX/4, and a distributed network of Apollo workstations. Both systems belong to the class of MIMD (i.e., multiple instruction stream, multiple data stream) architectures. We give some background on the parallel computing environment for each case and discuss the implementation. For a general discussion of parallel computing and references to the literature we cite Zenios (1989). In Hutchinson and Zenios (1991) we discuss an alternative approach to parallel computing for financial simulations, based on massively parallel architectures of the SIMD (i.e., single instruction steam, multiple data stream) classification.

3.1 Parallel computing on an Alliant FX

The Alliant FX is a shared memory vector multiprocessor. The system we use is an FX/4 configured with four computational elements (CE). The Alliant architecture utilizes parallel processing in two forms: vector functional units in each CE with hardware pipelining for every functional unit and parallel operations of the multiple CE. Multiple CE can execute concurrently segments of code like, for example, iterations of a loop or program subroutines. Communication between the processors is achieved through the shared memory. In addition FORTRAN programs may fork processes for parallel asynchronous execution to the operating system.

Each computational element uses a five-stage pipeline and multiple functional units that can overlap to achieve high performance on scalar operations. Instruction fetch, address calculations, and floating point add and multiply are performed by separate units. Any of the units may operate in parallel with the others. In addition each CE contains hardware to support a full vector process set: eight vector registers each with thirty-two 64-bit elements, eight 64-bit floating point scalar registers, eight 32-bit integer, and same number of address registers. Each element can execute vector operation of length thirty-two 64-bit words in parallel with the other elements for an effective vector length of 256 data elements. Vector operations utilize a twelve-stage vector pipeline.

Four possible modes of execution are possible on the FX system. *Scalar*, when all operations are performed serially by one CE. *Vector*, when operations are performed in groups of up to thirty-two elements by special vector instructions on the hardware. *Scalar concurrent*, when scalar operations are performed by a number of CEs concurrently. *Vector concurrent*, when multiple CEs operate concurrently on groups of up to thirty-two elements with the vector hardware.

It is easy to implement the algorithm for parallel execution on the shared memory multiprocessor. The simulation program is controlled by an iterative loop that repeats the simulation for the desired number of runs. Using a compiler directive we may instruct the compiler to execute this iterative loop concurrently. There are two sources of inefficiency with this approach. First, the number of tasks that have to be processed in parallel is very large (e.g., 1,000 for that many simulation runs) and each task has very small granularity (it takes less than 0.5 seconds for a single simulation run). Second, the computation of average and standard deviations is computed incrementally as each simulation run is completed. This operation cannot be executed in concurrent mode; it would violate data integrity when two or more processors would need to update the statistic at the same time.

To avoid both of these problems we partition the number of desired simulations into blocks *nblock*. Each block consists of 1/*nblock* the number of desired simulation runs. Each block is then scheduled for concurrent execution on multiple processors. The statistics are computed by each processor separately for the simulations in its block. Upon termination of all blocks control returns to the main program. The intermediate statistics of each block are combined to compute overall average and standard deviations.

3.2 Distributed computing on a local area network of Apollo workstations

The parallel implementation on the network of Apollo workstations is similar to the implementation on the Alliant FX/4. We determine the number of simulations to be executed on each workstation and the number of workstations to be utilized. The main program at the client station estimates the treasury curve, and then it requests server workstations to generate interest-rate paths and cashflows, compute average, and standard deviation statistics and return results back to the client. The client will then proceed to compute aggregate statistics, combining the results from the servers, and terminate.

An important distinction from the Alliant implementation is the need to explicitly handle communications among the workstations. The implementation of the client-server system is based on three features of the Apollo

Network Computing System (NCS): (1) concurrent programming support (CPS), (2) remote procedure calls (RPC), and (3) location broker system (LB). Concurrent programming support enables the spawning of multiple tasks for asynchronous timeslicing execution on a single processor. Remote procedure calls extend the procedure call mechanism from a single computer to a distributed environment. The location broker registers client-server interfaces in its databases for use by RPC.

The main program uses CPS to create multiple tasks on the client workstation. Every task receives the estimated treasury curve data (in particular the monthly one-year forward rates), and a parameter that specifies the number of simulations to be executed remotely. Each task gets a handle to a server environment at a remote workstation and establishes communication links. Data from the local tasks are passed on to the respective servers. Before the Monte Carlo simulation can begin each server must load into its local memory the prepayment table. The table is stored at a file server that is accessible by all workstation nodes. At present, reading the prepayment table is a very time consuming process. Input and output (I/O) on the Apollos is not particularly efficient. At the same time multiple servers need to access the prepayment table from the central data base, hence creating contention at the I/O channels. The impact of I/O on the efficiency of the distributed implementation is demonstrated in the next section.

3.3 Computational results

The OAS model of section 2 was implemented in FORTRAN 77, and is used here to study the efficiency of parallel implementations. Monte Carlo simulation is used for the generation of interest-rate paths. While the computer times will change if the binomial lattice model had been used instead, the observed speedups from parallelism will remain unaltered.

3.3.1 Simulations on the Alliant FX/4

For a typical run of 1,000 simulations we would use eight blocks. Each block consists of 250 simulation runs. The results for a typical run are illustrated in table 14.1. We observe from this table that reading in the prepayment data is a time consuming operation. When the simulation program is parallelized the data-read dominates the computation. It would be preferable to call the prepayment model to generate the relevant rates for each run of the simulation model, than generate first a huge prepayment table, and then use it as a database in the simulation. In figure 14.3 we plot the speedup factor against the number of CE. The speedup factor is defined as the ratio of solution time with one CE to the solution time with p CE.

Table 14.1. *Results of the simulation program with parallel computing on the Alliant FX/4*

CE	Data read (seconds)	Simulation (seconds)
1	110.3	235.2
2	109.6	133.2
3	109.7	90.1
4	109.6	67.9

Ideally this ratio should be equal to p (i.e., linear speedup). We observe from the figure that the simulation model achieves very close to linear speedup factors.

3.3.2 Simulations on the Apollo network

We ran 1,200 simulation runs on the network of Apollo workstations. Our network consisted of both DN3000 and DN4000 nodes, with different processing power. We observed from some preliminary runs that the DN4000 nodes would execute all the simulations in their respective blocks and would then remain idle waiting for the DN3000 to terminate. This phenomenon, known in the parallel computing jargon as *uneven load balancing*, can be eliminated using one of two approaches: we could create very large numbers of blocks and have each workstation process the next available block once its current block has been completed. Thus, faster nodes would complete more work than slower nodes and this would result in even load balancing. An alternative approach would be to create *a priori* different size blocks for the DN3000 and DN4000 nodes. The first approach is easier to implement, but results in a larger communication overhead in communicating data between the client and the servers for each new block. We therefore chose the second approach. Having established that a DN3000 node would take about twice the time of a DN4000 to execute the same number of simulations we created blocks of appropriate sizes, and each DN4000 was asked by the client to execute twice as many simulation runs as the DN3000 servers.

The results of the simulation model with a different number of workstation servers on the network are given in table 14.2. Three speedup factors are summarized in figure 14.4: (1) the speedup factor of the CPU time for the simulation program (excluding data read time for the

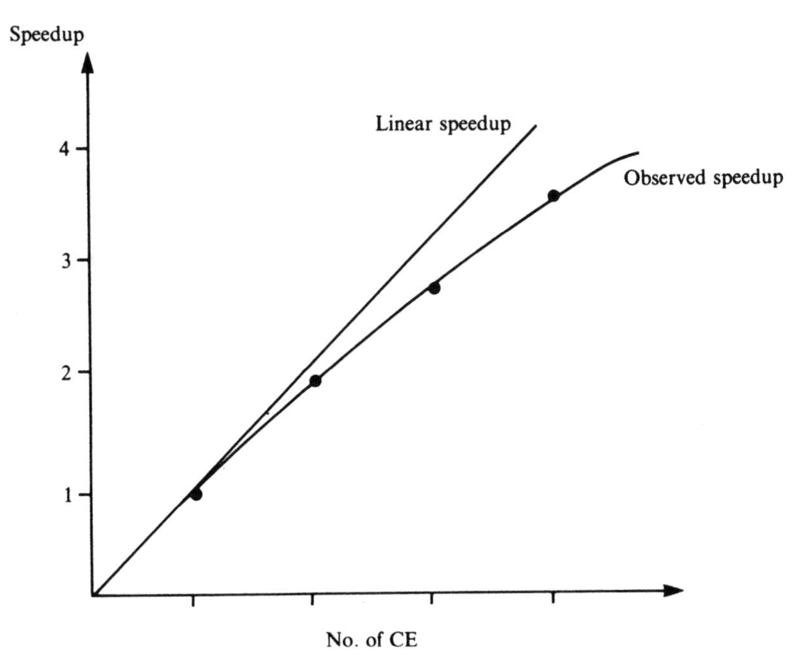

Figure 14.3 Speedup factors of the simulation model with parallel computing on the Alliant FX/4

prepayment table), (2) speedup factor of wall clock elapsed time as observed by the client (excluding data read time for the prepayment table), and (3) wall clock time as observed by the client workstation when the data read time for the prepayment table is included.

We observe from the results of the table and the figure that the simulation program achieves almost linear speedup factors. The difference between the CPU speedup and wall clock speedup factors is not significant. It is due to the control of the multiprocessing implementation by the operating system, and the time spent in communications between the client and the servers. Once more, however, the time spent in reading the prepayment tables dominates the computation as the number of servers increases. The comment made in the previous section about the need to parallelize the prepayment model holds true in the distributed environment as well.

3.3.3 Benchmark results

To conclude our discussion we present benchmark results of the complete OAS model on a wide range of computer systems; see table 14.3.

Table 14.2. *Results of the simulation program with parallel computing on the network of Apollo workstations*

	Data read		Simulation	
Servers	Wall clock	CPU	Wall clock	CPU
DN3000	375.8	319.9	2280.5	2147.9
DN3000	396.5	331.4	1219.1	1069.8
DN3000	390.3	329.7	854.1	713.2
DN3000	442.9	362.2	681.7	537.2
DN4000	440.2	367.6	482.4	360.2
DN4000	442.7	324.6	390.8	270.6
DN4000	455.7	360.8	308.8	216.7
DN4000	475.8	360.7	264.4	180.4

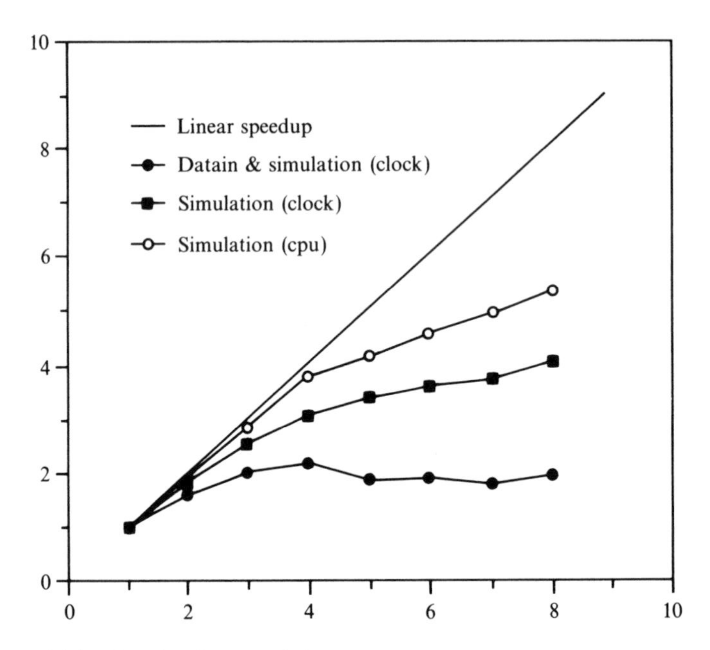

Figure 14.4 Speedup factors of the simulation model with distributed computing on the network of Apollo workstations

Table 14.3. *Benchmark results with the OAS model (CPU seconds)*

Computer	Monte Carlo simulation	Binomial lattice sampling	OAS for 3000 MBS with Monte Carlo
Apollo DN4000	480	440	16 days
VAXstation	330	320	11 days
DECstation 2100	120	130	4 days
DECstation3100	90	100	3 days
VAX 6400	60	60	2 days
CRAY X-MP	12	12	10 hours
Connection Machine CM-2a	1.1	0.7	1 hour

This experiment serves a dual purpose. First, it illustrates the computer intensive requirements of these models even when run on large mainframes. Second, it establishes that vector supercomputers (like the CRAY X-MP), or massively parallel architectures (like the Connection Machine CM-2a with 8192 processing elements) can have a significant impact on the way OAS models are being used. Models that would take several minutes or hours of computer time can now be solved in real time. To emphasize this point we estimate the time required to evaluate a portfolio of 3000 MBS – not a typical size for major investors in the MBS market.

4 Conclusions

Monte Carlo, and other forms of simulation, provide a very powerful analytic tool for the valuation of complex instruments like mortgage-backed securities. The need to repeat the simulation runs several hundred times, in order to reduce the statistical sampling error, makes these models computationally intensive. We have shown in this report that parallel and distributed computing technology can be employed successfully in this type of analysis.

The results presented here indicate that with parallel implementations the task of simulation can be made significantly more efficient. This will in turn imply ease of updating the results as new data become available, analysis of a large universe of securities without multiday runs and so on.

However, even with the parallel implementations described here, the solution times are not fast enough for a dynamically changing environment, like the one where mortgage-backed securities are traded. Massively parallel systems, like the Connection Machine CM-2 with thousands of processing elements, could be employed to carry out the simulations under real time conditions (i.e., a few seconds of wall clock time from the moment the button is pushed until the results are presented).

We expect to see parallel computers receive increasingly more attention in financial modeling applications in the near future.

ACKNOWLEDGMENTS

Research partially supported by NSF grant CCR-8811135 and AFOSR grant 89-0145 and by a research contract with Union Bank of Switzerland. The HERMES Laboratory was established under a grant from Digital Equipment Corporation. The simulation model was implemented by J. Garity and subsequently refined by P. Kang. Thanks are due to Mr H. Hill for providing references and feedback, and to Mr R. Ramanathan for providing the data.

REFERENCES

W. W. Bartlett (1989), *Mortgage-Backed Securities: Products, Analysis and Trading*, New York Institute of Finance.

F. Black, E. Derman, and W. Toy (1988), "A one-factor model of interest rates and its application to treasury bond options," Discussion Paper 1, Goldman, Sachs & Co., New York.

(1990), *Financial Analysts Journal*, January–February, 33–9.

R. Bookstaber, D. P. Jacob, and J. A. Langsam (1986), "Pitfalls in debt option models," Working Paper, Morgan Stanley, Fixed Income Analytical Research.

P. Boulay (1988), "The secondary market: planning strategy with the machine," *Mortgage Banking*, 48(7).

F. J. Fabozzi (ed.) (1985), *The Handbook of Mortgage Backed Securities*, Probus Publishing Company.

(1987), *Mortgage Backed Securities: New Strategies, Applications and Research*, Probus Publishing Company.

D. Heath, R. Jarrow, and A. Morton (1988), "Bond pricing and the term structure of interest rates," Technical Report, Cornell University.

T. S. Y. Ho and S-B. Lee (1986), "Term structure movements and pricing interest rate contingent claims," *The Journal of Finance*, 41: 1011–29.

J. Holm and T. LaMalfa (1988), "The secondary mortgage market: risky business – managing the pipeline," *Mortgage Banking*, 48: 30–7.

J. M. Hutchinson and S. A. Zenios (1991), "Financial simulations on a massively parallel connection machine," *International Journal of Supercomputer Applications*, 5(2): 27–45.

D. Kahaner, C. Moler, and S. Nash (1989), *Numerical Methods and Software*, Prentice Hall, New Jersey.

P. Kang and S. A. Zenios (1992), "Complete prepayment models for mortgage-backed securities," *Management Science*, Nov.

C. Pavel (1986), "Securitization," *Economic Perspectives*, 16–31.

S. M. Pinkus, S. M. Hunter, and R. Roll (1987), "An introduction to the mortgage market and mortgage analysis," Mortgage Securities Research Series, Goldman, Sachs & Co., New York.

W. F. Sharpe (1985), *Investments*, Prentice Hall, Englewood Cliffs, New Jersey.

O. A. Vasicek and H. G. Fong (1982), "Term structure modeling using exponential splines," *The Journal of Finance*, 37: 339–48.

S. A. Zenios (1989), "Parallel numerical optimization: current status and an annotated bibliography," *ORSA Journal on Computing*, 1: 20–43.

Index